Tetraonidae and Phasianidae of the USSR

Tetraonidae and Phasianidae of the USSR

Ecology and Morphology

M.A. Kuz'mina

Scientific Editor

Douglas Siegel-Causey

Smithsonian Institution Libraries
and
The National Science Foundation,
Washington, D.C.
1992

TT 81-52004

Teterevinye i Fazanovye SSSR: Ekologo-Morfologicheskaya
Kharakteristika. Nauka Publishers,
Alma-Ata, 1977

Translated from the Russian
Translators: P.M. Rao and D. Siegel-Causey
General Editor: Dr. V.S. Kothekar

Library of Congress Cataloging-in-Publication Data

Kuz'mina, M.A. (Mariia Alekseevna)
[Teterevinye i fazanovye SSSR. English]
Tetraonidae and phasianidae of the USSR: ecologo-morphological characteristics / M.A. Kuz'mina; scientific editor Douglas Siegel-Causey; [translated from Russian, translator P.M. Rao; general editor V.S. Kothekar].
p. cm.
Translation of: Teterevinye i fazanovye SSSR.
Bibliography: p.
Includes indexes.
1. Tetraonidae—Soviet Union—Ecology. 2. Tetraonidae—Soviet Union—Morphology. 3. Phasianidae—Soviet Union—Ecology. 4. Phasianidae—Soviet Union—Morphology. 5. Birds—Soviet Union—Ecology. 6. Birds—Soviet Union—Morphology.
I. Siegel-Causey, Douglas. II. Title.
QL696.G285K8913 1989
598'.616'0947—dc 19 89-4294
CIP

Translated and published for the Smithsonian Institution Libraries,
pursuant to an agreement with the National Science Foundation,
Washington, D.C., by Amerind Publishing Co. Pvt. Ltd.,
66 Janpath, New Delhi 110 001

Typeset by Systematic Graphic Services
and printed at Pauls Press, New Delhi, India

UDC 598.2/9–15 +598.2/90–14

This book reviews the ecological and morphological adaptations of Tetraonidae and Phasianidae of the Soviet Union, based on data available in the literature and many years of research by the author. The distribution, habitat, behavior, feeding habits, and the adaptation of these birds to climatic conditions are discussed in the first part. The second part of the book elucidates the structure of the locomotory organs and locomotion. The final chapter examines the ecological groups of galliform birds and suggests measures for conserving and increasing the numbers of some species of this economically important group of birds.

This book is addressed to zoologists and ornithologists, and to teachers, researchers, and students of biology.

Editor-in-Chief
A.F. Kovshar'

Foreword to the English Edition

The Smithsonian Institution Libraries, in cooperation with the National Science Foundation, has sponsored the translation into English of this and hundreds of other scientific and scholarly studies since 1960. The program, funded with Special Foreign Currency under the provisions of Public Law 480, represents an investment in the dissemination of knowledge to which the Smithsonian Institution is dedicated.

This volume reviews the natural history and morphology of the Tetraonidae and Phasianidae of the USSR, and presents a detailed survey of Soviet research up to 1977 on these economically important species. Maria Alekseevna Kuz'mina summarizes the results of her own morphological studies in the second part of this volume, and concludes with an analysis of the ecological and functional morphology of selected species. For many readers, this English-language translation presents for the first time a review of Soviet research on pheasants and grouse that would be inaccessible otherwise, and this edition is valuable for this reason.

The editors have taken great care to ensure that this translation is an accurate reflection of the original text and we have minimized interpretation and glosses as much as possible. Over a decade has passed since Kuz'mina finished this monograph, but her findings on morphology and natural history of these galliforms remain timely and relevant. We have included in this edition supplementary tables of common and Latin names of plants and animals mentioned, indexed to their location in the text. Common names are standardized as far as possible to North American usage; we have used regional names only where necessary.

Douglas Siegel-Causey
Museum of Natural History
University of Kansas
Lawrence, Kansas

The following Russian–English dictionaries and standard reference works were consulted in translating the Index of Animal and Plant Names as well as the List of Bird Names.

1. *Russian–English Biological Dictionary*
Compiled by C.W. Dumbleton, Oliver & Boyd, 1964

2. *Russian–English Agricultural Dictionary*
Compiled by B.N. Uspenskii, I.M. Podkamennych and E.I. Shkonde, Fizmatgiz, Moscow, 1960

3. *Russian–English Biological and Medical Dictionary*
E.A. Carpovich, Technical Dictionaries Co., 1963

4. *Russian–English Forestry and Wood Dictionary*
W. Linnard, Commonwealth Agricultural Bureaux, 1966

5. *Russian–English Bird Glossary*
J.M. Maclennan, Canadian Wildlife Service, 1958

6. *English–Russian Biological Dictionary*
Compiled by T.A. Kharisanova, Russkii Yazyk, Moscow, 1976

7. *Geobotanic Dictionary* (Russian–English–German–French)
O.S. Grebenshchikov, Nauka, Moscow, 1965

8. *Birds of the World: A Checklist*
James Clements, Croom Helm, London, 1976

9. *Checklist of Birds of the World*
Edward S. Gruson, Collins, London, 1976

10. *The Complete Birds of the World*
Michael Walters, David & Charles Inc., 1980

11. *Vocabularium Nominum Animalium Europae. Septem Linguis Redactum* (in 2 volumes)
Compiled by L. Gozmány, Akadémiai Kiadó, Budapest, 1979

List of Bird Names

English Name[1]	Russian Name	Latin Name[2]
Capercaillie		
Black-billed	Kamennii glukhar	*Tetrao urogalloides*
Western	Obyknovennyi glukhar	*T. urogallus*
Francolin		
Black	Turach	*Francolinus francolinus*
Clapperton's		*F. clappertoni*
Coqui		*F. coqui*
Crested		*F. sephaena*
Gray-winged		*F. africanus*
Latham's		*F. lathami*
Nahan's Forest		*F. nahani*
Scaly		*F. squamatus*
Shelley's		*F. shelleyi*
Grouse		
Caucasian Black	Kavkazii teterev	*Lyrurus mlokosiewiczi*
Northern Black	Obyknovennyi teterev	*L. tetrix*
Northern Hazel	Obyknovennyi ryabchik	*Tetrastes bonasia*
Ruffed	Vorotnichkovyi ryabchik	*Bonasa umbellus*
Severtzov's Hazel	Severtsova ryabchik	*Tetrastes sewerzowi*
Siberian Spruce	Dikusha	*Falcipennis falcipennis*
Partridge		
Barbary		*Alectoris barbara*
Chukar	Keklik	*A. chukar*
Gray	Seraya kuropatka	*Perdix perdix*
Daurian	Borodataya kuropatka	*P. dauurica*
Red-legged		*Alectoris rufa*
Rock	Obyknovennaya kuropatka	*A. graeca*
Ruppell's		*A. melanocephala*
Sand		*Ammoperdix heyi*
See-see	Pustynnaya kuropatka	*A. griseogularis*
Tibetan	Tibetskii kuropatka	*Perdix hodgsoniae*

[1]Common names follow *Birds of the USSR* (V.E. Flint, R.L. Boehme, Y.V. Kostin, A.A. Kuznetsov, 1984, Princeton University Press) where possible.

[2]Latin names follow the usage in this monograph.

English Name	Russian Name	Latin Name
Ptarmigan		
Rock	Tundryanaya kuropatka	*Lagopus mutus*
White-tailed	Belokhvostaya kuropatka	*L. leucurus*
Willow	Belaya kuropatka	*L. lagopus*
Peafowl		
Common	Pavlini	*Pavo cristatus*
Congo		*Afropavo congensis*
Pheasant		
Common	Fazan	*Phasianus colchicus*
Green		*P. versicolor*
Mongolian	Mongol'skii fazan	*P. mongolicus*
Pallas'	Manchzurskii fazan	*P. pallasii*
Quail		
African		*Coturnix africana*
Black-breasted		*C. coramandelica*
Common	Obyknovennyi perepel	*C. coturnix*
Harlequin		*C. delagorguei*
Japanese	Yaponskii perepel	*C. japonica*
Snowcock		
Altai	Altaiskii ular	*Tetraogallus altaicus*
Caspian	Kaspiiskii ular	*T. caspius*
Caucasian	Kavkazii ular	*T. caucasicus*
Himalayan	Temnobryukhnii ular	*T. himalayensis*
Tibetan	Tibetskii ular	*T. tibetanus*

Foreword

5 Of all the birds in the world, the Galliformes are by far the most important to man. Ease of taming, high fecundity, and excellent taste of the meat led to the domestication of a number of species and to the proliferation of a large number of breeds throughout the world for such diverse purposes as the production of eggs and meat, for sporting events, and for ornamental purposes.

At present, fowl hunting is practiced in all societies, and has not lost its importance as sporting overtakes it as a profitable enterprise.

It is highly probable that further progress in aviculture may require the domestication of new wild species for developing new breeds. The needs of conservation may also require raising new species for semicaptive and captive breeding.

The rational exploitation of wild Galliformes for commercial, scientific, and educational purposes requires conservation of all species of this valuable group of birds at optimum population levels.

Increasing changes of the natural environment (reduction of forested areas and expansion of human environments) affect the distribution and abundance of individual galliform species. The preservation of all species and the exploitation of some species for breeding, sport, and re-introduction call for different approaches for different species. A rational impact on faunas must be based on a thorough knowledge of the
6 ecology of each species, especially its ecological flexibility. The relationship between an organism and its environment is best understood by the ecomorphological approach, which means studying behavioral and morphological adaptations in their ecological context. This approach has been adopted in the present comparative study of two families of Galliformes (Tetraonidae and Phasianidae) in the USSR.

The ecological part of this book is based on the vast information available in literature and my own observations. The ecology of the galliform birds of the Soviet Union has not been well studied uniformly.

The most comprehensive information available on a number of widespread species of Tetraonidae (Capercaillie, Hazel Grouse, Black Grouse, and Willow Grouse) is in monographs for hunters, in works based on long-term studies conducted in sanctuaries and national parks, and in many contributions from different parts of the country. In addition, general monographs on the Tetraonidae of the Soviet Union have been published [Snigirevskii, 1946; Semenov-Tyan-Shanskii, 1959; and Teterevinye ptitsy (Tetraonid Birds), 1975]. Pheasants, on the whole, have not been studied as well as the tetraonids. No general monograph is available on this group of birds from the Soviet Union; information had to be gathered from individual, mostly short, articles.

My ecological study of Galliformes was conducted mainly in Kazakhstan where a diverse environment supports 12 species or 57.1% of all the Galliformes of the Soviet Union. The collection of ecological data commenced in the 1930's and continues to this day, and includes an intensive study of this group of birds over a shorter period, from 1948 to 1963. The data obtained filled gaps in our knowledge of the ecology of various species of phasianids, Data on tetraonids were collected from such regions as the Kazakh Altai, the Kalbinsk Altai, and northern Kazakhstan which have not been studied well so far. Only certain aspects of their ecology that are directly associated with a study of the morphology of locomotory organs are presented in this book. Analyzed in detail are the habitats, behavior, locomotion, mode of foraging, food composition and its seasonal changes, and nutritive value of species. Attention is paid to changes in life style in relation to food availability, emphasizing these changes among tetraonids and phasianids in winter.

The morphological part deals mostly with the locomotory organs. It is primarily based on data obtained from the osteological collection of the Institute of Zoology, Academy of Sciences of the USSR, and skin specimens kindly loaned to me by Yu.K. Gorelov (See-see Partridge) from Badkhyz Sanctuary, V.F. Gavrin (Black-billed Capercaillie) from eastern Siberia, and by Siberian hunters (Siberian Spruce Grouse and Altai Snowcock).

The format of this book is somewhat atypical. Genera rather than species are described, and interspecific differences within a genus are emphasized. Such an arrangement well represents both common features of a genus and differences between the species. The same rule applies for maps. The range of each species from the fauna of the Soviet Union is represented within the general range of the genus. In correlating the ranges, I used Peters' (1934) descriptions of the distribution of subspecies, and several other faunistic works on the distribution of the species [ptitsy Sovetskogo Soyuza Birds of the Soviet Union), Vol. IV, 1952; Johansen, 1956; Kozlova, 1960, 1970; Potapov, 1970; Izmailov and Pavlov, 1975]. This book is illustrated with original drawings by the late artist, Yu.B. Sofiev.

An examination of the ornithological collections of the Institute of Zoology, Academy of Sciences of the USSR, greatly assisted me in preparing this monograph. Consultations with the staff of this department, especially Konstantin Alekseevich Yudin and Aleksandr Ivanovich Ivanov, and several other ornithologists proved invaluable. I extend my sincere gratitude to each of them.

I am very much indebted to the late Elizaveta Vladimirovna Kozlova, who rendered invaluable advice and read the section on ecology, to Boris Karlovich Shtegman for sound counsel and suggestions in processing and presenting the morphological material, and to Aleksandr Nikolaevich Formozov who closely followed my work and provided me with a number of published works otherwise difficult to obtain. Finally, I express immense gratitude to Igor Aleksandrovich Dolgushin, the convener and director of ornithological studies in Kazakhstan, who took note of my interest in ecomorphological studies and extended constant encouragement to me.

Contents

Some Aspects of the Origin and Taxonomy of Galliform Birds

8 The order Galliformes is a very old, fairly well-differentiated, and large group of birds spread over a wide variety of environments in every continent except Antarctica.

The oldest fossil remains of Galliformes are known from the Middle Eocene. In North America, an almost complete skeleton of *Gallinuloides wyomingensis*, a species constituting an independent family (Lambrecht, 1933; Shul'pin, 1940; and Dement'ev, 1964), and skeletal remains of *Palaeophasianus meleagroides* (Brodkorb, 1964) were discovered from this epoch. Remains of some species of the genera *Palaeortyx* and *Filholornix* (Cracidae) found in Europe also date back to this epoch (Brodkorb, 1964). These discoveries provided a basis for assuming that the Galliformes arose probably at the end of the Cretaceous—the time of the first appearance of birds (Howard, 1950).

According to V.B. Dubinin (1956), who studied feather mites (Analgesoidea), the order Galliformes originated in the Australian zoogeographic region and Southeast Asia at the end of the Cretaceous or in the Early Eocene. Later, the land bridge between Australia and South America (through Antarctica) until the Middle Eocene facilitated the spread of the ancestors of galliform birds to South America, where the family of curassows, guans and chachalacas (Cracidae) and the suborder of hoatzins (Opisthocomi) originated, but the taxonomic status of the latter is not very clear yet. In Australia, the family of megapodes (Megapodiidae) separated from the ancient Galliformes. The close kinship between Megapodiidae and Cracidae is supported by a number of morphological features and their common feather mites.

The present-day discontinuous distribution of Cracidae and Megapo-
9 diidae is an important proof of the wide distribution of the ancestors of this order. However, there is no basis for assuming that Australia was the center of origin of these birds. It is equally probable that the ancestors of Galliformes originated in South America and later spread to the Australian zoogeographic region. At the same time, the theory proposed by V.B.

Dubinin that the further development of Galliformes took place in two widely separated areas, resulting in two totally different families—the small groups of arboreal Cracidae and hoatzins in South America and the terrestrial megapodiids in Australia—does not raise any doubt. It is quite possible that the large superfamily Phasianoidea, which appears to be the ancestor of all the remaining heterodactylous Galliformes, branched off from the megapodiids of Southeast Asia.

Even now there is no unanimity of opinion regarding the extent of the order Galliformes and its subdivisions into higher taxa, let alone its relationship with other orders.

According to Fürbringer's (1888) classification adopted in most subsequent publications, the order Galliformes contains two suborders—Galli and Opisthocomi (one species). Brodkorb, however, considered hoatzins as one of the families of this order. At the other extreme, hoatzins were either separated from the order Galliformes (Hudson and Lanzillotti, 1964), or were separated and hemipods included in it (Verheyen, 1961).

Marked differences exist also in the subdivision of the lower-order taxa of Galli. This suborder is often subdivided into two superfamilies —homodactyls and heterodactyls. In the first group, two families—the Cracidae and Megapodiidae—are well distinguished. Opinions differ, however, regarding the subdivision of heterodactyls. Most Soviet ornithologists favor Fürbringer's division of the superfamily Phasianoidea into four families: Phasianidae, Meleagrididae, Numididae, and Tetraonidae. This view is also reflected in a textbook of avian systematics (Kartashev, 1974). However, some ornithologists consider Tetraonidae merely as a subfamily of Phasianidae (Verheyen, 1961; Short, 1967).

A taxonomic review of galliform birds goes beyond the scope of this book, although an ecomorphological study of the two groups of this order may help elucidate their taxonomic position. Even though Galliformes are distributed almost throughout the world, their species composition in different climatic zones and environments varies widely. The species diversity is most high in the tropics with 143 species (or 55%) confined to
10 this zone (species taxonomy from Peters, 1934). The temperate zone of the northern hemisphere comes second with 56 species (21.5%). Thirty-seven species (14.1%) are common to the tropics and the temperate zone of the northern hemisphere. In the southern hemisphere, where the land area is far less, the number of species is limited. Here, only 12 species (4.9%) are common to the tropics and the temperate zone, and 3 species (1.1%) are confined to the temperate zone only. Finally, two species (0.8%)—*Numida meleagris* and *Excalfactoria chinensis*—while mainly distributed in the tropics, are also found in the temperate zone of both hemispheres.

Most tropical species have small ranges and the species diversity is high. This leads to diverse faunal communities typical of small territories. Moving toward the poles, the species diversity decreases, while the number and, as a rule, the population of widespread species increase.

More than half the representatives of the order Galliformes are associated with diverse forest environments ranging from tropical forests to the temperate taiga; only cracids and hoatzins, however, which spend their entire life cycle including nesting on trees, are truly arboreal. All the remaining Galliformes either lead a terrestrial life style or are terrestrial-arboreal birds, always nesting on the ground (except for Tragopans and possibly *Francolinus nachani*)*.

The diversity of tropical Galliformes and the close association of the Cracidae (undoubtedly, primitive gallinaceous birds) with tropical forests supports the view that the ancestors of Galliformes were birds of tropical forests. In the Soviet Union, Galliformes are represented by only 21 species in two families—Phasianidae and Tetraonidae.

Phasianidae is the largest family of the order (175 species), and is found predominantly in the tropics and subtropics. In the past, this family was represented by a wide variety of forms in southern Europe. Skeletal remains of 17 species of eight genera (*Miophasianus, Taoperdix, Paleoperdix, Pliogallus, Miogallus, Proalector, Paleocryptonix,* and *Plioperdix*) are known from the Miocene of southern Europe. In the Pliocene, when climatic conditions deteriorated the species composition of these birds changed sharply: only *Pliogallus* survived, and Recent genera emerged—*Ammoperdix, Alectoris, Gallus,* and *Phasianus.* In the Pleistocene, such thermophilic forms as *Gallus* (two species), *Phasianus* (two species), and *Francolinus capeki* survived in southern Europe, despite prolonged cooling of the climate.

11 Only the northernmost, widespread species of this family are found in the avifauna of the USSR. The homeland of the genus *Phasianus* is southeast Asia. It is inhabited by a large number of predominantly forest-dwelling species of pheasants. Compared to the other genera, Common Pheasants occupy the northernmost and most extensive geographic range, including desert oases, cultivated lands, and mountain broad-leaved forests. The center of origin of the genus *Tetraogallus* was in the high mountains of Central Asia, from where snowcocks later spread along the mountain ranges of the southern Palearctic in the west up to the Caucasus (Kozlova, 1952). The origin of other phasianids found in Russian fauna are associated with arid tropics and subtropics. The single species found in

*[sic]; should be *F. nahani*—Editors.

Trans-Caucasia and Turkmenia, Black Francolins (*Francolinus francolinus*), is the northernmost representative of a polymorphic genus widely found in southern Africa. Desert regions of western Asia are the homeland of Sand Partridges (*Ammoperdix* spp.). In the Soviet Union, one species of this genus inhabits Turkmenia, Tadzhikistan, and Uzbekistan, where snowfall does not occur every year and is of very short duration.

The origin of the genus *Alectoris* is associated with the Mediterranean region; all the recent species discussed here are found within this region. The species *A. kakelik*, widely distributed in the north, penetrated the mountains of the Caucasus, Soviet Central Asia*, and Kazakhstan to Mongolian Altai and the Central Asian mountains. The northernmost phasianids in Eurasia are quail and Gray Partridges, which breed even near the polar circle, although the quail migrate in winter to Africa and southern Asia. Gray Partridges from the northernmost parts of the range also migrate southward in winter.

The Tetraonidae are a family endemic to the Holarctic with 18–21 widely distributed species. The origin of this ecologically well-differentiated group is not yet fully known. Without doubt, the Tetraonidae originated from very early phasianids. This is supported by the ontogeny of the color of their plumage (Stegmann, 1932), as well as the presence of hybrids of pheasants with some genera of tetraonids obtained through artificial breeding (Kloiber and Rokitansky, 1954; Gray, 1958). However, there is no unanimity of opinion as to the place and time of origin of tetraonids and which group of phasianids is closer to their ancestral form.

12 Well-defined ecomorphological adaptations to severe climatic conditions indicate that tetraonids separated from thermophilic phasianids as a result of climate cooling and change in vegetation.

According to American researchers (Mayr, 1946; Short, 1967), North America is the homeland of tetraonids. This conclusion is supported by the discovery of early fossil tetraonids in the USA and the presence of a group characteristic of the steppe. This group is considered by these scientists as the most primitive among the tetraonids.

B.K. Shtegman (1938, 1966) suggested that tetraonids originated in the mixed forests of western America with a diversity of conifers. Presumably this region, along with the area presently submerged under the Bering Strait, was the birthplace of most of the avifauna of the taiga.

*The Russian literature frequently refers to two distinct geographical localities "Tsentral'naya Aziya" (Central Asia) and "Srednaya Aziya" (loosely translated as Middle Asia). The former term refers to Sinkiang Province in China (or the former Chinese Turkestan), while the latter refers to the former Russian Turkestan (the Kirgiz, Uzbek, Turkmen and Tadzhik SSR). Since there is no geographic locality by name Middle Asia, we prefer to call it Soviet Central Asia—Editors.

S.I. Snigirevskii (1946) suggests that tetraonids may have originated simultaneously with the development of the Turgai flora, widespread in Europe, Asia, and North America as early as the Tertiary period. Snigirevskii cites as evidence the correlation between the geographic range of the Turgai flora and the present-day distribution of tetraonids. During this period, tetraonids could have separated from partridges of the tropical forests in the northern part of the eastern hemisphere by changing their winter food to catkin trees. In the later work, Snigirevskii (1950) mentions Gray Partridges as a close relative of tetraonids.

V.F. Gavrin (1969) and R.L. Potapov (1970) are of the view that the family Tetraonidae originated in East Asia. Potapov notes that the ancestral form of tetraonids is to be expected in the mountains of this region.

A clear picture of the changes in the vegetation of the northern hemisphere during the Tertiary period has been presented by A.N. Krishtofovich (1959). According to him, the greater part of this region was covered by a temperate type of flora in the Early Tertiary. In the Oligocene, Turgai flora prevailed in northeastern Asia, the northern part of Kazakhstan, and northwestern America. The Turgai flora that served as the forerunner of broad-leaved forests in the eastern part of Asia appeared in the Eocene. By this time, catkin (poplar, oriental beech, sweet chestnut, elm, birch, and alder) were gaining dominance. Throughout the Pliocene, the Turgai flora gradually expanded westward and southward.

Considering the wide range of Turgai flora in the Oligocene in
13 northeast Asia and North America and the Miocene age of the earliest finds of tetraonids, one may conclude that this family appeared in the Late Miocene. During this epoch, conditions were favorable for the evolution of birds feeding on catkin trees during winter. It should be noted that both species of Hazel Grouse and Ruffed Grouse, which mainly feed on catkin trees, are considered the most primitive among the forest tetraonids on the basis of totality of characters.

Separation of the other group of tetraonids, those which feed on conifer needles, was associated with the development of coniferous forests and the emergence of taiga as a zonal type of vegetation over a considerable part of the middle [temperate] latitudes.

Opinions differ as to the origin of taiga flora. In the earlier paleontological works it was thought that the focus of origin of taiga was the coniferous mountain forests of East Asia, which gradually extended northward as the Turgai flora became extinct (Lavrenko, 1938; Tikhomirov, 1946; Tolmachev, 1954). According to more recent views, the development of taiga was simultaneous in the middle and higher latitudes of the Holarctic (Krishtofovich, 1959; Dorofeev, 1970). A change in the Turgai flora through the extinction of broad-leaved species and the development of taiga vegetation also occurred in the mountains of middle latitudes and the plains of the north. During the Miocene, taiga forests appeared throughout eastern Siberia and, in the Pliocene, became the characteristic element of the landscape of western Siberia and central Europe.

In the Neogene of East Asia and North America, Turgai flora dominated by deciduous forms prevailed initially. This was replaced by taiga flora as many broad-leaved species were displaced by conifers. These types of forests represent the "cradle" of the two well-differentiated ecological groups of tetraonids—the more primitive one using broad-leaved plants as winter food, and the more recent foraging on needles of conifers.

The uniform vegetation along the Pacific coast, in the Pliocene and Miocene, and the continental bridge between East Asia and North America make it difficult to decide about the center of origin of the Tetraonidae. This can be resolved only with sufficient paleontological material, presently lacking for northeastern Asia and very scarce for North America.

Review of Galliform Birds of the Soviet Union (Distribution, Habitat Preference, and Behavior)

Family TETRAONIDAE

Genus *Lagopus* Brisson—Ptarmigans

14 Of all the genera of galliform birds, *Lagopus* is the most northerly and is represented by three species, namely *L. lagopus* L. (Willow Grouse) and *L. mutus* Mont. (Rock Ptarmigan), having circumpolar and circumboreal distribution in the Northern Hemisphere, and *L. leucurus* Rich. (White-tailed Ptarmigan) which is endemic to the mountain ranges of western part of North America. The history of the genus is associated with the onset of abrupt cooling in the Northern Hemisphere. Fossil remains of ptarmigans are known from the Pleistocene of southern and central Europe (Lambrecht, 1933; Tugarinov, 1940), where these birds were pushed back during the glacial period (Fig. 1).

Opinions differ about the place and time of origin of the genus *Lagopus*. Some believe that it originated in north-western parts of America or in the Bering Strait region (Shtegman, 1938; Short, 1967), while Tugarinov (1929) and Johansen (1956) hold to its formation in eastern Siberia. Further, Johansen associates the formation of this genus with the alpine and subalpine belts of mountain ranges, and Short with the development of forests. Tugarinov believes that ptarmigans originated in the Quaternary but Johansen assigns their origin to the end of the Pliocene.

Forest birds were most likely ancestors of both ptarmigans and capercaillies. Even now, ptarmigans inhabit not only the tundra and taiga, but also forest bogs. It is possible that boggy forests were the typical habitats of the ancestral form. Food use of catkin trees (willow and birch) also supports the long association of these birds with broad-leaved trees and shrubs. Considering that bogs and lichens of pine forests held an
15 important place in the development of the tundra biocoenoses (Gorodkov, 1938), it can be said that the ancestors of ptarmigans were true forest birds,

and which led to a terrestrial-arboreal mode of life. Further progress to forest-free sites forced these birds to make a full transition to terrestrial life.

White-tailed Ptarmigan, the highly specialized present-day alpine species, inhabits the western alpine ranges of North America where it is confined to rocks and bare peaks with patches of alpine meadows (Godfrey, 1966).

Willow Grouse, the most variable species, is widely distributed in the tundra and forest-tundra, and even inhabits arctic deserts at some places. Willow Grouse are common in forest bogs and subalpine shrub thickets of mountain ranges, and at places of intrazonal vegetation penetrate into forest-steppes and steppes. However, their narrow food specialization, most clearly expressed in winter, strictly confines these birds to willow and birch shrubs.

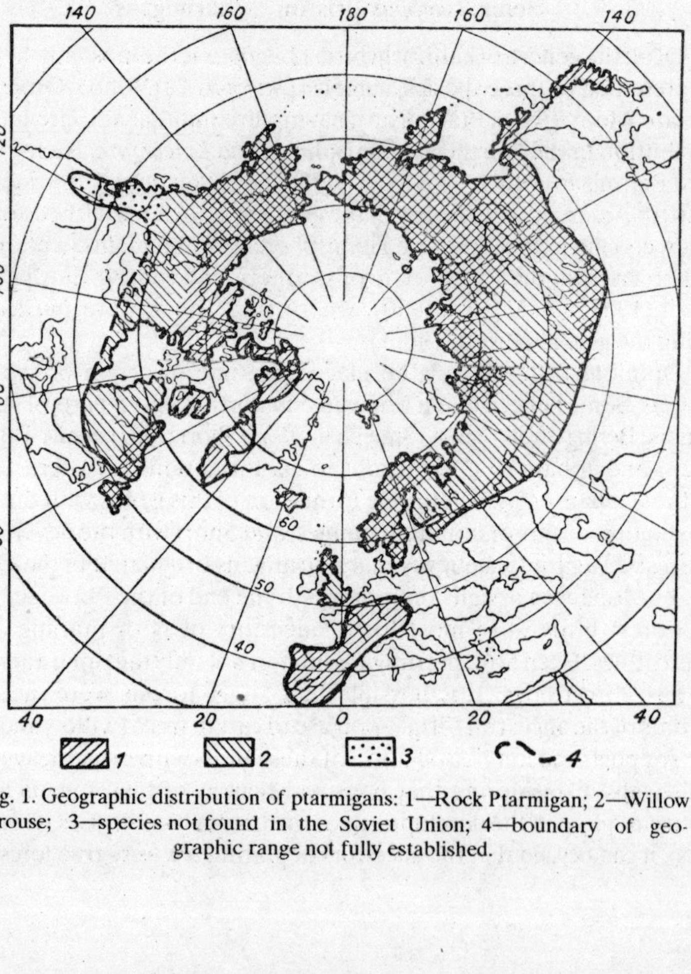

Fig. 1. Geographic distribution of ptarmigans: 1—Rock Ptarmigan; 2—Willow Grouse; 3—species not found in the Soviet Union; 4—boundary of geographic range not fully established.

On the tundra, the Willow Grouse is most abundant in hummocky, moderately moist areas with berry-bearing shrubs, dwarf birch trees, and shrubby brush, and distinctly avoids extensive areas covered with lichens and low, sparse grasses. It also inhabits sphagnum bogs with undergrowths of dwarf birch, marsh tea, sorrel, bog whortleberry, and marshwort.

In the forest belt, Willow Grouse inhabit moss bogs with bog whortleberry, bilberry, and marshwort, and in marshy river valleys; or it inhabits bogs overgrown with shrubs, or in patches of coniferous forests including birch. Its habitat differs particularly in the northern and southern boundaries of its range. On the Novosibirsk Islands, Siberian Willow Grouse inhabit arctic deserts devoid of shrubs, occupying the characteristic habitat of the now extinct Rock Ptarmigan (Kishchinskii, 1974).

In the southern part of its range, Steppe Willow Grouse inhabit the boreal relict vegetation of western Siberia and northern Kazakhstan. Birch logs, pine forests, moist aspen-willow or birch-willow groves ringed by northern willow, aspen, and birch are its principal habitats (Pavlov, 1948). Willow Grouse may also be found here in rocky pine forests with dwarf vegetation (Shtegman, 1934), in groves of dwarf birch with patches of sandy steppe complex, and on dune-like hillocks overgrown with spiraea and small-leaved birch (Livron, 1938; Ul'yanin, 1939). In the last ten years, extensive cultivation has led to a sharp reduction of Steppe Willow Grouse; where it has survived, it nests in grain fields.

Compared to the Willow Grouse, Rock Ptarmigan are a highly specialized species adapted to the severe conditions of flat and mountain tundra, forest-tundra, and zones of arctic desert. Its geographic distribution is more restricted and almost everywhere its density is considerably less than that of the Willow Grouse. Rock Ptarmigan inhabit mainly rocky tundra with sparse vegetation regardless of the altitude; it also lives in low, comparatively arid sections of rocky tundra in the northern parts of its
17 range and on plateaus of high mountain ranges, definitely avoiding sharply rugged terrain. In the southern parts of its range, Rock Ptarmigan are exclusively high-mountain birds, whereas in the far north it frequents both mountains and plains. The ranges of Willow Grouse and Rock Ptarmigan largely overlap but their habitats are usually distinct.

In Taimyr, Siberian Willow Grouse inhabit the northern part of the taiga in tundra with and without forests and its maximum numbers are found in shrubby tundra (Krechmar, 1966; Pavlov, 1974). Coastal arctic tundra is the primary habitat of Siberian Rock Ptarmigan. In moss-sedge and moss-cottongrass tussock tundras, it is much rarer than Willow Grouse (Birulya, 1907; Syroechkovskii, 1967; Pavlov, 1974). A similar division of the habitats of these ptarmigans is evident in the hilly regions. Rock Ptarmigan always live at higher altitudes, inhabiting predominantly mountainous or rocky arid tundra with berry bushes, while Willow Grouse are usually found in moister subalpine fields, and descend into valley tundras with willow and alder thickets.

In the mountainous southern parts of the range (the Sayan, Kuznetsk Alatai, and Altai Mountains, and the mountain ranges of northern Mongolia), the distribution of Altai Willow Grouse is closely associated with the subalpine belt, consisting of vast thickets of ground birch (*Betula rotundifolia*). Rock Ptarmigan in these ranges live higher, in rocky, much wetter, mountain tundra, where the grassy vegetation is represented by dwarf varieties of alpine grass, dryas, columbine, saxifrage, and buttercups (Sushkin, 1938; Irisov and Totunov, 1970). A clear-cut demarcation, however, between the habitats of these two species is seen only in regions where their ranges overlap. At places where only one of them is found, its habitat becomes more diverse and Willow Grouse is seen in habitats typical of Rock Ptarmigan, and vice versa. One example is the habitat of Willow Grouse in the Novosibirsk Islands and the mountains of eastern Siberia where Rock Ptarmigans are absent. Here, Willow Grouse mainly inhabit bald peaks with sparse growths of mountain pine, lichens, mosses, and low grasses (Afanas'ev, 1934; Pavlov, 1948; Izmailov, 1967). Habitats of Rock Ptarmigan are quite diverse on the Komandorskii Islands where Willow Grouse are absent. Rock Ptarmigans occupy tall grass meadows on
18 lower slopes, and in valleys of rivers and brooks upstream. They also inhabit sparsely vegetated peaks, and bogs on plains and wet areas (Andreev, 1971), utilizing the typical habitats of both species. The great diversity of the habitats of Willow Grouse and Rock Ptarmigan in these cases can be explained by the absence of interspecific competition. In such conditions, even the highly specialized Rock Ptarmigan adapts to more diverse biotopes, but is limited when both species coexist.

Differences in the biotopic distribution of these ptarmigans are observed not only in the snow-free period but also in winter; Rock Ptarmigan occupy less snowy regions while Willow Grouse usually inhabit snow-covered areas. However, during migration, these birds may be seen together, sometimes even forming mixed flocks.

In winter, Willow Grouse move in large flocks, sometimes hundreds of birds, and migrates from the open tundra on islands close to the mainland to the forest-tundra and into the river valleys of the forest zone rich in willow thickets. Food availability is restricted to a small area for the winter life of ptarmigans (Mikheev, 1948; Voronin, 1969); hence, migration could be triggered by food shortage. In some years, mass migrations of Willow Grouse have occurred. For example, in 1954, mass migration in eastern Siberia involved an area of 2,500,000 km^2, but the majority of these migrants perished (Gagina, 1958).

Both species of ptarmigans are more active in winter than the forest tetraonids. They spend considerable time and energy foraging and rest very little. They do, however, exhibit some differences. In Taimyr, Willow Grouse feed five–six times a day in winter (Pavlov, 1971) during the day and at twilight; in the Kolyma Basin, Rock Ptarmigans feed only in mornings and evenings, resting for three to four hours during the day in open holes in the snow or under snow (Andreev, 1975). Rock Ptarmigan are often found at places with uneven snow cover, and prefer exposed, wind-blown rocks where they forage and rest in the nearby, snow-filled depression. Willow Grouse frequent the snow-covered areas where it feeds on twigs of shrubs.

According to Andreev, in the Kolyma Basin, Rock Ptarmigans inhabit
19 the upper valleys where the atmospheric temperature is 5 to 10°C higher than that in the habitats of *L. lagopus.* Snow depths may vary from 5 to 10 cm. Willow Grouse frequent holes in the snow cover more often than do Rock Ptarmigans (Uspenskii, 1969).

It was observed that in the Lapland National Park, "... in the windy season, Rock Ptarmigan always sit with their head facing the wind, in the position of best air flow around the body. In a blizzard, they always sit near rocks; if it sits on the windward side it would soon be covered with snow, as in a show hole" (Semenov-Tyan-Shanskii, 1959).

Willow Grouse move very fast on the ground, their speed is comparable to the average trot of a horse. Running undisturbed, this species takes long strides with its neck out-stretched, somewhat bent toward the ground, back hunched, and tail down. Alarmed birds usually run with their tails raised and somewhat spread out, shaking them like a hen hatching chicks (Sushkin, 1938).

The mobility of Rock Ptarmigans while foraging can be judged from the report of Kapitonov (1962). He observed flocks of birds running ahead of harnessed reindeer and breaking willow buds while running. Moving on the ground, they often gather on small hills or other high ground and, stretching up almost vertically, survey their surroundings. Sensing danger, they run away or hide but rarely fly. Apparently, with their protective coloration the birds feel safer on the ground than in air.

Willow Grouse and Rock Ptarmigans fly much more than other tetraonids. Willow Grouse take off from the ground with much noise and, after rising 3–4 m, gain speed. During flight, they alternate rapid wing beats with smooth noiseless gliding. After 300 to 500 paces, they often rise steeply and then dive to the ground. Such a flight is characteristic of the male during courtship, when the aerial display terminates a steep climb followed by a downward swoop accompanied by loud calls. The rest of the year, they fly only occasionally (except during migration) and silently. The

main differences between the flights of Willow Grouse and Rock Ptarmigans are the great speed, ease, duration, and quicker wing beats of the latter. According to Bent (1932), the faster flight of Rock Ptarmigans more resembles that of a falcon than a ptarmigan. The courtship flight is particularly varied and males display on the ground, in the air, and, at
20 times, even perched on trees. The display (mating) flight may extend sometimes for 400 to 500 m (Kapitonov, 1962). The nature of the flight may vary sharply. Fast, straight flight with maneuvered landing, or quick alternation of soaring and diving, when the bird rises from the ground flapping its wings, gathers height and speed, then soars sharply upward with steady wings followed by a nearly vertical landing. Sometimes, these maneuvers proceed in a sequence, one after the other (Andreev, 1971).

In the autumn–winter period, Willow Grouse sometimes migrate, later appearing 200–300 km away from their nesting sites. Their range of flight without rest, however, is rather limited.

Seasonal migrations of Willow Grouse between the Novosibirsk Islands and the mainland (120 km) and between Kolguev Island and the mainland (80 km) are not to be considered as the range of their nonstop flights. There is evidence (Birulya, 1907) that they depart from the Kotel'ny Islands (Anzhu Island) only when the straits are covered with thick ice on which the migrants can rest. Rock Ptarmigans can fly without rest for longer distances, as confirmed by their flight across unfrozen straits from islands on the coast of North America.

Of these two closely related species of ptarmigans, Rock Ptarmigans are more highly specialized, adapted to less snowy regions and to mainly dry, rocky tundra.

Genus *Tetrao* Linnaeus—Capercaillies

This genus includes two species, *Tetrao urogallus* L. (Western Capercaillie) and *T. parvirostris* (Bonap.) (Black-billed Capercaillie), which are distributed in the boreal forests of Eurasia (Fig. 2).

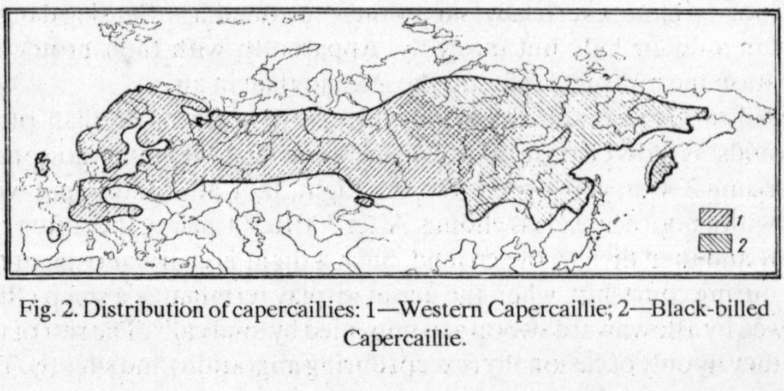

21 Fig. 2. Distribution of capercaillies: 1—Western Capercaillie; 2—Black-billed Capercaillie.

20 Based on present-day distribution, habitats in specific types of coniferous forests, coloration, and some morphological features, Snigirevskii (1946) associated the origin of this genus with the Pliocene, when much of northern Eurasia and North America was still covered by Turgai flora. According to this view, until early Quaternary glaciation, capercaillies probably inhabited the entire northern Palearctic, from Great Britain to Kamchatka. The geographic range of these birds was reduced to isolated patches with the onset of glaciation. As the glacial front receded, these
21 birds once again spread out to form a continuous range. However, the prolonged isolation in two geographic pockets led to the formation of two species of capercaillies well-differentiated in colour, morphological features, and behavior (courtship display).

Study of courtship calls, colour of plumage of adult birds, molting, and relative length of the tail revealed *T. urogallus uralensis* as the most primitive form of this species (Johansen, 1957; Potapov, 1971).

Capercaillies are phylogenetically closest to the genus *Lyrurus*. In his description of the new species, Linnaeus combined Black Grouse, Capercaillies, and Willow Grouse into a single genus, *Tetrao*; only much later Swainson separated Black Grouse into an independent genus. There is no doubt about the phylogenetic proximity of these genera, and is confirmed by commonly encountered hybrids. Recently, Short (1967), while revising the genera, reverted to the old concept and combined *Tetrao* and *Lyrurus* into a single genus. Though the Capercaillie and Black Grouse often live together in the same forest, their ecological differences are quite distinct, particularly in winter when they take to different habitats, and their food is sharply different. It could thus be assumed that although *Tetrao* and *Lyrurus* originated from extremely proximate ancestral forms, they evolved under different conditions—Capercaillies in coniferous forests, and Black Grouse in deciduous forests and forest-steppes—resulting in fairly well-differentiated groups that can rightly be considered as independent genera.

Capercaillies stand closest to the taiga genus of Spruce Grouse
22 (*Falcipennis*) in some ecological features, namely, similarity of courtship behavior and adaptation to coniferous forests that provide them winter food.

At present, the ranges of Western and Black-billed Capercaillies overlap in a large part of Trans-Yenisey Siberia, creating conditions for the frequent appearance of hybrids known as the Dark-gray Capercaillies. The fertility of at last the hybrid females is confirmed (Kirpichev 1961, 1965), which emphasizes the phylogenetic affinity of these well-differentiated species.

The distribution of capercaillies is closely associated with various types of pine and cedar forests. Bogdanov (1874) briefly, but accurately, described its typical habitat: "Old, shady pine forests with damp soil overgrown with ferns, young woody underbrush, and sphagnum and peat bogs—is the true habitat of capercaillies." The author has given a very accurate general picture of their habitats, but in terms of the composition of vegetation, terrain, and other conditions these could describe distinctly different forest types*. It has been noted that, being strictly resident birds, capercaillies inhabit rather diverse forests where, in changing biotopes in different seasons, they can find both food and safety. In winter, these birds forage on trees and prefer open forest, while, in summer, on the other hand, they take to denser areas of the forest with abundant fruit bushes. In forest types having the favorable nesting and wintering habitats widely separated, Capercaillies resort to seasonal or short migrations.

In most regions of the European part of the USSR, distribution of Western Capercaillies is associated with different types of pine forests. Marshy, thin pine forests bordering swamps are their favorite haunts in winter, as well as serving as nesting areas.

From summer to autumn, broods and single males are often found in forest sections offering good protection and an abundance of fruit bushes. These may be pine-spruce patches with shrubs and underbrush or pure spruce and birch groves with tall grass cover. Confinement to pine forests for the greater part of the year is explained by the fact that in this part of the range pine needles are the primary food item of these birds throughout the snow season. In summer and autumn, the denser sections with
23 abundant fruit bushes provide the broods and molting birds with both safety and food. In the Urals and western Siberia, they inhabit pine as well as cedar forests with glades, marshes, clearings, and recovering burned-out patches (Danilov, 1975).

In the mountain regions of the eastern part of their territory, cedar forests play an important role in the life of Western Capercaillies. Here, cedar needles and cones are as important in their winter food as pine needles in pine forests. In regions where cedar forests are small (e.g. south-western Altai), the nesting and wintering sites are often quite separated and birds resort to seasonal migrations extending 20 to 50 km. They winter in cedar forests, sometimes up to an altitude of 2,200 m, and descend in the breeding season to spruce-fir and larch forests mixed with aspen, birch, and other deciduous species. The tall grass cover in such forests affords safety for the brood and berries provide abundant food. Western Capercaillies are found less often in deciduous forests.

* *Forest massif* in original—Editors.

In the southern extremity of the Ural Mountains, these birds inhabit not only the mountain pine-birch-larch forests, but also in isolated broad-leaved forests that are less suitable for winter life. Hence, seasonal migrations are more pronounced in this part of the range. In winter, only solitary birds remain in deciduous forests, and that only in areas where there are occasional pine trees (Kirikov, 1952, 1975). If coniferous forests are destroyed over large areas, capercaillies sometimes move to nearby deciduous forests and settle, an action which reflects their winter diet.

The distribution of Black-billed Capercaillies over most of eastern Siberia and on Sakhalin Island is associated with plains and mountain-larch taiga. More often, this species inhabits sparse larch forests bordering broad floodplain terraces and water divides alternating with extensive dwarf birch and fire-damaged areas.

Larch is as important to Black-billed Capercaillies as pine or cedar is to Western Capercaillies. Hence, their adaptation to vast tracts of larch taiga is wholly understandable. In some regions, this species inhabits not only the larch forests but, as in the northern regions of Buryat and in Irkutsk, also cedar and pine forests (Kirpichev, 1960).

In southwestern Kentei (eastern Mongolia) where larch and pine cover 60–80 km at a stretch, these birds inhabit pine forests, but prefer
24 cedar to larch in the eastern part of the mountains (Kozlova, 1932). Kamchatka Black-billed Capercaillie inhabit both birch and coniferous forests. Along the eastern coast of Kronotskii Bay, this species is found only in deciduous forests where the primary species is Erman's Birch and the undergrowth consists of dwarf mountain ash, honeysuckle, and dense grass (Averin, 1948; Markov, 1968).

Though the natural history and behavior of Western and Black-billed Capercaillies have much in common, there are fairly distinct differences too. Both species are ground dwellers in the snow-free period of the year, but in winter and breeding season they become terrestrial-arboreal. For much of the year, they remain in flocks with males and females separate. Flocks may comprise from only a few to 20–30 birds.

In winter, Western Capercaillies are less active; feeding and short rests on tree branches alternate with prolonged rest on the snow while food is digested. They usually feed twice a day but during extremely severe frosts they leave their snow holes only once. For the most part, they remain in very limited areas.

Western Capercaillies rarely walk over snow. Only toward spring, when the snow crust is well formed, do these birds venture out on the snow surface. Black-billed Capercaillies are far more active during winter than Capercaillies and walk over the snow to nibble at shoots of young larch. Winter in eastern Siberia is particularly severe, but is less snowy. In the early winter months, when there is very little snow and the temperature falls to −40°C to −50°C Black-billed Capercaillies, unable to dig in the snow, sit immobile on low larch trees, forsaking its natural caution.

The free movement of this bird over snow is explained both by its less snowy habitat as well as its light weight. The toes of Black-billed Capercaillies are better feathered than those of Western Capercaillies; these feathers not only keep the toes warm, but also enlarge markedly the supporting area of their feet.

Black-billed Capercaillies do not always dig fresh holes in the snow. In Kentei, where there was little snow, Kozlova (1930) discovered holes with large quantities of excreta, and far more than could have been deposited in a single night.

Western Capercaillies move quite fast on the ground, though not as fast as Ptarmigan or Hazel Grouse. The gait of a walking bird is often clumsy because of its large size and low-slung body. While walking, the torso is almost horizontal with only its front part slightly raised and the neck stretched. In hilly regions, birds walk uphill, on slopes, or along them, but invariably take to the wing when descending. It is interesting to note that Western Capercaillies always bypass obstructions like felled trees, stumps, and dead wood, and never venture to climb them. This points to some restriction in their movement, which is exploited by Altai hunters to trap them.

Black-billed Capercaillies are distinctly more active and spend more time on land. On trees they obviously feel more comfortable, though their movements are restricted to hopping. Feeding and resting birds usually perch halfway up the tree, but in autumn they are often seen on tree tops. Their low-slung body, which somewhat restricts its movement on ground, helps these heavy birds maintain a steady perch on tree branches. It is further aided by their long, broad tail that ensures its balance while moving among the branches, and in their resting position (Teplov, 1947; Efremov, 1968).

Perching on a tree, which provides a wide field of vision, this alert bird can spot danger at a distance, hear any suspicious rustling, and take off quickly when needed. In summer, living on the ground, Western Capercaillies try to keep to very dense forest areas where they can hide in case of danger and take to flight only when absolutely necessary.

Western Capercaillies usually fly a little and do not rise above the forest canopy. Vertical take-off is very noisy and the airborne bird either perches on the nearest tree or with skillful maneuvering glides to another part of the forest. Flight involves rapid wing beats alternating with gliding and is often almost silent. They fly more often during the breeding season, especially in autumn. According to Semenov-Tyan-Shanskii (1959): "Single capercaillies often fly in autumn across lakes a kilometer or more wide and sometimes over hilly tundra at such a height that it is difficult to identify them, especially when they fly on a steady wing." Such autumn migrations of Western Capercaillies help disperse these birds, which otherwise live in restricted areas. There are no perceptible differences in the flight of Black-billed Capercaillies.

Behavioral differences between the two species of capercaillies are particularly distinct during courtship. The Western Capercaillie spends the first part of its breeding season on trees, descending later to the
26 ground, where the males chase each other and fight fiercely. Black-billed Capercaillie display only on the ground (Dorogostaiskii, 1913; Kozlova, 1930; Kirpichev, 1965). The male, with his tail raised high and wings dropped, struts about rapidly in small circles shaking his entire body. Singing, he leaps up to a height of 1–2 m with a loud rustling of the wings, and then descends as if on a parachute, unfurling his tail and dropping the wings. Breeding males often engage in "tournament bouts" in which the rivals do not advance, but instead fall back and whirl about each other; violent fights have not been observed in these bouts.

The reason for the adaptation of Black-billed and Western Capercaillies to different types of coniferous forests and even their habitation in the deciduous forests (e.g., Black-billed Capercaillies in Kamchatka) must be sought in the conditions of evolution of these species. I feel that the ancestral form of capercaillies, which inhabited in past the eastern part of the range of this genus, was presumably less specialized than the recent species. The diverse winter foods of Black-billed Capercaillies characterize them as more adaptable species of the genus, apparently closer to the ancestral form. In different parts of its habitat, Black-billed Capercaillies adapted to life in various kinds of forests. It is probably a very young, more arboreal and highly specialized species. This view is supported, to some extent, by a study of the courtship song, plumage, and tail length of subspecies of the Capercaillie, which show that the young males of the eastern subspecies, *T. urogallus uralensis,* should be regarded as closest to the ancestral form (Potapov, 1971).

Genus *Lyrurus* Swainson—Black Grouse

This genus comprises two species—*Lyrurus tetrix* L. (Black Grouse) and *L. mlokosiewiczi* Tacz. (Caucasian Black Grouse).

Black Grouse are without doubt the closest relatives of Capercaillies and probably originated from a common ancestor; however, they evolved not in the coniferous but sparse, deciduous forests. The "cradle habitat" of this genus, unlike that of capercaillies, was apparently the deciduous forests with Turgai flora, which were widely prevalent in the Pliocene of eastern Siberia and Kazakhstan, and could have been the homeland of Black Grouse. A detailed systematic review of Black Grouse (*Lyrurus*) by Stegmann (1932) revealed that the more primitive forms (judging from the color of adult males and plumage stages) inhabit the eastern part of its geographic range, in which lies the center of origin of this genus (Fig. 3).

27 At the beginning of the glacial period, Black Grouse probably inhabited the southern part of its present-day range and were more widespread there. Many skeletal remains of this species were discovered in Punic era funeral sites on Malta (Lambrecht, 1933); in Odessa District skeletal remains were discovered dating back to the beginning of our era,* while those discovered in the lower reaches of the Amu-Darya River belong to the seventh—eighth centuries (Burchak-Abramovich and Tsalkin, 1971).

Over a century ago, Bogdanov (1868), in his work on the biology of the Black Grouse in the European part of the range showed convincingly that the original habitats of this species were open country, small deciduous forests, and steppes overgrown with small shrubs and tall grass. This species appeared in dark pine forests only when large forests were destroyed by felling and burning. Subsequent study of its extensive distribution has confirmed this viewpoint, i.e., in all the ecogeographic zones, Black Grouse inhabit only places where open environments alternate with forest areas or brushwood.

Colonization of the forest zone, which is observed even now, was forced by expanding agriculture in the forest-steppe zone and the reduction of the birch groves (Menzbir, 1900). This phenomenon was aided by large-scale felling of northern forests and cultivation of cleared land.

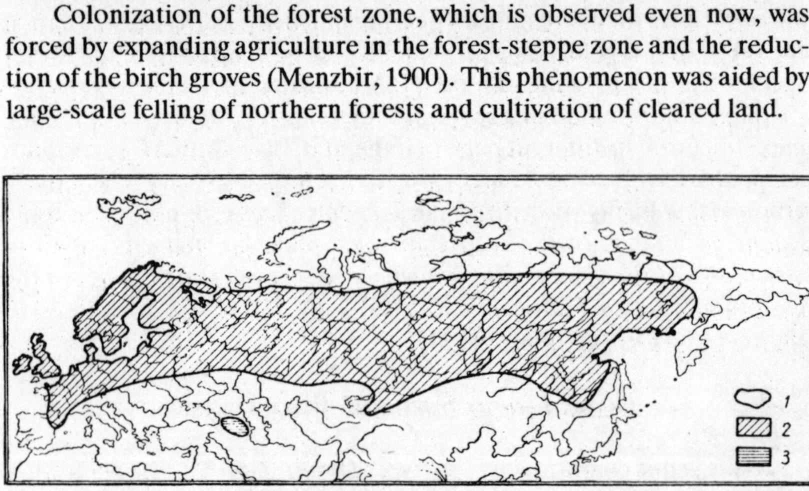

Fig. 3. Geographic range of Black Grouse (*Lyrurus*): 1—boundary of distribution; 2—Black Grouse; 3—Caucasian Black Grouse.

*2000 years before present—Editors.

The exceptional ecological adaptability of Black Grouse enables it to utilize diverse environmental and climatic zones. It inhabits both forest-steppe and forest zones, and penetrates along open areas almost up to the northern border of forest zones. It inhabits not only the brushwood steppe
28 zone but even semideserts. This species also inhabits both plains and mountains, but is confined invariably to brushwoods or forests with open areas.

Humidity, apparently, is only of secondary importance to the Black Grouse, since it inhabits marshy places among taiga as well as brush-steppes in the south and brushes and insular forests in sandy areas.

The habitats of North European Black Grouse are varied. It lives in both plains and mountain ranges; in the Carpathians, it lives in a zone of dwarf Swiss mountain pine, speckled alder, and common juniper thickets (Strautman, 1963). In the plains of the European part of the USSR, North European Black Grouse inhabit sparse deciduous and mixed forests. A necessary condition during summer, however, is the availability of fruit bushes and reliable protection for the brood. An even close relationship between the distribution of the nominal species of Black Grouse and birch groves has been observed in the Urals in the Pechora-Ilychsk Sanctuary (Teplov, 1947) and in central Yakutia (Larionov, 1962).

Baikal Black Grouse live in similar conditions, preferring sparse forests of birch, pine, alder, aspen, and larch, and particularly areas with small knolls between creek valleys and brooks (Fetisov, 1934).

Ussuri Black Grouse experience much sharper seasonal changes of biotopes than do the other subspecies. In the snow-free period, Ussuri Black Grouse are confined to birch groves, oak copses, marshes and wet hummocky meadows. In winter, these birds gather on islands and banks of large rivers overgrown with willow shrubs and feed on their buds. In other regions, they winter in fields along the edge of taiga and in mountains (Shul'pin, 1936).

Steppe Black Grouse inhabit the steppe regions of the European part of the USSR, forest-steppes, and steppe regions of Kazakhstan, western Siberia, Altai foothills, and, at places, semidesert areas. Its unusually high population density in forest-steppe and steppe zones of the recent past suggests that this terrain offers optimal conditions for the existence of this species.

In the forest-steppe, Black Grouse inhabit predominantly sparse birch forests (located often in marshy soil), interspersed with cultivated fields and meadows. Though more widespread in the plains, it does not avoid hilly forests. In the steppes, Black Grouse prefer sparse island forests, fallen birch or aspen in the steppes, sand or pine groves on granitic
29 soils. In the mountain ranges, this species lives on slopes with thick

overgrowths of pea-shrubs, dog rose, and chokecherry, but only where these are interspersed with mixed-grass glades.

In semideserts, Black Grouse penetrate arboreal and shrubby intrazonal vegetation, and colonize river flood-plains, mountain ranges with distinct vegetative zones, and, only as an exception, desert oases.

In the Ural River valley, this species inhabits flood-plain forests of black poplar, white poplar and elm with dog rose and black thorn undergrowth, and in places where forests alternate with floodplain meadows (Dubinin, 1953). On the Tusumsk sands (lower reaches of Turgai River), Black Grouse inhabit oases composed of tall Russian wild olive and willow, and in particular places with thick underbrush entangled with bindweed. Junipers grow here in small numbers (Sushkin, 1909).

In vegetation of the highland Kazakhian small-leaf forests that are populated by steppe Black Grouse, their characteristically distinctive northern coloration contrasts sharply with that of the surrounding semidesert. The mountain slopes in this region are often overgrown with common juniper creepers; sparse aspen and birch groves, spiraea, and dog rose cover large areas. Sparse pine groves are not uncommon. Bottomland, deciduous forests of willow, chokecherry, aspen, birch, and black currant are found along mountain river valleys. Thus, in the highland regions of the semiarid zone, the living conditions for Black Grouse differ little from those in other mountain regions, not only of the steppes but also of the forest zone.

The southern part of the geographic range of the species, i.e., the Altai, Saur, Tarbagatai, Dzhungarian Alatau mountain ranges, and the northern ranges of Tien-Shan, are occupied by the ecologically distinct hill subspecies, Tien-Shan Black Grouse. It inhabits mainly the forest-meadow-steppe belt of mountain ranges containing diverse types of trees and shrubs.

In the northern Tien-Shan ranges, this subspecies lives mostly in the upper part of the spruce zone and juniper groves to 2,800 m and, at places, even to 3,000 m. It is rarely found in the deciduous forest zone.

In the Dzhungarian Alatau range in summer, the birds keep to deciduous forests, i.e., apple groves with hawthorn, thick overgrowths of dog rose, and birch-aspen groves. In winter, the birds are confined mainly to the upper boundary of spruce and juniper groves.

In the Tarbagatai range, Tien-Shan Black Grouse inhabit a characteristic zone of pea shrub and spiraea thickets, scattered aspen groves, some wild apple, and patches of creeping juniper. In the snow-free period, these birds keep close to cultivated areas where shrubs are not as dense and
30 strawberries are plentiful. After harvest, they feed in clearings (Bibikov and Korelov, 1961). In south-western and southern Altai, these birds live

in mixed forests where deciduous species predominate. Deciduous forests and especially different types of birch, represent the principal habitats of the Black Grouse over much of the range. Apparently, the species originated in these areas, a fact supported by its food associations, mainly in the winter.

The isolated habitat of Black Grouse in the mountain ranges of the Caucasus led to the evolution of a separate mountain species, Caucasian Black Grouse, which nevertheless preserved the main ecological characteristics typical of this genus in association with open country (although alpine) and adaptation to deciduous vegetation in winter. This species inhabits the upper part of the forest zone, subalpine meadows overgrown with Caucasian rhododendron (*Rhododendron caucasicum*), and alpine meadows. The range of its altitudinal distribution is 1,200 to 2,600 m. Habitats of Caucasian Blackcocks differ little from those of Black Grouse, in particular the subspecies in the Carpathians, northern part of Tien-Shan, and Dzhungarian Alatau mountain ranges.

For the greater part of the year, Black Grouse live in flocks, males often segregated from females; mixed groups are rare and their number, depending on the population density in a given area, may vary from a few to 300–500 birds. In the autumn-winter period, they migrate considerably in search of areas with abundant food, avidly visiting harvested fields to consume the fallen grain. During this period, Black Grouse can be sighted far beyond their nesting sites. Winter migrations are more prominent in Ussuri Black Grouse. In breeding sites in Primor'e*, they are very rare or sporadic, but in winter they appear in flocks large enough for hunting in the willow groves along the lower reaches of the Amur and Sungari rivers (Shul'pin, 1936). During winter, these Grouse spend much of the day in snow burrows, but during brief feeding periods these birds are far more active than are Capercaillies. This is because of their lighter weight. Black Grouse dig into snow, dropping onto it from trees; rarely do they dig while walking.

In very severe frosts, Black Grouse become much more inactive. During active periods, birds usually keep to trees where they forage and rest. Only in regions where their chief winter food are berries of juniper or some other low vegetation, do they come out on the snow, nibbling at the berries in thickets or foraging on the ground. In the less snowy northern
31 Kazakhstan (Naurzumsk pine forest), they prefer to feed in the steppe clearings, flying there in the morning and returning to the birch groves at night, where the ground is thickly covered with snow (Ul'yanin, 1949).

*Maritime Provinces, Far-Eastern Siberia—Editors.

The winter life of Black Grouse in the mountains of northern Mongolia is somewhat unusual. Autumn breeding of the Black Grouse in many areas of the range is well known; in Kentei, however, breeding commences from mid-October and continues into late January. The birds remain very active during winter and their diurnal behavior changes completely. During this period they rest in snow burrows only at night. They spend the morning hours on the snow-covered lek and within two or three days, narrow, tampled pathways become visible, along which the males run and hop with springtime exuberance. They leave the lek around nine in the morning, fly to the fields and then to the forest, and forage until evening. Before sunset, these birds return to the trees near the lek and reach their roosts by dusk (Kozlova, 1930). Because of the protracted autumn breeding at these sites, the period of low activity of the Black Grouse is only 1.5 to 2 months.

In their behavior, Black Grouse fall midway between the terrestrial ptarmigans and the largely arboreal species like Hazel Grouse, Spruce Grouse and capercaillies. Unlike the ptarmigans, Black Grouse do not attempt to escape by running when chased, but almost always take to the air. Apparently, they are inferior to ptarmigans in speed and endurance. Considering that Black Grouse predominantly roost on the ground and not on trees (a characteristic of other forest tetraonids even during the snowfree period), breed on the ground, and lead an exclusively terrestrial mode of life in the open, its poor adaptation to an arboreal life style, compared to other forest tetraonids, becomes quite clear.

Black Grouse are mobile birds. In spite of a low-slung body, it runs very fast by stretching the neck forward and slightly inclining the hind part of the body. During the snow-free period it rarely perches on trees; only in late autumn does it forage on trees, but until the ground is completely covered with snow it prefers the berries of herbaceous plants. In winter flocks of Black Grouse are often seen on trees where they feed and rest for short spells.

32 Compared to other galliform birds of our avifauna, Black Grouse are a much better flier and can cover longer distances, and are inferior only to quail and ptarmigan in this respect. They take to flight with a loud noise but with apparent ease.

The fast and straight, less-maneuverable flight of Northern Black Grouse is accompanied by frequent noisy wing beats. In the autumn-winter period, flocks of Black Grouse and Western Capercaillies fly well above the forest and often migrate to distant places. Northern Black Grouse, flying at a greater height, differ from Western Capercaillies only in their swifter flight (Semenov-Tyan-Shanskii, 1959). Regular flights from roosting sites to feeding grounds and back, sometimes as much as 30 km each way, have been observed in northern Kazakhstan (Ul'yanin, 1949).

Caucasian Black Grouse move rather fast on the ground, and walk only up slope but always descend by flying. When chased, males often run away and rarely fly with a characteristic flapping of wings. Unlike males, flushed females often fly away to perch on the crowns of trees (Noska et al., 1895). Caucasian Black Grouse fly for rapid descent or when migrating over long distances to other valleys. These birds usually fly in a smooth arc, which reduces its speed to some extent. Flushed birds may fly a distance of up to 5 km at a stretch and, when chased by hawks, "rapidly gain height, trying to fly over the predator" (Averin, 1938). All of this shows that Caucasian Black Grouse are endowed with high flying qualities, although while descending, it glides passively.

Unlike the well-known display flights of Black Grouse, courtship among Caucasian Black Grouse is a very sedate affair. The courting male simply arches its head, raises its tail without spreading, periodically leaps upward, and flaps his wings. Courtship is silent and only rarely do birds trill with the beak. Fights between courting males have never been observed (Lorents, 1884; Averin, 1938).

The two species of Black Grouse are quite similar. Distinct differences in plumage, shape of tail, and the ritual of courtship display are the result of the complete isolation of the Caucasian Black Grouse. On the other hand, with respect to habitat preference, Caucasian Black Grouse do not differ any more than do different subspecies of Black Grouse (especially the mountain Tien-Shan Black Grouse) compared to the other species of the plains.

Genus *Falcipennis* Elliot—Spruce Grouse

33 Opinions differ regarding the size of this genus. Until the end of the 1930s it was considered monotypic, consisting of only the species *Falcipennis falcipennis* Hartlaub (Siberian Spruce Grouse). The two other related species of North American Spruce Grouse were placed in an independent genus, *Canachites* Stejneger. Considering the small differences between the two, Yamashina (1939) proposed that North American and Siberian Spruce Grouse be combined into a single genus. Stepanyan (1962) shared this view but suggested that the generic name *Falcipennis* be retained and *Canachites* be treated as its synonym. Short went further by combining Spruce Grouse with the genus *Dendragapus*, and retaining the latter name for the enlarged genus. In my opinion, it is most convincing to combine the Asiatic species and the two North American species of Spruce Grouse into one genus. Opinions differ also on the relationship of Spruce Grouse with

other tetraonids. Short considers the genus *Dendragapus* (together with Spruce Grouse) as the main trunk, of which *Tetrao* (together with *Lyrurus*), *Timpanuchus**, *Centrocercus,* and Bonasa are the divergent branches. Potapov (1970) considered the genus *Falcipennis* as the most primitive group of present-day tetraonids (with the exception of the Hazel Grouse and Ruffed Grouse), since it, apparently, is the closest relative of the ancestral form from which the majority of the tetraonid genera originated.

The genus *Falcipennis* is found on every continent of the Old and New Worlds. There is no doubt that in the past the ancestral form of this genus inhabited Beringia, eastern part of the Asian continent, and western part of North America, and probably was the connecting link with the other conifer feeders such as *Tetrao* and *Dendragapus* (Fig. 4).

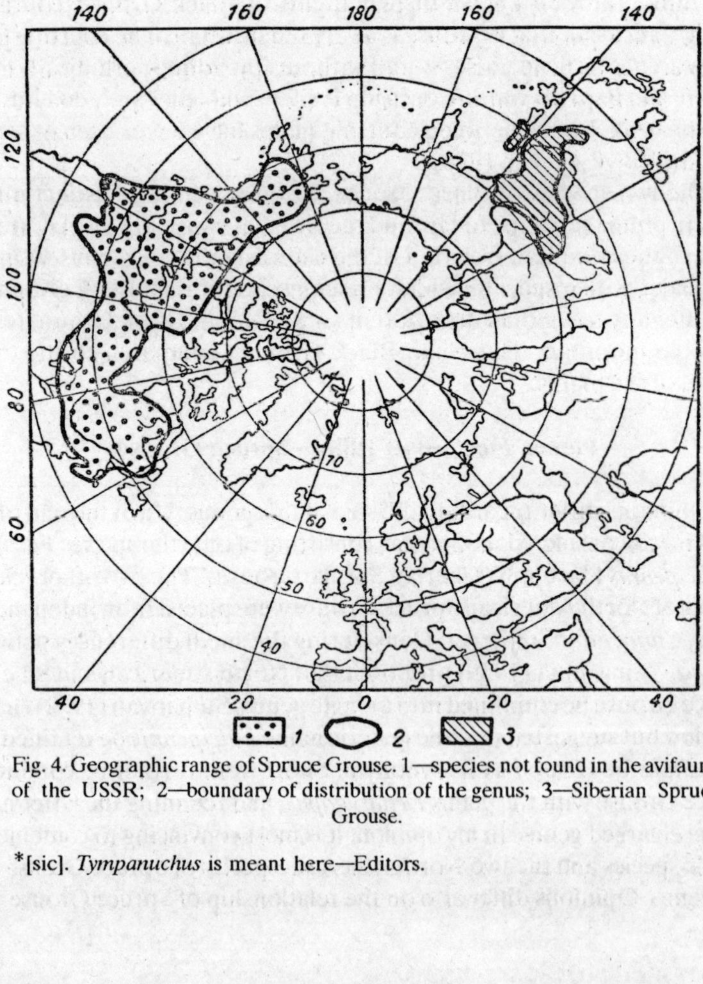

Fig. 4. Geographic range of Spruce Grouse. 1—species not found in the avifauna of the USSR; 2—boundary of distribution of the genus; 3—Siberian Spruce Grouse.

*[sic]. *Tympanuchus* is meant here—Editors.

Spruce Grouse are typical representatives of the taiga fauna, and its distribution is closely associated with several types of coniferous forests which provide food. The distribution of Asiatic Spruce Grouse is confined to boreal forests where the dominant species are Yeddo spruce and Manchurian fir. During the snow-free period, its habitats are more varied; but in winter, the Spruce Grouse is not found outside fir and spruce forests.

Shady areas in forests with meadows, thick sphagnum moss covers, and places of abundant berries of bog whortleberry, crowberry, cloudberry, and wild currant form the typical habitat of Spruce Grouse (Shul'pin,
34 1936). In dense boreal forests, it is often found in the upper reaches of streams and on the northern slopes of water-sheds (Yudakov, 1967). On Sakhalin Island, Spruce Grouse inhabit spruce-fir forests bordered by an underbrush of cowberry, bilberry, and scattered currant bushes. In the mountains these birds are found up to the limits of distribution of cedar and Erman's birch (Gizenko, 1955).

In the snow-free period, these birds readily prefer high-mountains above the tree-line, or sparse forest. In the Sikhote-Alin Reserve, they have been found mostly at higher elevations (1,300–1,500 m) and very rarely below 700 m, on bald peaks of high-mountain tundra which abound in cranberry or in old, burned-out areas of sparse larch forests (Kaplanov, 1938).

During the autumn-winter period, Siberian Spruce Grouse remain in
35 flocks. Flocks of American Spruce Grouse generally consist of 6–12 chicks and females; old cocks lead a solitary life. The data indicate that small flocks are also characteristic of Siberian Spruce Grouse.

In winter these birds are less active and feed mainly on one tree and roost in snow burrows. In the southern part of its range, however, Canadian Spruce Grouse roost even on tree crowns. Bent (1932) cites some interesting data relating to observations made in Canada by Walter (1907). He came across birds which had gathered in snow storms on thick branches of spruce and fir. Undisturbed, they remained on trees for several days since finding food was no problem.

Spruce Grouse lead a sedentary life, performing only short seasonal migrations associated with food change. In the summer months it is found at times even in the open, but never far from forests.

A common characteristic of Spruce Grouse is their extremely trusting nature, which many researchers have labeled "stupidity". In summer, Spruce Grouse are less active, hide in the thickest part of the taiga, and hence are rarely seen. Their trusting nature can be judged from the regularity with which these birds visit traps in their wintering grounds laid by Sakhalin hunters (Lisenka, 1929).

It has been observed that Siberian and American Spruce Grouse readily hop away from man rather than fly. In flight, they rise vertically, loudly flapping their wings, and land on tree branches 1.5–4 m from the ground, where they quietly hide themselves. Repeated shots do not frighten the hidden birds; neither are they disturbed by the practice of catching by throwing a noose around their neck. The apparent fearlessness of these grouse is more a defensive adaptation since an immobile bird is difficult to detect (Afanas'ev, 1934).

There is no record in literature on the flight of Siberian Spruce Grouse, except for descriptions of its short flights from the ground to nearby trees. Apparently, American Spruce Grouse fly more often. When flushed, it also rises vertically with much noise and lands on tree branches; however, it can fly short distances almost quietly even in a dense forest. Flushed in the open, it takes off with a deafening noise and flies into the nearby forests, somewhat like the Ruffed Grouse (Bent, 1932).

During the breeding season, these birds are more lively and differences in flight characteristics of species are more distinct. According to Kaplanov (1938), Siberian Spruce Grouse, like Capercaillies, commence courtship on trees, then descend to the ground, but, here again, use their wings only for hopping. Potapov (1969) observed the courtship of Siberian
36 Spruce Grouse on the ground. The cock pivots in circles in a small area or moves back and forth, often stopping and listening. Occasionally, his pace gathers speed and the vertical tail alternately fans out and closes. The bird then pauses, emits a rumbling sound, performs two short vertical leaps not higher than 30 cm, flapping its wings both in the air and on the ground. When Potapov attempted to catch a courting bird, the male rose with ease, fluttering its wings, flew a few meters, and resumed the courtship display.

The peculiar nuptial flight of American Spruce Grouse, as well as their short flights in the forest, suggest their high flight capacity. During breeding, Canadian Spruce Grouse perform nearly vertical flight 2–6 m high and then descend to the ground (Lumsden, 1961).

Mountain Spruce Grouse males fly more often during breeding. They court even on trees, rising there vertically from the ground, and perform a horizontal courtship flight with frequent wing beats, which culminate in a powerful beating of the wings on the back (McDonald, 1968). Considering the more complex courtship displays of American Spruce Grouse it may be suggested that the Siberian Spruce Grouse is a more primitive species.

Genus *Tetrastes* Keyserling and Blasius—Hazel Grouse

This genus includes two Palearctic species: *Tetrastes bonasia* L. (Hazel Grouse) and *T. sewertzowi* Przewalski (Severtzov's Hazel Grouse). It is true that in Short's work this genus is even larger and includes the monotypic genus of American Ruffed Grouse (*Bonasa umbellus* Doyg.). According to the principle of priority, he named the enlarged genus *Bonasa* (Stephens, 1819). In my view, combining these two indisputably proximate groups of Hazel Grouse of the Old and New Worlds into a single genus is not justified. It is more likely that Hazel Grouse and Ruffed Grouse originated from the same common ancestor; nevertheless, through evolution they became quite distinct geographically in a number of morphological features and in courtship behavior.

Apparently, Severtzov's Hazel Grouse, with its restricted geographic distribution in the mountain forests of Kansu and Szechwan provinces and Tibet, is closest to the ancestral form in plumage color. Hazel Grouse, widespread in the boreal forests of Eurasia, fall midway between the
37 species discussed before and the most advanced and quite distinct Ruffed Grouse—endemic to North America (Fig. 5).

There is no doubt that East Asia is the homeland of Old World *Tetrastes*; it is the center of origin of deciduous forms of Turgai flora (Vul'f, 1944) where the small habitat of the more ancient Severtzov's Hazel Grouse was preserved.

The higher population density of the Hazel Grouse in mixed forests in the southern part of its range, its adaptation to deciduous forests in the taiga zone, and the close association of winter feeding with alder, birch, and other deciduous species, serve as direct proof of its speciation in the deciduous forests. Weak sexual dimorphism, incompletely feathered tarsometatarsus, and the absence of ritual courtship allow us to consider Hazel Grouse as the most primitive group of tetraonids of Eurasia. Both species of Hazel Grouse inhabit predominantly mixed coniferous-deciduous forests.

The distribution of Severtzov's Hazel Grouse is restricted to the mountain forests of central China, where it inhabits birch and coniferous

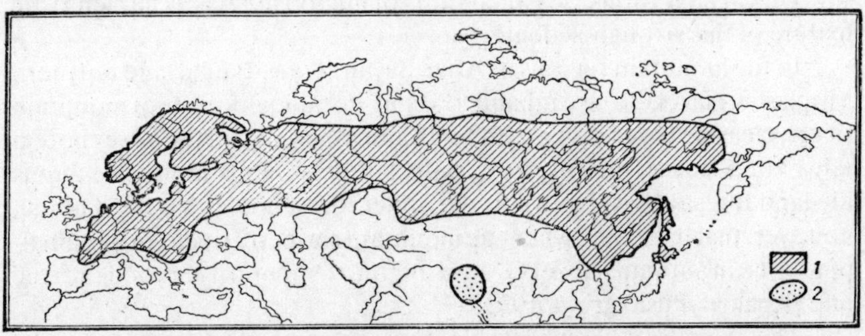

Fig. 5. Geographic distribution of Hazel Grouse. 1—Hazel Grouse; 2—species not found in the avifauna of the USSR.

forests and, by means of willow groves may reach the tree-line; as an exception it is found in summer even among rhododendrons in the alpine belt (Przewalskii, 1875–1876; Berezovskii and Bianki, 1891; Bianki, 1907).

Hazel Grouse inhabit fairly dense, damp forests with well-developed underbrush and abundant fruit bushes. Its habitats vary in different parts of its range. It lives in forests in the plains as well as mountains. At places it is adapted to various species of alder, which is of great importance in its
38 winter food. It can be assumed that the evolution of this species is also associated with alder; supporting this view, the plumage of this bird effectively blends with alder bark. Distribution of the different subspecies of Hazel Grouse almost coincides with the boundaries of various species of alder (Gavrin, 1956). At places where alder is absent, Hazel Grouse is found in mixed forests with other deciduous trees, settling more readily in birch groves. If deciduous species offer these birds their winter food source, conifers offer protection. Moving northward, as conifers begin to dominate in mixed forests, Hazel Grouse are adapted to areas of mixed forest or patches of deciduous forests among taiga—lowland deciduous forests along rivers, clearings overgrown with birch, aspen, and alder. Similar preference for deciduous forests has been observed in mountain taiga where along lowland forests of rivers, these birds may reach at some places the upper limit of coniferous forests.

Under favorable conditions of food and protection, Hazel Grouse are found throughout the boreal forest zone—from Manchurian taiga through the broad-leaved forests in the south to the deep spruce forests in the north. However, being a distinctly sedentary (nonmigrant) species, Hazel Grouse is seen only in those parts of the forest where there is a diverse combination of biotopes (microhabitats) to provide its life requirements year-round within a small area.

Depending upon the state of the vegetative cover while in search of more abundant food and better protected sites, Hazel Grouse undergo short, local migrations. Seasonal shifts of microhabitat is a characteristic feature of this strongly sedentary species.

In the mountain regions of Altai, Sayan, Trans-Baikal, and northern Mongolia, Hazel Grouse inhabit lowland, deciduous forests of mountain rivers, clearings overgrown with birch and aspen and, in the lower boreal taiga zone, dense, mixed forests with tall grass. Only along deciduous lowland forests does it move to the upper forest line. It has been noted, however, that in Kentei, when the mountain cowberry ripens in the subalpine zone in autumn, Hazel Grouse migrate temporarily beyond the tree line (Kozlova-Pushkareva, 1933).

The habitats of Ussuri Hazel Grouse are quite diverse. In Primor'e, this species inhabits dense brushwoods of Manchurian taiga rich in woody

39 and scrub vegetation with luxuriant growths of vines and ferns. It lives in sparse oak forests with a rich undergrowth of different shrubs. Apparently, in this part of its range, oak is of great importance to this bird since acorns are its main diet in autumn. In mountain regions, this species inhabits spruce-fir forests with thick sphagnum moss cover, and, in the lower reaches of the Amur, prefer open areas of taiga, fringes of clearings, and river banks where conifers alternate with birch, aspen, alder and poplar (Shul'pin, 1936; Vorob'ev, 1954). In the islands of Japan and North Korea, it inhabits mountain forests, but often descends in the autumn-winter period to fields that otherwise are not its typical habitat (Taka-Tsukasa, 1932). For much of the year, Hazel Grouse live in pairs or with broods, but do not form flocks. In Japan and North Korea, however, in winter and during severe frosts, birds gather in flocks of 20 to 30 and wander along the mountain slopes (Austin, 1972); for this reason, Taka-Tsukasa considered it to be even a gregarious bird.

Marking and ringing Hazel Grouse in Kirov District have shown their exceptionally sedentary life style in this region. Distances between the banding and recovery sites (sometimes several years later) were usually no more than 500 m, in a majority of instances even less, and only rarely reached 1.5 km (Gaidar, 1974). The strictly sedentary life of Hazel Grouse is particularly evident in winter, when pairs or small groups of birds stick together in a small section of rather dense forest.

In early winter and late autumn, Hazel Grouse feed mainly in trees, spending the day on the ground, and roam over thin snow or fly to fruit bushes. They roost on trees in pairs or small groups, and perch on spruce boughs at heights of 2–6 m above the ground.

In midwinter, when the snow cover grows sufficiently deep and severe frosts set in, Hazel Grouse spend much of the day in snow holes, flying out only briefly to nearby trees for feeding. Only on warmer days do they at times wander over the compact snow. The winter distribution of Hazel Grouse depends not so much on the depth of snow as on food availability.

From the end of December through March, in regions with severe frosts and fairly thick snow, Hazel Grouse stay under snow for up to 20 hours a day, flying away for food only in the morning and evening (Donaurov, 1947). When the crust of frozen snow is hard and difficult to break, they roost on snow and in thickets of bushes, but not in trees.

40 Hazel Grouse lead a secluded life in dense forest and are rarely sighted. When alarmed they crouch on the ground among bushes, or more often, fly to the closest tree and remain motionless.

The annual cycle of Hazel Grouse can be divided into two periods based on behavior: a shorter but more active breeding period devoted to nesting and brood care, during which the birds remain exclusively on the ground, and a longer period of 7–8 months when they rightfully can be called arboreal.

Inhabiting forests which are often stewn with dead wood, fallen trees, and projecting stumps, Hazel Grouse can negotiate complex obstacles on the ground during the snow-free period. They climb small elevations with ease, maneuver skillfully through dense grass and bushes, and only rarely run on foot. When running away from danger, they hop into hummocks or stumps, briefly survey the scene, and run farther. In winter, they do not avoid the frozen snow crust and, as judged from footprints on fresh snow deposited over the hard crust, sometimes run for hundreds of meters (Formozov, in litt.). Running birds stoop and stretch their heads somewhat like a ptarmigan on the run. They can leap to a height of 1.5 m, and are as free in trees as they are on the ground. From late autumn to spring, they feed almost exclusively on tree tops, and, except when nesting also roost there during the snow-free period.

The flight of Hazel Grouse is limited both in speed and range, but they possess the maneuverability required of birds living in thick undergrowth. The bird makes quite a noise while taking off and perches on the nearest tree halfway up its height, or a few meters above the ground. Then it changes to a horizontal flight through the trees, but not rising over the forest canopy. Flying among dense branches, Hazel Grouse change their body angle so that their outstretched wings become vertical and they glide effortlessly through obstacles. At times, such a flight has been observed in sparse forest areas as well. Its takeoff from a tree and the flight is noiseless. The flight range apparently does not exceed 150 to 200 m, judging by birds flying over open stretches from one mountain slope to the other or open areas in a forest.

The entire behavior of Hazel Grouse confirms the excellent adaptation of this species to living in dense forests.

Family PHASIANIDAE

Genus *Coturnix* Bonnaterre—Quail

41 This genus comprises six species of galliform birds widely distributed in the open environments of the Old World (Fig. 6). Two species—*Coturnix coturnix* L. (Common Quail) and *Coturnix japonica* Temm. and Schleg. (Japanese Quail)—inhabit the Soviet Union.

Judging from the present-day extensive distribution of this genus, the ancestors of quail were apparently widely distributed in the tropics and subtropics of the Old World and possibly, like other Galliformes, were sedentary birds. The present-day breeding range of quail includes open lands in different geographic zones up to the northern forest boundary.

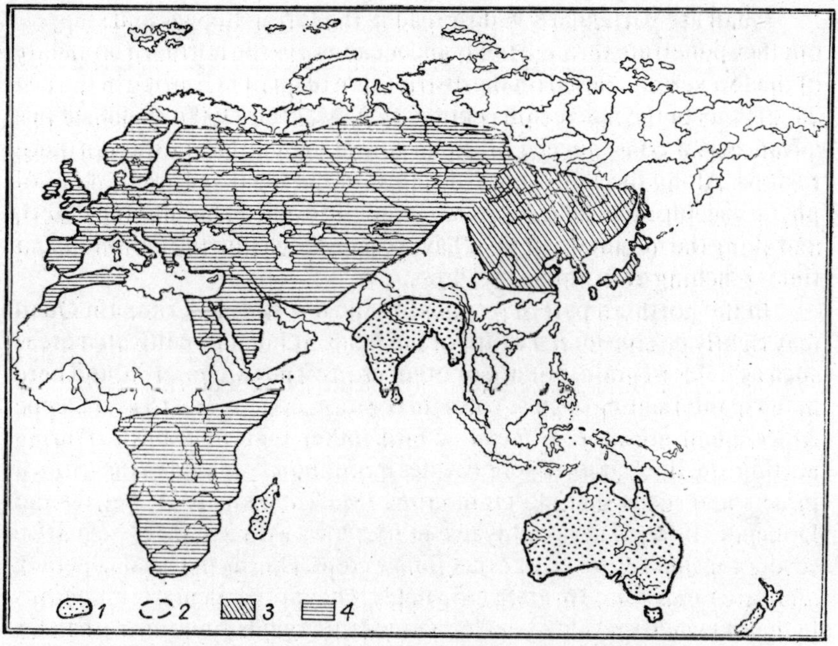

Fig. 6. Geographic range of Quail. 1—species not found in the avifauna of the USSR; 2—partially established ranges of distribution; 3—Japanese Quail; 4—Common Quail.

Their penetration into regions with severe climatic conditions after the glacial period were possible because of long seasonal migrations. This
42 ability for long-distance migration is exhibited only by quail found in avifauna of the USSR—Common Quail and Japanese Quail. Even the subspecies *C. c. africana* Temm. and Schleg. (African Quail) found throughout South Africa is a sedentary (nonmigratory) bird. Locomotion in quail, although better than of other Galliformes, is far from perfect. In my opinion, this proves that migration in quail is a comparatively recent adaptation evolved after the glacial period simultaneous with colonization of the northern part of its range.

Fossil quail are known only from the Pliocene of southern Europe. *Plioperdix* sp. Gailard has been described from France, and *Plyogallus coturnoides* Tug. from the vicinity of Odessa. These two finds were later combined in a single genus, *Plioperdix* (Kretzoi, 1940). No other evidence is available concerning the antiquity of this group of birds.

All quail are birds of open landscapes having a fairly thick grass cover. Naturally, their habitats vary considerably in different climatic belts; but in all cases, a dense stand of grass is essential as shelter for these small birds.

Quail are particularly widespread in the forest-steppes and steppes, but they penetrate through fields and meadows to the northern boundary of the forest zone. Based on the distribution of quail in Europe in the last 25–30 years of the nineteenth century, M.A. Menzbir (1900) explained this advance as a consequence of expansion of agriculture in the northern regions. Along the short-grass meadows of the tugai and areas of mesophytic vegetation in lake depressions, quail advanced far into the desert, and along the mountain steppes have reached high into the mountains, at times reaching even alpine meadows.

In the northern part of its range, in the forest zone, Common Quail may rightly be considered a bird of the fields. It inhabits cultivated areas such as fields of grain, potato, or other crops. They are rarely found here in the natural setting of grass cover. In contrast, in steppe and forest-steppe zones, quail are more a meadow bird rather than field birds. During nesting, the thick grass cover provides more abundant food in the form of insects and seeds than do grain crops (Mal'chevskii, 1961; Petrov and Donchev, 1969). They readily live in meadows with a natural vegetative cover or in alfalfa, clover, or other forage crops. During the nesting period, quail are rarely found in grain crop fields. They are particularly numerous in regions where meadows or steppe and fallow lands alternate with fields of grain crops. Here, the birds find extremely favorable conditions for food
43 and shelter, not only during nesting but right up to migration, since they can feed on stubble left behind after the harvest.

Common Quail occur throughout the forest-steppes of western Siberia and Kazakhstan; in the more southern steppes, it is common only in more mesophytic areas such as river valleys, meadows around lakes, and basins with meadow flora, sometimes even penetrating glades of insular pine forests (Dolgushin, 1947). In southern Kazakhstan and Soviet Central Asia, they inhabit foothill zones with irrigated agriculture and meadow-steppe slopes of mountain ranges. Elevation is not of great importance to these birds; their vertical distribution depends only on vegetation. In the Pamir-Alai mountain ranges, Common Quail apparently nest at elevations not lower than 1,600–1,800 m, the upper limit being 2,700 m in Zeravshan and 3,000 m in Pyandzh (Ivanov, 1969). In the valley regions of Tadzhikistan, their main habitats are irrigated land supporting crops of alfalfa, cotton, rice, and melons. In mountain regions, they are confined to fields of cereal, flax, millet, and alfalfa. At still higher altitudes where there is no cultivation, they inhabit subalpine and alpine meadows with sparse dogrose and juniper bushes (Abdusalyamov, 1971). The upper limit of distribution of Common Quail in the Tien-Shan range is 2,800–3,000 m (Shnitnikov, 1949; Korelov, 1956; Kovshar', 1966). In the Dzhungarian Alatau, it is found up to 1,400–1,800 m. In Altai, where there is hardly any cultivation in the mountains, the Common Quail inhabits rich grassy meadows, often with isolated willow bushes, from the foothills to heights of 2,000–2,200 m (Sushkin, 1938; Irisov and Totunov, 1970).

In different parts of their range Common Quail prefer places with a definite humidity, avoiding extremely wet or dry areas. In the northern parts of their range, Common Quail nest in cultivated land, specifically avoiding extremely wet areas; in the forest-steppe zone, on the other hand, it prefers natural vegetation.

In the steppe and desert regions, Common Quail inhabit river valleys, basins, and fields near shrubs or areas with more mesophytic vegetation; finally, in the southern regions of the country, it becomes a predominantly "mountain" species. It is characteristic that in the more humid Tien-Shan ranges, Common Quail are found from the foothills to the subalpine meadows; in the more arid Pamir-Alai and Kopet Dag ranges its distribution is largely restricted to the higher altitudes. It is true that Common Quail can nowhere be called a true mountain bird, yet they move high into
44 mountains only in search of its favorite vegetation, meadow-steppe flora, specifically avoiding extremely rugged terrain and keeping to mountain plateaus, terraces, or gentle slopes.

The habitats of Japanese Quail are generally more humid than those of Common Quail. Based on observations in Primor'e, Shul'pin (1936) assumed that Japanese Quail were initially birds of humid meadows. Later, as this species started spreading, it began colonizing dry meadows with various kinds of shrubbery and partly cultivated fields, although it is not too fond of the latter. It avoids dense tall-grass meadows such as the prairies. In marshy areas, it prefers high grounds, at times among small willow bushes. Another preferred habitat is the fringes of burned-out areas where it forages. In spring, it stays in wet, even marshy, areas that are quagmires. During spring flight, it is found along the shores of in-lets, among small clumps of reed and sedge. In the upper reaches of the Chita River and on the Vitim plateau, Japanese Quail occupy glades with dwarf birch and intervening meadows, as well as open areas near scrubs of birch and dwarf willow (Pavlov, 1948). In eastern Kentei, it occupies dry bottom areas of the steppe and humid, hummocky river valleys. In Kerulen River valley, this type of habitat is the most common (Kozlova-Pushkareva, 1933).

The nonmigratory species of this genus are birds of the open grassland. In Burma, *C. coramandelica* (Black-breasted Quail) inhabits natural grass stands as well as cultivated land near villages (Smythies, 1953). The African Harlequin Quail (*C. delegorgiei*), widespread in Africa, is found only in the open country, and its area of distribution is thus restricted by forests (Bennerman, 1953).

Life in dense grass cover has left its imprint on the morphology of Common Quail and has shaped its entire behavior. For these small birds, grass stands provide a reliable protection from winged predators and Common Quail are not prepared to leave this shelter even for a short while. It is difficult to flush these birds as they usually crouch and only rarely fly, and that for a short distance, and hide again in the grass. Considering that during migration, Common Quail fly tens and even hundreds of kilometers without rest, this behavior cannot be attributed to their poor flying capabilities. They simply do not want to abandon the safety of their shelter. They run very fast, skillfully maneuvering through the grassy tangles. Their laterally compressed body aids free movement in the dense vegetative cover.

45 Moving on foot, Common Quail stretch their necks somewhat and stoop, appearing like a miniature Ptarmigan. At the same time, their gait is peculiar; at each step the bird swings its head rhythmically.

Common Quail are nocturnal birds and rest during the day. At twilight, they become more active and forage; during migration, they fly only at night.

The flight style of Common Quail differs from the typical flapping flight of other Galliformes. A flushed bird flaps its wings incessantly and takes off not 'in a flash' but fairly gently, not rising very high above the ground; after flying for some distance, it hurriedly descends to the ground. When necessary, the bird rises above tree tops (10 to 15 m), and having crossed the obstacle, descends to the grass. In daylight quail fly only for short distances. In August and September, they leave their nesting sites singly or in small groups and gather in large flocks as they proceed southward. They fly at night, resting in the day in fields or places with a dense grass cover.

Mass congregations of these birds on the Black Sea coast are well known to occur before migration across the sea. Based on rather limited banding data, Shevareva (1965) identified the following migration routes of quail. The birds of the western Ukraine fly westward in autumn across Switzerland, France, Spain and Morocco, and return in spring via Italy. The more northern populations from the Leningrad District possibly follow the same route. The bulk of quail from the European part of the USSR, the Urals, Bashkiria, western Kazakhstan, and Stavropol', congregate on the Black Sea coast and, after crossing the sea, fly along the Nile River valley and reach Equatorial Africa. Quail from western Siberia probably migrate to Pakistan or India; one specimen banded in the vicinity of Tashkent in autumn and recaptured the next year in Altai territory supports this view.

The majority of quail winter in the northern part of tropical Africa (South Sahara, Gambia, Chad, and hills of Ethiopia); some fly as far as the Equator and to southern Africa for wintering, while others enter Arabia, Iran, Afghanistan and India.

In the southern parts of their breeding range, quail regularly winter in
46 snow-free regions, although in very small numbers. They have been sighted in the Mediterranean, southwestern Europe, the Caucasus, Turkmenia, Tadzhikistan, and southern Uzbekistan. During very mild winters, isolated quail have been found even in the more northern, less snowy regions: Don steppes, Astrakhan' territory, Volga-Ural steppes, and Balkhash and, according to Sushkin (1938), even in the Altai in the Chulyshman valley.

The speed, height, and range of flight without rest (during migrations) can be estimated only from indirect data. Flight height directly depends on weather conditions. On clear windless nights, the noise of birds flying at great heights is audible; in fog on the other hand, these birds fly low as evidenced by carcasses of birds that perished by hitting power transmission lines. Mountain ranges do not obstruct the migrants, which soar gradually along gorges. The most difficult part of their journey is flying over the Black and Mediterranean seas. It has been observed that quail returning in spring from their wintering quarters are so exhausted that, on landing on the coast, they lie absolutely motionless for some time. Apparently, such long flights tax their endurance to the limit.

Flights across the Black Sea are performed only on calm and rather clear nights. During storms on the Crimean coast, birds which have flown in from the north gather in large numbers and then disappear on the very first calm night. The report of quail resting on water, during migration, is supported by just one observation of a "splash down" in the Red Sea on November 10, 1957 (Kost, 1967). A bird flying over the sea landed on the water surface, and remained there for over half a minute rocking in the waves. Apparently its feathers were not very wet; the bird rose from the water with ease and flew away. One can hardly evaluate the ability of quail to rest on water during migration from this single observation.

The behavior of quail in wintering areas in Arabia, where these birds are found from the middle of September to early May, differs very little from that during breeding season. Here, they remain close to crop fields or in short grass and hide during the day, at times using bushes or even rock outcrops on level ground as shelter. At this time of the year, food consists of small seeds (75%), snails and insects (5%), and plant material. In September, the migrating birds fly through towns at a height of about one meter and are sometimes caught like butterflies with nets (Meinertzhagen, 1954).

Genus *Perdix* Brisson—Partridges

47 Three species of partridges represent the northernmost phasianids (only migratory quails are an exception, which penetrate even farther north for nesting in certain regions). Two species of partridges are found in the Soviet Union. The western and central parts of the range of this genus in the east, up to Sayan and Altai, are occupied by the widespread species *Perdix perdix* L. (Gray Partridge) and the eastern part is colonized by *Perdix daurica** Pall. (Daurian Partridge); the mountain ranges of Central Asia include a small range of *Perdix hodgsoniae* (Tibetan Partridge). The range of Daurian Partridge overlaps that of the other two species (Fig. 7).

Among fossil remains *Paleoperdix* is considered the closest to the *Perdix*. Four species have been described in *Paleoperdix* from fragmentary remains of bones of the hind limbs, from the Lower and Upper Miocene of France (Milne-Edwards, 1869–1871; Brodkorb, 1964).

Like other members of the family Phasianidae, ancestors of partridges were more thermophilic. Further evolution of the genus *Perdix* occurred, however, in areas of gradual cooling and resulted in the separa-
48 tion of a group which, in ecological characteristics, was intermediate between tetraonids and phasianids. The earliest fossil of *Perdix* (*cinerea*?) in Europe dates back to the Pliocene (Lambrecht, 1933).

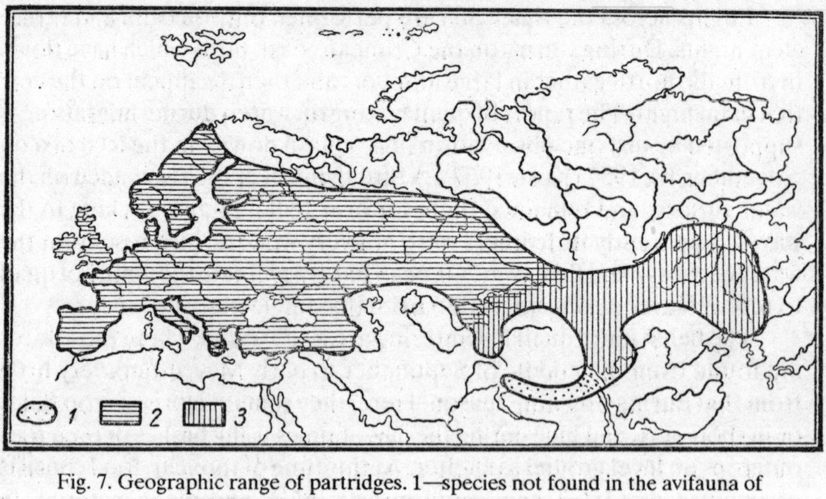

Fig. 7. Geographic range of partridges. 1—species not found in the avifauna of the USSR; 2—Gray Partridge; 3—Daurian Partridge.

*[sic]; should be *Perdix dauurica*—Editors.

In his list, Brodkorb (1964) includes only *Perdix juresaki* Kretzoi from the Lower Pleistocene of Romania; possibly, this is the same specimen but identified differently. In the Upper Quaternary deposits of Apsheron Peninsula, more fossil remains of Common Partridges have been found than that of the Chukar Partridge (Burchak-Abramovich, 1962). Considering the less dense plumage coloration and the high ecological flexibility of the Gray Partridge compared to other species of this genus, it may be assumed that this form is closest to the ancestral.

All species of this genus inhabit open country with grass cover and shrubs and positively avoid dense forests but inhabit their fringes or glades. Gray Partridge mainly inhabit steppes, meadows, cultivated lands interspersed with overgrowths of weeds, brushwood, and windbreaks which provide reliable shelter. From the open country they penetrate the forest zone, at places reaching the northern boundary of forests, and in the south, enter far into the desert using mesophytic vegetation.

In western Europe, the optimal conditions for Gray Partridges are plains at elevations of 200–300 m with fertile, rather moist soils under cultivation, and uncultivated lands covered with rich stands of grass.

Distribution of this species in mountains is determined by wintering conditions and the depth of the snow cover. In the southern Carpathians, Gray Partridges move up to the lower part of alpine zone, i.e., to 2,000 m (Niethammer, 1942), while in the more snowy Tatry and eastern Carpathians, these birds do not nest above 700–800 m (Strautman, 1954).

In the forest zone of the European part of the USSR, Gray Partridges follow the advancing agriculture, along the edge of fields, fallows, downed timber, and large glades.

In the steppe regions, they are most numerous at places where fields alternate with dense network of windbreaks or where the terrain is rugged, with large ravines and gullies (Osmolovskaya, 1966). In the forest-steppe of western Siberia and northern Kazakhstan, Gray Partridges live on the edges of fields and, at times, are found in virgin steppes or old fallows, but only close to shrubs, or forest groves, fallen timber, and windbreaks.

49 In the floodplains of the Ural River, Gray Partridges are found on open elevated patches where grass meadows are interspersed with reed, sedge, and groves of black or other poplars (Dubinin, 1953). On the outskirts of the Muyunkum desert and in the Chu valley, this species inhabits sand dunes with saxaul bushes or in mixed scrubs of salt tree, tamarisk, and saxaul (Gladkov and Grinberg, 1932). In Caucasia, in the Tien-Shan mountain ranges, Gray Partridges, though typical birds of meadows and steppes, move to high altitudes along grassy slopes (Dal', 1949; Korelov, 1956; Kovshar', 1966; Khanmamedov, 1969). Maximum numbers are found in the foothills and middle ranges, but at places their

range extends higher into juniper groves (Ugamsk range) and even up to subalpine meadows (Talassk Alatau).

Although Gray Partridges inhabit various biotopes in the open country, they are still typical of steppe and forest-steppe zones where their population is higher. They exhibit great preference for cultivated areas; however, intense cultivation deprives this species of the necessary cover (brushes, grassy patches) resulting in a steep fall in its population.

In comparison with Gray Partridges, Daurian Partridges are found more frequently in mountain regions. They inhabit foothills, mountain slopes, river plains and valleys and comparatively rarely move up to the alpine belt. At the southwestern boundary of their range in the Semirech'e of Kirgizia, Siberian Daurian Partridges inhabit the area from the foothills and intermontane undulating valleys up to elevations of 3,000–3,500 m (Shnitnikov, 1949; Yanushevich, et al., 1959). As a rule, Daurian Partridges in these places occupy higher elevations than Gray Partridge. At the lower boundary of its distribution, this species occupies edges of fields and old fallow fields with sallow thorn, dog rose, and barberry bushes; at higher elevations, they may be found among the dwarf vegetation of watershed uplands. At some places, especially in autumn, this species almost reaches the permafrost zone (Severtsov, 1873).

In northern Mongolia, Daurian Partridges are common in the river valley steppes in bottom-land forests, meadows, and foothills. In Khangai, they often inhabit marshes near relict lakes or sources of mountain brooks at the upper boundary of cedar forests (Kozlova, 1930). In the mountain ranges of Central Asia, this species of partridge was caught in the Suok and Kobdo river valleys among cheegrass overgrowth (Bianki, 1907).

In Trans-Baikal, Manchurian Daurian Partridges live along the edges of fields of grain crops, near shrubs, or in small forests; they also inhabit
50 dry river valleys with overgrowths of sallow horn, willow, and chokecherry, grass meadows among sparse forests on mountain slopes, elm groves, and various clumps of bushes on the steppe slopes of river terraces (Khankhasaev, 1969).

The third species of this genus, *P. hodgsoniae* [Tibetan Partridge], are typical representatives of the alpine fauna. This species inhabits the Kansu province, Tibet, and the Himalayas mainly in the alpine zone, rarely descending to the high-mountain valleys. The upper limit of its range is 5,000–5,600 m (Bianki, 1907; Salim Ali and Ripley, 1962) and the lower limit 2,800–3,600 m (Przhewalski, 1875–1876; Schäfer, 1938).

Typical habitats of Tibetan Partridges are high-mountain plateaus ordry slopes with bushes of *Caragana spinosa, C. jubata,* and overgrowths of rhododendron and dwarf juniper. In the high-mountain valleys, they are found among overgrowths of *Potentilla tenuifolia* and grasses, are an avid visitor of crops in high-mountain areas. This species sometimes is found in the most mesophytic areas in meadows and shrubs of the alpine region (Berezovskii and Bianki, 1891).

The extensive utilization by Gray Partridges of the natural habitat in various geographic zones, the gradual extension of their range northward (to Arkhangel'sk) and southward (valley of the Urals, sand dunes of Sam on Ustyurt), and their remarkably successful acclimatization in several regions of the USA and Canada (Porter, 1955; Westerskov, 1965, 1966), reveal the exceptionally high ecological flexibility of this species. On the other hand, Daurian Partridges differ very little in appearance and behavior from Gray Partridges, and are highly stenobiontic as revealed by its confinement to less snowy regions. Unsuccessful attempts to introduce Daurian Partridges in the outskirts of Moscow in 1957 (Gabuzov, 1958; Sergeeva and Sumina, 1963) and in the central and southern Urals in 1961 (Skripchinskii, 1962) confirm this assumption.

All partridges of this genus show preference for cultivated areas. Even Tibetan Partridges, which are high-mountain birds, readily inhabit high-mountain cultivated area. Daurian Partridges are found more at the edges of fields than in natural biotopes, while Gray Partridges are birds of anthropogenic environments for much of their range. Such an attraction to cultivated areas may be explained by their high food requirement, especially in winter.

While inhabiting open habitats, partridges select areas which offer
51 appropriate protection from enemies and inclement weather. After harvesting, birds nesting in fields migrate to the underbrush of weeds, bushes, and forests where they remain throughout the winter.

Winter is the most difficult period in the life of Gray Partridges. Insufficient food and difficulty in negotiating loose snow compel these birds to migrate to settlements, roads, and, at some places, to even migrate to less snowy regions for wintering. During this period, the ability of Gray Partridges to change their behavior and adapt to severe conditions becomes manifest. Birds occupying regions of deep snow cover have developed the ability of regular migration to south and southwest, where the wintering conditions are more favorable.

Information on the migration of Gray Partridges was first summarized by Zhitkov and Buturlin (1906), who emphasized the sedentary life of partridges in the northern, central, and western parts of European Russia, and the regular, annual migrations of large number of the southern and southeastern partridges. These authors attribute such migrations to the severe continental climate in this part of the range. Later observations showed that migrations of Gray Partridges also involve the northern regions of Kazakhstan (Shaposhnikov, 1936) and the northern and northeastern districts of the European part of the USSR (Osmolovskaya and Formozov, 1950; Rodionov and Nemtsev, 1954).

In the northeastern regions of Leningrad District, already in late August to early September, partridges begin to break into numerous flocks, sometimes of merely 100 to 150 birds. By the beginning of winter the populations thin down, leaving behind small wintering groups of up to ten birds. Toward spring, the number of over wintered birds is only 20 to 30% of the original population and is supplemented only by the migrants from the south.

These observations lead one to assume that migrations of Gray Partridges cover a considerably larger area than was earlier thought of and, apparently, evolved in the not so distant past as the partridges colonized regions with abundant snow in winter.

In southeastern Europe, partridge flocks could well be aggregations of migrants from the more northern regions which local birds possibly joined. However, there is other evidence to show that, from some places where Gray Partridges regularly migrated, these birds have now become
52 sedentary. In the past, Gray Partridges migrated almost *en bloc* and returned to the breeding grounds in the Cis-Urals only in spring. In 1955, in the forest-steppes of Bashkiria, partridges did not migrate until the snow was very deep, while in the winter of 1956–57 many wintered there (Fedorako, 1957). Presumably, migrations of these birds depend on weather conditions and, in some years, the intensity of migration changes markedly.

In winter, Gray Partridges remain in small flocks of a few broods and, sometimes unbroken families. Their behavior and biotopic distribution depend to a great extent on the severity of winter and weather conditions. Partridges select areas where the snow is often blown away and food is easy to obtain; they are found on roads, in fields, and windbreaks. On clear, relatively warm days, they forage throughout the day, pecking at seeds on the snow or vigorously shoveling snow with their feet. In Finland (Pulliainen, 1966), the activity of partridges in the feeding grounds rises as the temperature drops, reaching their peak at −20°C.

Gray Partridges roost under cover of bushes or brushwood, under haystacks, or in pits dug in shallow snow, usually in flocks pressed close to each other; at times, they use solitary open holes. In central Kazakhstan, they dig individual holes of up to 10 cm deep and group holes up to 12 cm deep (Krivitskii, in litt.). During severe, windy winters in the Trans-Volga region, they congregate in the forest zones where there is considerable snow accumulation which enables them to roost in snowdrifts (Mal'chevskii, 1947). Roosting in forests is of particular importance to Gray Partridges during severe frosts. Here they roost with broods pressed close to each other (Poyarkov, 1955). Gray Partridges introduced into Canada roost in snow pits, but, when the snow cover is more than 30 cm deep, they roost like tetraonids in true burrows dug into the snow (Westerskov, 1965).

In the Chulyshman valley of the Altai, Daurian Partridges congregate in winter in flocks of several tens of birds. Autumn onward, these birds are seen near fields where grain was harvested. In winter, they feed on the stubble by raking the shallow snow, but when the snow cover is deep, they come to the threshing barns where grains are being threshed (Folitarek and Dement'ev, 1938). They roost in flocks in open snow pits, but for Daurian Partridges inhabiting less snowy regions, roosting in snowdrifts is not known.

53 With respect to winter roosting, Gray Partridges occupy a somewhat intermediate position between members of the family Tetraonidae which spend much of the day in snow holes, and Phasianidae which roost in flocks under various types of cover and do not use to advantage the favorable microclimatic conditions of the snow cover.

Unfavorable conditions in wintering sites usually lead to high mortality among these birds. Partridge mortality is particularly high in heavy snow winters with frequent thaws often accompanied by icing. Deprived of food, birds die of starvation. The magnitude of this phenomenon in some years may be judged from the work of Dement'ev and Shimbareva (1941), who present data on Gray Partridge mortality during the severe snowy winter of 1939–40 in the Ukraine. In that winter, the depth of the snow cover reached 70 cm, even over 2 m deep in lowlying areas, and partridges were nearly wiped out. The birds also suffered in Trans-Caucasia. In western Siberia, only a negligible number survived along the Tyumen River; their number was reduced to almost half in the Om' valley. Bird mortality was high north of Rybinsk.

In the Il'mensk Sanctuary (southern Urals) no fewer than 1,500 partridges were counted at the beginning of winter, but the following summer and autumn not a single Gray Partridge was sighted within a radius of 25–30 km of the sanctuary (Ushakov, 1949). High mortality among Gray Partridges was recorded the same winter in Germany. However, if a severe winter with high partridge mortality is followed by climatically favorable years, their population is rapidly restored due to their high fecundity. According to published reports, a normal clutch consists of 12–24 eggs but may even contain 26 eggs (Ul'yanin, 1949). The high fecundity of these birds emphasizes their exceptional vigor and ability to adapt to extreme conditions, not only by changing behavior (migration, partial use of snow cover as a protection from cold) but also as a result of their physiological features.

All species of partridge are exclusively ground birds: they forage, nest, and rest on the ground without ever flying onto trees or bushes. In regions of sympatry, Gray and Daurian partridges are difficult to distinguish apart, and no separate accounts of the behavior of Daurian Partridges are available in the literature.

According to observations reported from England where the population density of Gray Partridges is very high (one pair per 1.6 and even 0.8 ha), they keep to a very limited territory during the nesting period and
54 defend against encroachment by other birds. They very rarely fly. If these birds are not disturbed they may not be airborne for more than a minute per day (Blank and Ash, 1956). They feel safer in the grass, among bushes, and in fields than in the air; in case of danger, they try to hide or run but rarely fly. Gray Partridges hide very cleverly using all kinds of cover; in extreme cases, they crouche close to the ground where they are barely noticeable because of their protective coloration.

Gray Partridges run fast and with ease. Undisturbed, they walk with a stoop and an outstretched neck, and look rather clumsy. As soon as they begin to run at speed, they rise fully on their legs with the neck held high and stride with ease.

Flight of Gray Partridges is fast and powerful. This species rises from the ground with noise and rapid flapping of the wings. It usually flies at a height of 2 to 5 m from the ground. Their frequent wing strokes produce a characteristic sound alternated with silent gliding, and they land with outstretched, immobile wings. When chased, it flies for 100–200 m, lands on the ground, and hides in the grass. Flights to feeding and roosting sites may be performed at great heights and cover long distances. Flights of Gray Partridges from one shore of a large lake to the other 2–3 km away have been recorded. It is not known how far partridges can fly without rest during migration, but the route undoubtedly extends for not only hundreds, but thousands of kilometers.

According to Sokolov's (1895) observations, along the lower reaches of the Volga and the steppes of southeastern Europe, partridges fly for not less than 400–500 versts.* The migratory route of partridges is considerably longer than this and covers not only the lower Volga but also the coasts of the Black and Azov seas, penetrating even farther into Germany and France (Alferaki, 1882). Considering that partridges are also known to migrate from northeastern Europe, it is quite likely that some populations travel one thousand kilometers.

In the lower reaches of the Volga, the main migratory flight commences after the first frosts and is particularly noticeable in October; in some years, however, migrating birds may be held back until the end of December (Vorob'ev, 1896).

*A verst is 1.067 km—Editors.

The migration of partridges, according to my observations, starts
55 before sunset. In the early twilight, the distant chirping of birds is heard, followed by the sighting of flocks of 4–18 birds. During mass flights, hundreds of flocks fly one after another. Partridges fly very low over land as in short flights; this feature has been exploited in the past by hunters who trapped hundreds and thousands of Gray Partridges in nets tied to poles along the lower reaches of the Volga.

Genus *Ammoperdix* Gould—Sand Partridges

According to Menzbir (1904–1909), the genus *Ammoperdix*, comprising only two species, *Ammoperdix griseogularis* Brandt [See-see Partridge] and *A. heyi* [Sand Partridge], may be placed between Francolins and Rock Partridges. The origin of this genus, like that of the rock partridges, dates back to the Pliocene when the Mediterranean climate was drier and xerophytic vegetation was widespread.

Judging by the distribution of the two species of *Ammoperdix* and a lone fossil metatarsus of *A. ponticus* Tug. (Tugarinov, 1940) from the vicinity of Odessa, the distribution of Sand Partridges in the past was considerably wider and covered the southern parts of Europe. Discoveries of extinct members of genera *Alectoris* and *Ammoperdix* (*Alectoris pliocenica* and *Ammoperdix ponticus*) in the same fossil bed near Odessa indicate that these two genera originated in the same territory. Later, the

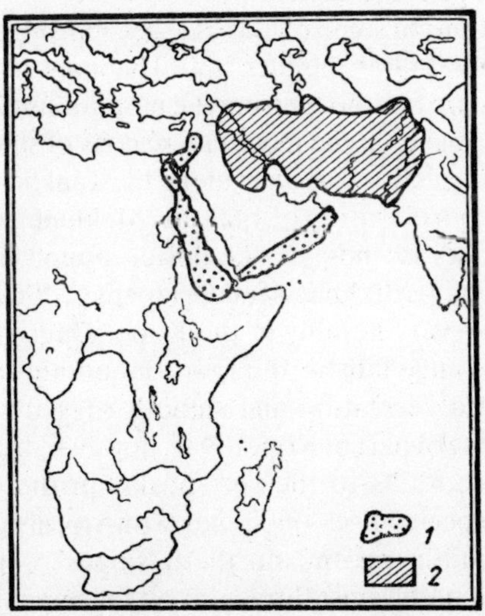

Fig. 8. Geographic range of Sand Partridges. 1—species not found in the avifauna of the USSR; 2—See-see Partridge.

more adaptable *Alectoris* spread extensively in the northeast, while *Ammoperdix* contracted its range and held to a limited territory with a distinctly arid climate. The present range of this genus coincides only with the southernmost part of the range of the Rock Partridge (arid mountains of Soviet Central Asia, West Asia and the Arabian Peninsula) and extends southward to the deserts of Egypt. The northern boundary of the range of this species passes through regions where there is no permanent winter snow cover (Fig. 8).

Both species of Sand Partridge ecologically are very close. *A. griseogularis* occupies predominantly desert slopes, low knolls, and low hills of mountain ranges, mainly up to 1,000 m above sea level; during the nesting season it is seen up to 1,200–1,350 m. It often inhabits the slopes of loess and conglomerates with talus separated by deep ravines. Vegetation on such slopes is sparse and fades early. A necessary condition for the existence of this species is the availability of water. In the absence of fresh
56 water these birds use brackish water. Sand Partridges often live on the same mountain ranges as Chukar Partridges, but this species keeps to gentler and less overgrown slopes, and avoids rocky gorges preferentially inhabited by Rock Partridges. Only rarely is it found on rocky slopes with sparse vegetation. Hence these two biotopically separated species are often found very close together. Compared to Chukar Partridges, See-see Partridges are stricter stenobionts with greater affinity to arid mountain regions. This seems to explain the relatively limited range not only of this species but of the genus as a whole.

At the same time, in some regions See-see Partridges live under rather different conditions. In Iran, besides typical habitats, it sometimes nests in sand dunes if they are not too far from the mountains (Zarudnyi, 1903). It lives considerably higher in the mountain regions of southern Uzbekistan: in the Kashka-Darya River basin and along the Tankhai River, this species has been found at an elevation of 1,600 m (Meklenburtsev, 1958). In the Kugitang range, it ascends to the lower boundary of the juniper belt—1,200–1,600 m (Salikhbaev and Ostapenko, 1964). The habitats of See-see Partridges vary notably in the Kopet-Dag. The greater part of their geographic range falls in the lower mountain ranges, either with steppe or semiarid vegetation and smooth rugged relief (Dement'ev, 1952). In the northern part of Kopet-Dag, however, this species has been found from the foothills to the very peaks. In the lower part of the mountains, this species lives on slopes with wormwood—emphemeral groups of vegetation, on the northern slopes of the range among wormwood—ephemeral-grass floras, and on the upper part, in an atypical
57 habitat—among tugai meadow vegetation (Kolesnikov, 1956). As an exception, it has been found on rocks with juniper groves. Zarudnyi (1896)

came across a large flock at such a place in Kopet-Dag; during the warm nours of the day he saw birds perched on thick branches in the shade.

From the brief description available of the habitat of *A. heyi* [Sand Partridge] (Meinertzhagen, 1954; Mackworth Captain, 1952)*, it is evident that it lives in conditions similar to those of See-see Partridges, namely, dry, gentle slopes of rocky hills with sparse scrub and scree.

Information on the life and behavior of See-see Partridges is scanty. A more complete picture of the behavior of this bird is available from Shlyakhtin (1896). See-see Partridges are very similar to the Chukar Partridge in behavior but differs in other respects. This species is more cautious, leads a more secretive life, rarely raises its voice, and hence is considerably difficult to detect.

When chased from bushes, the brood does not fly far off; birds flutter onto a stone, and sit quietly until the hunter comes close. Sensing that the danger is imminent, they run farther up the slope, hopping and jumping higher and higher. After a gunshot, unlike the Chukar Partridge, the brood does not scatter but flies together.

In summer, See-see Partridges stay close to water sources; when alarmed, they do not fly very far and return in a short while and take cover against the intense summer heat in tamarisk and blackberry bushes. Both adult birds and freshly hatched chicks require water. The female takes the tiny chicks to a water source but not right up to the water and carries water to them in her beak.

During the autumn-winter period, See-see Partridges become even more cautious. Alarmed birds fly far off, thrice the distance covered by Chukar Partridge. The same flock encountering a hunter a second time does not wait for him to approach, but flies 100–150 paces away with a barely audible, melodious whistle, then flying to a distance at least as far as on their first encounter. Hunting See-see Partridges is much more difficult and less rewarding than hunting Chukar Partridges.

Unlike Chukar Partridges, See-see Partridges do not collect in large flocks in winter, but remain in small groups and do not perform vertical migrations. However, during snowfall, it disperses from the northern to the southern slopes where the snow melts faster and food is more readily available. Such a dispersal has been recorded in the Kugitang range from December to February (Salikhbaev and Ostapenko, 1964).

Genus *Alectoris* Kaup—Rock Partridges

58 The geographic distribution of rock partridges, a small, well-differen-

*[sic]; should be Mackworth-Praed and Grant, 1952—Editors.

tiated group, is associated with the mountainous relief of the southern part of the Palearctic. The genus *Alectoris* include 4–5 species. According to some, there is one species, *A. graeca* Maisner, from the Pyrenees to the eastern boundary of the range of the genus. Others consider that the range of this particular species covers only the Mediterranean region; the territory east of Balkan Peninsula is inhabited instead by another distinct species (*Alectoris kakelik* Falk.). Most Soviet ornithologists support this view. All the species of this genus are inhabitants of rocky mountain slopes. Judging from paleontological evidence, the present distribution of these birds, and their great preference for rugged terrain, the Mediterranean region and West Asia seems to be the homeland of this group (Fig. 9).

This genus apparently originated in the Neogene. According to Vul'f (1944), the Tertiary flora of the Mediterranean region was distinctly tropical. By the end of the Tertiary, under the influence of falling humidity and lowered temperature, the moist Mediterranean vegetation vanished. Elements of the temperate flora and xerophytes slowly began to spread. The Pliocene flora was already quite similar to the present-day vegetation. During this period, conditions prevailed for the origin of Rock Partridges
59 inhabiting dry terrains. The only fossil of the extinct *A. pliocenica* Tug., collected from the vicinity of Odessa, dates to the Pliocene (Tugarinov, 1940). Fossils of *Alectoris* sp. were found in the upper Pliocene deposits of southern France (Lambrecht, 1933). Fossils of *A. graeca* were discovered in Pleistocene formations in the vicinity of Baku, Odessa, and the Upper Paleolithic horizons of western Siberia (Tugarinov, 1940; Burchak-Abramovich, 1962), and of *Alectoris kakelik* Falk. From the Paleolithic sites of southern Kazakhstan (Aubekerova, 1965). The present-day distribution of rock partridges inhabiting the subtropical parts of the Mediterranean and

58 Fig. 9. Geographic range of Rock Partridges. 1—Chukar Partridge; 2—species not found in the avifauna of the USSR.

West Asia is very limited: *A rufa* L. inhabits the Pyrenees peninsula and southwestern Europe (France); *A. barbara* Bonn. occupies only the mountain of North Africa; *A. graeca* lives in southern Europe—in the Alps and Appenines, on the mountains of the Balkan Peninsula, and the Mediterranean islands; and *A. melanocephala* Rüppel lives in the western and southern mountain regions of the Arabian Peninsula.

The range of Chukar, the only species of the group found within the Soviet Union, is considerably wider.

The distribution of this species over a large territory with varied climatic conditions and marked differences in vegetation is explainable only by the great adaptability of this species. Much of its range falls in regions with continuous and rather severe winters; only in the south do Chukar inhabit the dry subtropics, which is closer in climatic conditions to those of other species of Rock Partridges.

The distribution of Rock Partridges is associated with the nature and extent of ruggedness of the terrain. They inhabit small and often isolated groups of low hills as well as mountain ranges from the foothills to the alpine belt, preferring rocky slopes with patches of grass and shrubs alternating with screes and rock outcrops. In terms of topography, the vegetation is evidently of secondary importance to Rock Partridges. The majority of these birds prefer sparse grass and shrubbery and they are very rarely seen among dense bushes and woods. Red-legged Partridges (*Alectoris rufa*), which prefer sparse pine forests and parks on mountain slopes, are an exception (Bodenstein and Neubauer, 1954). The close association of this species with tall arboreal vegetation is also evident from the fact that it usually perches on trees.

The widest range of vertical [elevational] distribution is characteristic
60 of Chukar. At places, this species inhabits low hills almost at sea level, while some subspecies in Soviet Central Asia inhabit alpine meadows 3,400–4,000 m above sea level, and ascend even higher, up to 4,500 m in Central Asia.

Chukar like other species of Rock Partridges, inhabit arid slopes of ravines where talus and outcrops are interspersed with areas of grass-cover and clumps of shrubs. One of the essential conditions is the availability of water. Judging from the very large abundance of Chukar, the most favorable habitats for this species are mountains and low ranges of less snowy and arid regions, as well as the lower regions (1,500–1,800 m) of the large mountain ranges. Furthermore, a number of subspecies live only at great heights in the high mountain regions of Central and Soviet Central Asia.

Caucasian Chukar are found on the rocky slopes of foothills with sparse grass to a height of 2,000 m (Azerbaidzhan). This species does not avoid foothills with cereal cultivation when there are rocky sections nearby.

In central Kopet-Dag, Chukar are found from the foothills to the peak (2,000–2,200 m). Here their distribution is associated with semidesert ephemerals (500–1,200 m), steppe couchgrass (700–1,700 m), feather grasssheep's fescue steppes (1,000–1,800 m), and mountain xerophytes (1,700–2,000 m). It also inhabits areas of intrazonal vegetation in juniper groves, tragacanth, milk vetches, shrub undergrowth, and various fruit trees (Fedorov, 1949).

In Turkmenian deserts Chukar inhabit not only isolated mountain ranges—the Greater and Lesser Balkans—but also habitats atypical of the species—sandy coasts of the Uzboi—in the region of freshwater lakes, among saxaul and sand acacia. It is also found on sand dunes along the middle reaches of the Murgab and Tedzhen rivers.

The distribution of Tien-Shan Chukar is associated with meadow steppe and subalpine belts. Their breeding habitats are rocky slopes with rather thin grass vegetation and scattered shrubs. In the mountains of Turkestan, they are found from 1,600 to 3,200 m above sea level; in Talassk Alatau 1,300–3,000 m; in Narynsk range, Kirgiz, and Tersk- and Kungei-Alatau slopes from the foothills to 2,500—3,000 m; and in the Carpathians from the foothills to 1,500–1,700 m.

61 In the Pamir-Alai mountain range, Chukar inhabit desert hills at 500 m and rise to 4,000 m. The topmost areas are occupied by the high-mountain Karakorum subspecies (Ivanov, 1969), while Tien-Shan Chukar inhabit the lower regions.

The northern ranges of Tien-Shan, Dzhungarsk Alatau, Tarbagatai, Saur, and Zaisan basin are inhabited by Dzhungarian Chukar. Distribution of this species is associated with isolated hills in desert areas and with mountain ranges, where they inhabit open slopes from desert foothills to alpine meadows. The upper boundary for breeding is at an elevation of 2,500–3,000 m, with the main abundance up to 1,500–1,800 m. In desert mountains, it is numerous in ravines with a sparse grass cover, in which ephemerals with thickets of scrub saltwort, steppe cherry, ephedra, and spiraea predominate. It readily inhabits slopes with mixed grass and sparse shrubs. At some places, it also lives on slopes with tall shrubs and even tall arboreal vegetation (prune, crab apple, and hawthorn) if wide expanses of talus and outcrops are available in the vicinity.

Thus the living conditions of Chukar may vary considerably in different parts of its range, although this species never crosses the limits of the mountain regions. Life in mountain regions with varied climatic conditions, distinctive types of vegetation at different heights, and slopes of different exposure have made it possible for this thermophilic species to penetrate northward and inhabit even alpine regions.

A snow cover more than 10 cm deep restricts movement of Chukar on the ground and deprives them of food. That the spread of snow cover is the main factor limiting its distribution is distinctly seen at the northern boundary of its range. Within the Soviet Altai, where the snow cover is very deep almost everywhere and where the snowfree areas on the slopes are very limited, Chukar inhabit only the southwestern regions in very small numbers. At the same time, they are quite common in the snow-free Mongolian Altai and penetrate far north into the Tanny-Ola range and some northern ranges of Mongolia, but are absent in the highly snow-covered Kentei and Khamar-Daban. The existence of Chukar in regions with prolonged and severe winters is perhaps possible only because of the uneven distribution of the snow cover and complex mountain topography: on well-warmed southern and southwestern slopes and wind-blown areas, the large snow-free areas provide sufficient food.

62 As soon as winter sets in and snow covers the mountain ravines, Chukar move to the nearby less snowy regions. In some cases, it undertakes even longer seasonal migrations in the same direction, starting in autumn before snowfall. In the Gissar-Karateginsk range, autumn migration has been traced to a distance of not less than 37 km (Popov, 1960).

In winter, as for much of the year, Chukar live in flocks. The size of the flock usually depends on the population density in a given area. The greater part of the short winter day is spent foraging and flocks are formed only occasionally while basking on rocks under the sun. Chukar feed only right under the sun, at times even descending to the foothills to feed along the ice edge. In blizzards and heavy snowfalls, this species does not venture out for foraging, but hides under the safety of rocks or stones; with prolonged bad weather, it starves. I have often observed that Chukar feed voraciously just before inclement weather sets in and snow fully covers the slopes. They roost and find cover in rocky niches, usually on the upper part of slopes, and gather there in dense flocks pressed close together.

In very severe snowy winters, when the slopes are entirely covered with a thick layer of snow, absence of food and their total helplessness in the snow lead to large-scale mortality among these birds. Their numbers are reduced to one-tenth and, at times, even to almost total extinction. If the next two to three years are favorable, however, the population is restored to its earlier level due to the high fecundity of this species. A clutch usually contains 7–23 eggs, more often 10–18; sterile birds are very rare. High fecundity, which ensures its distribution in regions with very severe climatic conditions, is an important feature of the biology of this species.

The behavior of Chukar and their anatomical peculiarities underscore its distinct adaptation to life in the hilly areas. Unlike Gray Partridges,

Black Francolins, and Common Quail, which crouch in the grass, Chukar can be seen almost everywhere on rocky slopes with sparse vegetation. In the open, they escape from enemies by protective coloration, flocking, ability to conceal themselves, and their speed in downhill flights or uphill runs along slopes in case of danger. It is extremely difficult for a hunter or
63 predator to sneak up on a resting or feeding flock of birds since one or several of them always stand "guard". They escape from predatory birds by hiding among stones, under rocks, and in bushes. The color of the bird blends so well with its surroundings that a potential "kill" is not readily detected even when fully exposed. Chukar usually escape from enemies on ground either by running or flying away.

Chukar spend their entire life on the ground among stones, rocks, and bushes, and only rarely perch on trees. They move very fast and skillfully on the ground, run freely and gracefully both on the plain and along steep slopes, and climb steep rocks with ease. They even take short runs downhill, especially during courtship. This species is always on the move except when resting. While moving undisturbed, or when clambering uphill, its front is held upright. On the other hand, when running fast along a slope, downhill, or on the plains (especially on sighting a predatory bird), it tilts its front downward and moves with a stoop. Overhanging rock cliffs and talus present no barriers to its movement. Their long thin toes with soft pads and fine curved claws help them cling to even the smallest rock projection and climb absolutely inaccessible precipices. If rock indents and footholds are wide spaced, the bird leaps, flapping its wings during ascent, to reach the next foothold, which it grasps with its prehensile toes.

Chukar rarely choose flying over locomotion on the ground. This species flies mainly while shifting to another slope or moving downhill along ravines. Upward flight over the slope is more strenuous and very rarely resorted to, and even then only for short distances. During take-off from slopes or rocks, it propels itself with its legs and hurtles downward, often flapping its wings, and then glides on a steady wing. Going down, it seldom flies in a straight line but more often describes a wide arc. It flies freely and rapidly, gliding for about 200 m and sometimes even a longer, rarely flapping its wings. The speed of flight and the distance covered increase with the height of the drop. The flight of Chukar differs from Gray
64 Partridges in being less labored but more intense. Like Gray Partridges, however, Chukar flushed from cover take to wing rapidly with noise and fly low above the ground.

Sudden vertical flight is resorted to only when the bird has no escape. In such cases, this species may rise abruptly upward ten or more meters from the rocky foothills. Active flight greatly exhausts this bird, a fact exploited in the barbaric method of "hunting" on freshly fallen snow.

After three or four forced flights, the bird is so exhausted that it can be picked up by hand.

It is very difficult for Chukar to fly long distances without rest. During migratory flights between the Gissar and Karateginsk ranges, this species has to cross the Ilyak River valley, which is 2 km wide with a height difference of 200 m from the top of the range to the foot of another. Not all birds are able to negotiate this distance; some drop into the valley after crossing the river, while others land on the same side of the river and return on foot to attempt the difficult flight once again on the following evening (Popov, 1960). Apparently, this distance of 2 km with small drop in height is the limit of the flight capabilities of this species, since it is accomplished with active flight.

Genus *Francolinus* Stephens—Francolins

This is the most polymorphic genus of galliform birds. According to Peters (1934), it includes 32 species, while Hall (1963) recognizes 40

Fig. 10. Geographic range of Francolins. 1—Black Francolin; 2—species not found in the avifauna of the USSR.

species. The present-day range of francolins comprises two isolated areas. In Africa, 26 species are known south of 17° N (if we accept Peters' number of species) and constitute the nucleus of the order Galliformes in this area. Only six species are known in West Asia, southern Asia and Trans-Caucasia (Fig. 10).

Until the middle of the nineteenth century, the northernmost species of this genus, *F. francolinus* L. [Black Francolin] inhabited southern Europe; it was introduced into southeastern Spain, Sicily, and the islands of the Aegean Sea. Fossil *F. capeki* (Lambrecht, 1933) were discovered in Hungary and Czechoslovakia and date back to the Pleistocene.

The majority of francolins inhabit dry grass undergrowth and scrub thickets. Considering that other genera of quail of the family Phasianidae
65 living in Africa, e.g., Painted Quails (*Ptilopachus* spp.), inhabit similar terrains, there is little doubt that the ancestors of the present-day African genera of phasianids were birds of the open country. It is quite probable that the other ecological "niches" in tropical Africa were not inhabited in the past by galliform birds, which provided other habitats for the more abundant and adaptable genus *Francolinus*.

Mountain species such as *F. africanus* Stephens, *F. sephanea* Smith, and *F. clappertoni* Children may be distinguished, and by their habitat are reminiscent of Palearctic rock partridges.

A small group of francolins inhabiting tropical forests became separated; some of these species (*F. shelleyi* Ogilve-Grant, *F. cogui* Smith, and *F. squmatus** Gassin) occupy glades and fringes of large forests. Some had already become ground-arboreal species (*F. lathami* Hartlaub and *F. nachani*** Dubois). Both species roost in trees, and the latter [Nahan's Forest Francolin] is known to nest in the hollows of trees and not on the ground (Mackworth-Praed and Grant, 1952).

Thus, the genus *Francolinus* constitutes a large group of galliform birds, and is undoubtedly associated in its origin with the open, arid terrain of Africa.

66 A very small northernmost part of the geographic range of one species of francolins extends into the Soviet Union. This species inhabits Trans-Caucasia and the south-western regions of Turkmenia. The possibility of its extension farther north is restricted by climatic factors, primarily the depth and duration of the snow cover.

The francolins in Trans-Caucasia are recent immigrants that penetrated from the south during the warm and arid epoch of the postglacial period (Vereshchaigin, 1960). The absence of fossil remains of francolins

*[sic]; should be *F. squamatus*—Editors.

**[sic]; should be *F. nahani*—Editors.

in the Middle Quaternary "Binagadinsk fauna" of the Apsheron Peninsula and the evolution of the Quaternary environment of Trans-Caucasia support this assumption. Had francolins been present in Trans-Caucasia in the preglacial period, they would have been forced out of this area by heavy snowfalls in the period of maximum cooling and glaciation.

The habitats of francolins are somewhat diverse, and localities are preferred where vegetation affords good protection and water is readily available.

In Trans-Caucasia, the most preferred natural biotopes of francolins are the scrubby tugai vegetation consisting of tamarisk, dewberry, Russian olive, sallow-thorn, and patches of reed. Extensive cultivation in these areas, however, has reduced the natural biotopes to small patches and francolins have now turned into birds of "cultivated" land. During the nesting period, a very high population of francolins is found in orchards, in cotton, alfalfa, and grain fields. All the same, in winter these birds prefer natural habitats.

In southern parts of Turkmenia, the range of Black Francolins is confined to lake shores ringed with reeds and overgrowth along irrigation channels consisting of various rushes, camel's thorn, wormwood, and saltwort, with scattered bushes of *Nitraria schoberi* and tamarisk. The height of these plants normally does not exceed a meter and is often considerably less.

In Turkmenia, Black Francolins live among weeds and shrubs without marked seasonal migrations, while in the Caucasus, migrations are associated with changes of habitats are well defined. In October-November, a large number of birds migrate from nesting habitats such as cotton, alfalfa, and grain fields to tugai, reeds, orchards, and brushwood where they remain until spring. The winter habitats of birds during this period are more abundant in food and also protect them better from wind, snow, and a variety of enemies.

67 Black Francolins live in pairs, almost always among dense vegetation where they move freely among tangled grass and branches of shrubs. Their medially compressed body helps in such movement. These birds are found in open areas, not far from overgrowth, and only during feeding or breeding when the males lose their guard and permit approach within 20–30 m.

Black Francolins are ground birds; they run exceptionally fast and skillfully among dense vegetation as well as on open land. Usually even a dog finds it difficult to pursue a running bird. The speed of a running francolin in open areas has been likened to that of a flying stone. The running bird tilts the front part of its body toward the ground, stretches its neck, and moves its legs so fast that they are difficult to perceive. These cautious birds find maximum protection among dense grass or shrubs and therefore resort to flight only in extreme cases.

Having noticed a hunter, francolins remain in reeds or shrub undergrowth and can conceal themselves well and take to the wing at the very last moment. In areas with sparse vegetation, they escape by running, but at times crouch even here. An immobile bird is difficult to spot even on bare ground, to say nothing of a bird in cover.

From an open area or short grass, these birds rise low, almost noiselessly, at an angle and then fly in a straight line alternating wing strokes and gliding. Their flight is almost silent, and slower than that of Gray and Rock partridges. From tall undergrowth or solitary bushes, the birds take off almost vertically, but, after rising slightly above the surrounding vegetation, fly in a straight line. Such a flight is noisy, but much less than in the case of Gray Partridges and more so than the pheasants. An undisturbed bird flies quietly while a frightened bird rises with a low cry. The flight range as well as takeoff depend on the surrounding area. A bird on the wing tries to remain in the air for as short a time as possible, descends hurriedly and hides in bushes or grass. It generally flies only a few tens of meters, but in some cases can negotiate up to 250–300 m. It has been observed that the take-off exhausts the bird greatly. If forced to rise repeatedly in quick succession, its energy is so drained that the bird can be picked up by hand. Francolins were "hunted" in this manner by the Turkmenians after a snowfall.

68 Considering the restricted flying capacity of francolins and their preference for dense bushes, Starosivil'skii's (1897) observation about their regular flights to fields 2 to 3 km away for foraging raises doubts.

The northward distribution of francolins is determined by the depth and duration of the snow cover. In the Trans-Caucasia, their range boundary passes through areas where the snow cover persists for about 30 days and its depth does not exceed 20 cm.

Low temperatures are less important to these southern birds than is snow cover. There are reports of francolins satisfactorily enduring frosts of even —20 to —25°C hiding in bushes in flocks. During heavy snowfall, however, they die *en masse.* During snowfall, francolins gather in flocks under large bushes and in warmer shelters, waiting for the inclement weather to pass (Vereshchagin and Dyunin, 1949). But when the snow cover is deep mortality due to starvation is high for they cannot search food under the deep snow.

The extent of mortality in frosty winters can be judged from Khanmamedov's (1956) data: in the severe winter of 1949–50 in the region of a Kurdamirsk coypu farm in Azerbaidzhan, 75% of the population died (300 bird carcasses were gathered). Almost all the birds dissected had no food in their gizzards except a few which contained the remains of green sea scirpus.

The continuing survival of francolins in the Trans-Caucasia, in spite of the periodic decimation of the greater part of their population, is ensured by their high fecundity, much higher than among African species.

Genus *Tetraogallus* Gray—Snowcocks

This genus comprises five alpine, highly specialized species, which are ecologically quite distinct from other members of the family. All species of snowcocks are found in the Soviet Union, of which Caucasian Snowcocks (*T. caucasicus* Pall.) are endemic to our fauna (Fig. 11). Two differing views exist concerning their alpine origin.

Kozlova (1952, 1960) considers that this genus originated in the high mountains of Central Asia. Later it dispersed to the north along mountain ranges, and the original form split into two branches: the Mongol-Tibetan branch (with two species—Tibetan Snowcocks, *T. tibetanus* Gould and Altai Snowcocks, *T. altaicus* Gebler) and the Caucasian-Himalayan branch
69 (with three species—Himalayan Snowcocks, *T. himalayensis* Gray, Caucasian Snowcocks, *T. caucasicus* Pall., and Caspian Snowcocks, *T. caspius* Gmelin). This view is based on a study of age-related plumages and present-day distribution of the species.

Many researchers (Shtegman, 1938, 1946; Vereshchagin, 1947; Bannikov, 1958; Stepanyan, 1958, 1960a) have pointed to the origin of several species and genera of birds and mammals in the mountains of Central Asia with subsequent dispersal westward. Thus, Kozlova's view agrees with the established opinion that the center of origin of a number of terrestrial vertebrates lay in the high mountains of Central Asia, with their subsequent dispersal westward through the mountain ranges.

Fig. 11. Range of Snowcocks. 1—Caucasian Snowcock; 2—Caspian Snowcock; 3—Tibetan Snowcock; 4—Himalayan Snowcock; 5—Altai Snowcock.

The second view holds that the ancestor of snowcocks lived in the territory presently inhabited by snowcocks until mountain ridges began to form. Speciation in snowcocks progressed concomitantly with orogenesis (Beme, 1960; Baziev, 1965, 1972). In my opinion, this point of view is less convincing.

Notwithstanding diverse views on the evolution of snowcocks, the prolonged evolution of this group in high mountain ranges is undisputable. This also explains the narrow specialization of snowcocks for mountain life, which influenced their life style, feeding, behavior, and morphology.

70 The habitats of individual species of snowcocks in generally similar alpine conditions are sufficiently different that one species is strictly confined to alpine regions while others have a fairly large vertical distribution.

Altai Snowcocks occupy the northernmost part of the range of the genus, and live in mountain massifs with extremely severe and protracted winters. In the Altai, this species inhabits the rocky tundra of cirques and the near-peak part of mountains with rocky talus and sparse vegetation of tundra grasses, dryas, columbine, primrose, and rarely dwarf birch and bergenia. Patches of snow are found in the upper boundary of such areas even in midsummer (Irisov and Totunov, 1970). In the upper reaches of the Bukhtarma River, nesting sites of this snowcock have been found on escarpments 1,300 m above sea level, among rocks with bushes of spiraea and in grassy patches, where grass and sedge predominate. The upper boundary of its range passes above 3,000 m through steep, steppe-like, grass-covered slopes and wide plateaus of rocky moss and lichen tundra.

In western Sayan, Altai Snowcocks inhabit rocky slopes of mountains in the alpine zone at somewhat lower levels; the maximum height of the range is only 2,500 m (Sushkin, 1914). In the Mongolian Altai, where the terrain is more even, this snowcock occupies more gentle slopes. In the Khangai high mountains, it inhabits screes in the rocky tundra close to the lower limits of the permanent snowline and on outcrops in the subalpine belt. In the Ikhe-Bogdo region, the most preferred habitats of this bird are the sloping passes between ravines covered with mountain-steppe vegetation or the meadow slopes of ravines (Kozlova, 1930, 1932).

In winter, well-defined migrations of Altai Snowcocks are observed in many regions; these migrations are caused because of the very deep snow on the peaks of mountain ranges. At this time, in the Sayan ranges, the bird is commonly seen in the foothills on low rocky mounds. In the Mongolian Altai, it descends to passes above the forest boundary and, during particularly severe winters, even to river valleys where it remains in flocks among herds of domestic cattle.

Tibetan Snowcocks inhabit the highest mountain ranges, where they keep to the steep, rocky slopes of deep ravines in the upper part of the alpine belt. In the Pamir range, it lives at heights of 4,300–5,000 m (Ivanov, 1969), only slightly below the snowline. At places, however, it ascends even higher after the chicks have hatched; these birds have been found to roost
71 at heights of even 5,200–5,400 m (Abdusalyamov, 1957). On the Tibetan hills, it lives above 4,500 m, at places with rocks outcrops or talus (Kozlova, 1952). In western Himalayas, the upper limit of its range in summer reaches 5,800 m. Potapov (1966) ascribes the Tibetan Snowcock's preference for particularly great heights to its hygrophilous nature.

Seasonal vertical migrations of Tibetan Snowcocks are very distinct. In summer, after the chicks hatch, the birds ascend to the highest altitude; in winter, when the snow becomes very deep, they descend to the lowest areas. In the western Himalayas, in winter, they live at heights of 3,000–4,000 m. In the Pamirs, in very snowy winters when crossing the valleys is possible, according to hunters' reports, they descend to the foothills; in less snowy winters, they remain at much higher altitudes (Meklenburtsev, 1946). Vertical diurnal wanderings, associated with foraging, have also been observed. Bianki (1907) notes: "At Dza-Chu (Mekong) in September, when the Tanguts harvested barley, Tibetan Snowcocks often descended from the hills to feed on the grain. My (laboratory) assistant shot birds in the mornings in these fields without having to climb the hills."

The range of Himalayan Snowcocks covers mountain ridges differing widely in height, geomorphology, climate, and plant cover. The limits of altitudinal distribution depend considerably on the latitude of the area. In the Pamirs, Turkestan Snowcocks are not found below 3,800–4,000 m during the breeding season; after the chicks have hatched, footprints of these birds have been found up to 5,200 m (Ivanov, 1969). In the Gissar, Zeravshan, and Turkestan ranges, this snowcock descends to 2,400–2,600 m, the upper limit of its summer range being 3,800 m (Popov, 1959).

In Boro-Khoro, *T. h. sewerzowi* Zar. nests at 900 1,500 m (Shestoperov, 1929) and the lower limit of its nesting range in some ravines of the northern part of the Dzhungarian Alatau has been recorded as 1,200 m. The vertical distribution of this bird apparently depends on a number of factors, of which topography and vegetation are very important. Hence, even in the close-set ravines of a single hill, its vertical distribution may vary markedly.

Habitats of Himalayan Snowcocks are more diverse than those of other species, but one essential condition is the presence of steep mountain slopes with rocky outcrops over a large expanse, which allow these cautious birds a broad view of the area.

72 During the breeding season, these snowcocks are found even in alpine meadows and in the subalpine zone where they nest under juniper bushes and in the meadow-steppe slopes of the forest zone. In the Kul'dzhinsk hills of Boro-Khoro, they are found at 900—1,500 m on rocky slopes overgrown with small shrubs, and in the ravines of the northern part of the Dzhungarian Alatau, they occupy steep slopes with rocky outcrops and vast areas of large rock screes. Here, the vegetation consists of grasses, the main species being sheep's fescue, xerophytic herbs, bushes of ephedra, spiraea, and patches of juniper. The steppe environment of these places is characterized by its typical inhabitant, the souslik*, which lives alongside the snowcock on the same slopes, but in less steep localities and with less rocky ground.

The wide vertical distribution range of Himalayan Snowcocks and the habitat diversity in various vegetation zones emphasize the high ecological adaptability of this species compared to other snowcocks.

The range of Caspian Snowcocks is associated with subalpine and alpine zones. Its vertical distribution in the Caucasus lies between 1,800 and 3,500 m, or may be slightly higher at certain places. Caspian Snowcocks inhabit mostly steep slopes, where patches of xerophytic meadows alternate with screes and precipices, and reaches rhododendron bushes. Many of the mountain ranges inhabited by snowcocks lack perpetual snow; however, these birds exhibit a special preference for places where patches of snow remain longer. Snowfall in mid-October forces these birds to descend to their nesting grounds, but even in the most severe winters they do not come up to the upper boundary of the forests (Spangenberg, 1948).

The life of Caucasian Snowcocks is associated with the alpine zones. In spring, with the beginning of vegetative growth, snowcocks descend to the upper part of the subalpine zone where the vegetation comes up earlier, but they never nest there.

The most preferred habitats of this species are the rocky, barely accessible alpine areas with large rock outcrops interspersed with alpine herbage meadows. Bare, flat rocks hold no attraction to these snowcocks. The upper limit of their distribution ranges from 2,600 m in winter to 3,400-3,700 m in summer (Baziev, 1962). In winter, they descend to nesting grounds on slopes with a southern exposure, where snowfree areas are always available due to solar radiation. In addition to the usual short seasonal movements of individual populations, Caucasian Snowcocks are
73 also known to undertake regular seasonal migrations (Baziev, 1968). Unlike the usual migrations forced by heavy snowfall, migrations of transitory populations commence long before the onset of unfavorable

*Ground squirrel *Citellus* spp.—Editors.

winter conditions and are performed along the old established routes for up to 20 km. Such migrations of Caucasian Snowcocks, according to Baziev, are the result of the former glaciation of some ravines.

The severity of climatic conditions in high mountains is comparable to that of high latitudes—prolonged cold winters and cool summers. The fairly long winter day in the mountain regions of the southern Palearctic, however, and the uneven distribution of the snow cover on slopes of various degrees and exposures create, more favorable wintering conditions for animals foraging on the ground despite low temperatures.

Winter life of Himalayan Snowcocks of Tersk-Alatau was studied by Zimina (1956). Below I include some data from this interesting work. In winter when the alpine ranges are covered with deep snow, this bird descends to 300–400 m and often migrates 5 km or more from its summer grounds. At this time, these birds remain in flocks of 8–10, but when feeding, many such flocks gather sometimes in somewhat larger flocks.

Himalayan Snowcocks spend the greater part of the daylight hours foraging. Between 7:00 and 8:00 a.m., they fly out of their permanent roosting site in rock niches, under overhanging ledges, or in juniper bushes on the upper part of the slopes and go to the much lower parts of south-facing slopes. They forage until almost midday, gradually climbing along the slope, then rest from 2:00 to 3:00 p.m. on rocks or on the ground and readily dust bath on dry patches. After resting, they resume their downward flight, usually not going far and while feeding gradually return to their roosting sites, which they reach in total darkness. During such diurnal movements, they descend about 50–100 m; flights of 200 m from the roosting places are exceptional.

During cloudy, snowy weather, when foraging is difficult, these birds do not rest but forage throughout the daylight hours. Deep snow cover not only deprives them of food but also makes their movement difficult. In spite of being heavy and short-legged, these snowcocks move freely on soft 5–10 cm deep snow; they move with difficulty on 10–15 cm deep snow, and their movements are halted on 15 cm deep snow. On soft, deep snow,
74 snowcocks usually move on foot, gradually making a wide, well-trod path.

After snowfall, these birds dig in shallow snow at their feeding sites like hens scratching the ground. Digging the soil in search of food is characteristic of all phasianids found in the Soviet Union, but digging in snow for this purpose is known only for snowcocks, Gray Partridges, and pheasants. This ability of birds should be regarded as a very important adaptation facilitating the procurement of necessary food in difficult winter conditions. Snowcocks are greatly helped in their food procurement by ibexes and boars in Tien-Shan and by turs and chamois in the Caucasus (Nasimovich, 1947).

The co-occurrence of snowcocks and ungulates and their ability to dig in snow enables them to obtain food even during severe winters. It is for this reason that snowcock populations do not decrease appreciably even in years with unfavorable winters. Life in alpine regions with rugged terrain calls for special skills in these large ground birds. In snowcocks, rapid dispersal to long distances at lower heights, or flight at the same level to an adjacent ravine are accomplished through active flight when descending. They climb the slopes, however, simply by hopping.

More complete observations are available on the locomotion of Himalayan Snowcocks. Whether walking or running, these birds lift their legs high at each step, but do not place the foot far away. Even though their strides are not long, these birds move with ease along steep slopes. Because of such a gait on level land, a bird with a wing injury can be chased and caught without difficulty (Shtegman, 1954). The advantage of such a gait is distinctly revealed when chasing a bird uphill. Even when the bird does not climb particularly fast hopping from one rock to another, catching it is almost impossible. Himalayan Snowcocks run with ease but not rapidly; a running bird holds its head high somewhat tossed back, with its tail pulled up on the back so that the lower coverts are distinctly visible. Moving downhill, it does not stand on its toes, but walks very slowly, as though creeping, with the entire metatarsus resting on the ground. In moving uphill, it easily negotiates even steep rocks by flapping its wings and hopping.

This unique behavior of snowcocks, as well as their flight specializa-
75 tion, allows them to move very freely on steep slopes. This explains why snowcocks rarely stay in the valleys of ravines or on plateaus. Only Altai Snowcocks venture onto level plateaus to forage for about 1 km from the nearest rocks.

The unidirectional flight specialization of all snowcocks, revealed in their rapid glides downhill, made these birds almost completely lose their ability for active flight and for takeoff from level ground. When trying to rise from a flat surface, Himalayan Snowcocks often flap their wings and run, and only then are they able to takeoff with some difficulty. Often birds when surprised on a level surface, run to the nearest slope and only then get into a gliding flight. Flight from steep slopes or rocks is quite effortless. Pushing back with its legs, a bird swoops downward and soon changes over to a rapid glide. It flies either in a slightly inclined straight line or, during descent, describe a wide arc and cease flight only on landing. In the last lap of such a flight, the bird bounces sharply upward due to inertia.

The flight of Altai Snowcocks is more powerful. A running bird may take off even from a level surface. Besides downward passive flight, snowcocks, like partridges, are capable of flying short distances by frequent, almost vibrating wing strokes alternated with gliding on spread wings (Kozlova, 1930; Sushkin, 1938). Tibetan Snowcocks also have not lost their ability to take off from flat level surfaces.

Observations of the flight of a Caucasian Snowcock escaping from a predator revealed that, when necessary, this species can fly very fast and with considerable maneuverability. In such situations, the bird seeking protection makes sharps turns during fast flight, making long loops both horizontally and vertically (Noska et al., 1896). The takeoff of Caucasian and Caspian snowcocks even from gentle slopes close to the peak is very difficult; these birds are not found at all on completely level surfaces.

Rapid gliding with loss of height, which requires minimal muscular power, is characteristic of all species of snowcocks and is most economical in the highly rugged terrain.

The evolution of snowcocks in alpine regions led to the formation of
76 a group of galliform birds with a very narrow adaptation to severe climatic conditions and rugged terrain. This is clearly manifested in their behavior (flight, gait), composition of food, and development of markedly thick plumage.

Genus *Phasianus* Linnaeus—Pheasants

The discontinuous range of two species of this genus, extending from the islands of Japan to Bulgaria, is populated by a large number of distinct forms. These forms represent a continuous series of variation in morphological features and have been assigned varying taxonomic ranks (Fig. 12). According to Buturlin (1935), the genus *Phasianus* includes 20 species with 46 subspecies. Such a detailed division was also accepted by Zarudnyi

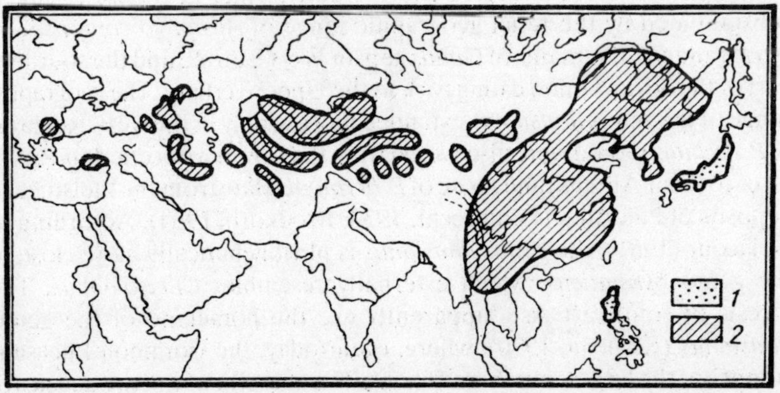

Fig. 12. Distribution of Pheasants. 1—Species not found in the avifauna of the USSR; 2—Common Pheasant.

(1896). Hartert (1905) and Peters (1934) held a diametrically opposite view, considering this genus to be monotypic. According to the prevailing view, the genus has been subdivided into two well-defined species—the Green Pheasant (*Phasianus versicolor* Vieill.) and the Common Pheasant (*P. colchicus* L.); the latter includes 31-37 subspecies, according to different authors (Ivanov, 1951; Delacour, 1951).

The numerous subspecies of pheasants have been broadly divided into two groups by Kozlova (1970) on the basis of coloration and plumage pattern: eastern (type *P. c. pallasii—P. c. torguatus**) and western (type *P. c. colchicus—P. c. mongolicus*). In males of the western group, the color of the rump and upper tail coverts differs slightly from the rest of the back
77 and the plumage lacks a grayish-blue tinge. The distribution of this group of subspecies extends from the western part of the range of the species of Pamir-Alai and the northern mountains of Tien-Shan in the east. The eastern group of subspecies comprises males with the tail coverts distinctly differing from the rest of the back because of a grayish-blue tinge. This group includes the subspecies of pheasants inhabiting the eastern part of the range extending west to the western border of the Kobdosk valley and the western border of the Nan'-Shan mountains.

Both groups of pheasants are represented in the avifauna of the USSR. Manchurian Pheasants (*P. c. Pallasii* Rothsch.), occupying the northernmost part of the range of the eastern group, live in the Far East, while the subspecies of the western group inhabit southern Europe, Soviet Central Asia, and Kazakhstan.

Southeast Asia is the homeland of the entire subfamily Phasianinae. besides the 12 genera of Pheasants, the subfamily includes Argus Pheasants, Peafowls, Jungle Fowl, and *Afropavo congensis.* In the past, members of this group were distributed considerably farther in the west. This is substantiated by the relict geographic range of *Afropavo congensis* in Africa and fossil remains of *Gallus* spp. in West Europe and the southern part of the Soviet Union dating back to the Upper Tertiary. The geographic range of the genus *Phasianus* extended more widely in the west. Remains of *P. archiaci* Gould were discovered from the Lower Miocene deposits of Greece (near Athens) and those of *P. hermonis* Bate from the Pleistocene deposits of Palestine (Lambrecht, 1933; Brodkorb, 1964). According to Delacour (1951), the genus *Phasianus* is phylogenetically more close to the genus *Syrmaticus,* which externally resembles *Chrysolophus*. The forests of southeast Asia apparently are the homeland of the genus *Phasianus* (Kozlova, 1970), where, even today, the Common Pheasant comprises the largest number of forms. The ancestor of the two species of

*[sic]; should be *P. c. torquatus*—Editors.

this genus probably resembled the group of eastern subspecies (Kozlova, 1970), which have dove-colored and green-tinged back and tail coverts.

In his description of the habitats of the Common Pheasant, Delacour (1951) states that, unlike other genera of pheasants whose distribution is chiefly associated with forests, this species predominantly occupies open terrain. In my opinion, it is not quite correct to regard this species as a bird of the open country. Due to its exceptional adaptability it inhabits the most diverse biotopes and open lands and impenetrable tugai, always selecting well-protected areas close to water. Reliable shelter, presence of water holes, and depth of snow cover are the main factors determining the
78 distribution of this group of birds. The absence of fairly reliable shelter leads to mass mortality among these birds especially in winter when many carcasses have been found on snow.

Green Pheasants live mainly in low-lying areas, valleys, and plateaus, as well as in rice fields and tea plantations. It rarely reaches a height of 1,000 m, inhabiting sparse deciduous forests with meadows and scrub undergrowth.

Natural conditions differ greatly in different parts of the range of the Common Pheasant. These include the seacoasts with temperate climates, southern regions with short winters and temporary snow covers, and desert oases with hot summers and protracted but less snowy cold winters where temperature drops to —35 and —40°C (Mongolia and Balkhash). This species inhabits the northeastern limits of its range in particularly severe conditions, in the Primor'e, where deep snow and severe frosts create very difficult wintering conditions for birds. Common Pheasants live on plains and mountains, often up to 2,800 m above sea level. In much of its range, it inhabits "cultivated" regions and spreads from there to new territories. In the Primor'e, the chief habitat of Manchurian Pheasants is shrub and wooded vegetation in the broad river valleys. With the felling of forests, clearing of areas for agriculture, and the growth of bushes in such clearings, these birds moved farther north of their original range and began inhabiting oax forests with thick undergrowth, which offered them good protection and abundant food (Shul'pin, 1936).

The habitats of Common Pheasants are extremely diverse in the western part of its range (Soviet Central Asia, Central Asia, Kazakhstan, and the Caucasus). They are found in oases, deciduous mountain forests, and valleys brought under cultivation. They live in inaccessible tugai and thickets of rushes along the shores of lakes, banks of rivers, and especially in places overgrown with Russian wild olive, willow groves, Asiatic poplar, and bushes of salt tree, dog rose, sallow-thorn, and other shrubs interspersed with patches of meadows. They also inhabit foothill shrubs, the deciduous forest zone, and Nan'-Shan mixed forests. Vertical distribution

79 in the various mountain ranges varies quite considerably. In Nan'-Shan, they are found up to the scrub vegetation of the subalpine zone (2,800 m); in northern Tien-Shan, they do not cross the limits of deciduous forests (1,500–1,600 m), while in the northern Caucasus the birds are more restricted (700 m above sea level).

Their undemanding habitat selection, utilization of agricultural land and forest clearings to expand their territory, and life in regions of continuous and rather severe winters point to the extreme endurance and high adaptability of Common Pheasants. The extension of the northern boundary of the range of these pheasants in the Far East due to the felling of forests and expanding agriculture is particularly noticeable (Shul'pin, 1936; Nasimovich, 1950). Until the 1920's, pheasants were encountered along the Zee River to 51°N, 300 km northward of the earlier known limit. By 1954, the boundary of distribution of pheasants along the Zee River had reached 53°40' N and along the Amur Valley 53°10' N (Barancheev, 1964).

Agriculture in Soviet Central Asia also proved favorable for pheasants. Irrigation in arid regions, construction of irrigation channels, and expansion of agricultural areas favor penetration of Common Pheasants into the previous desert lands. Here it inhabits weed-infested areas and scrub near fields of corn, millet, clover, sesame, cotton, and other crops. Construction of large dams on the Syr-Darya and Ili, as well as increased flow in the Chu and Talas rivers, have led to the desertification of the lower reaches of these rivers and adjacent areas. This has resulted in a steep drop in the pheasant population in the southern regions of Kazakhstan.

The great ecological adaptability of pheasants has helped in their successful and wide introduction in western Europe and the USA. In these countries they mainly occupy cultivated regions; at places their population density is higher than in the natural environment of their homeland. They readily inhabit irrigated areas rich in food, and which provide good shelter (Ward, 1955).

Winter and deep snow cover are inimical to pheasants. The northern boundary of distribution for much of their range coincides with the less snowy plains, foothills, and continental dry plateaus where the depth of snow cover usually does not exceed 10 to 20 cm (Formozov, 1946).

Heavy winter snow deprives these birds of food under the snow cover, hinders movement, and, in most cases, leads to mass mortality. In such years the pheasants suffer considerably less in tugais where they feed on berries of Russian wild olive and sallow-thorn; mass mortality of these birds has been recorded at places where they feed solely on the ground (Stroganov, 1946).

80 Voluminous data on the population dynamics of pheasants from 1875 on are presented in Naimovich's (1950) work; it is clearly shown that deep

snow cover is the primary factor in bird mortality during severe winters. Winter with frequent icing are also unfavorable.

Deep snow not only deprives pheasants of the opportunity of forage on the ground, but impedes their movement as well. The bulky birds readily sink in the snow and cannot move through it. When the snow cover is 10—15 cm deep, pheasants usually do not peck in it for food and, when there is a frozen snow crust, even a thin layer of snow hinders foraging.

Pheasants are usually resident birds of small territories and migrate from the biotope to another only in search of food and shelter. In some regions, however, they do undertake distinct migrations. On the Trans-Ili Alatau, vertical migrations to less snowy foothills have been recorded.

In the lower reaches of the Ili River (Balkhash region), pheasant migrations (based on the presence of gastroliths in their gizzards) of at least 150 km have been recorded (Shtegman, 1946).

In the Far East, mass migrations of pheasants from China have been recorded, especially in heavy-snow winters. In 1960, in a stretch of 300 km in Amur area, pheasants appeared as early as in September while mass migrations were observed in November. They moved on foot mostly in mixed flocks with males predominating. Despite the availability of food, mortality was very high. One hunter found 106 bird carcasses, 86 of which were males (Barancheev, 1965).

In winter, pheasants roost on snow in reed beds or in thick bushes well-protected from the wind. They feed for the greater part of the day depending on food availability in the area, foraging for weed seeds on the ground or nibbling at berries on trees and bushes. At times, they dig into the snow in search of food. In between feedings, they rest in open snow holes near the roosting sites. However, in the Far East, like tetraonids they forage in the snow for food. According to Sleptsov and Gorchakovskaya (1952), in winter when the snow is not very deep, pheasants escape danger by running, and if they take to wings, fly for barely 200 to 300 paces and then land. If the snow is deep and loose, they dive into it and sometimes slide for 2-3 m in it.

81 Over much of their range, pheasants inhabit reed beds, fields, and bushes; in fact, their entire life is spent on the ground. In autumn and winter, they lead an arboreal life in the tugai and deciduous mountain forests, feeding on berries of Russian olive, sallow-thorn, and acorns; sometimes they roost in trees.

These birds mostly forage among trees, especially in heavy-snow winters when food on the ground is inaccessible. Also, roosting in trees provides protection from ground predators. Even when leading an arboreal life, pheasants spend most of their time on the ground where they feed and rely on their fast legs to escape predators. On the other hand, they are less adept at hopping from branch to branch, as shown by their awkward movements.

A detailed description of pheasant behavior has been given by Tugarinov and Kozlova (1945). Salient features from their work, supplemented with observations from other sources, are given below in describing the behavior of this bird.

Pheasants are cautious birds. They are always alert while feeding in the open or moving through bushes. The thick undergrowth, however, offers a greater sense of security to it.

Pheasant movements are light and free. Undisturbed, they are rather slow; they hold the head erect and immobile while moving forward. In the breeding season, males move along definite paths, halting from time to time and giving a mating call, and then fluttering their wings so rapidly that the eye cannot follow the strokes. In this period, the birds are more active, and often walk hurriedly or even run.

Pheasants run very fast not only in the open, but also in dense undergrowth, maneuvering deftly among bushes and leaping easily across irrigation channels without using their wings. Sensing danger, they run away or crouch; only when the enemy is very close (and even then, only when absolutely necessary), they take off vertically, rise above the bushes, fly a short distance, and then land in a thicket. The purpose of such a flight is to quickly reach another cover avoiding flight among dense vegetation.

The flight of pheasants out of thickets is unique. It consists of a very powerful takeoff followed by short maneuvering. A great deal of energy is required for such takeoffs; a strong push with the legs with simultaneous strong wing-strokes imparts the required momentum for further flight.
82 With wings folded and slightly turned back, they pull out of thickets with a noise and, rising 4–5 m above vegetation, change to a horizontal flight along a straight line or gentle arc, and alternate gliding on steady wings and flying with frequent wing strokes. On approaching the landing site, birds descend slightly and land almost vertically, braking with a lowered tail and balancing with wing strokes. When a flying pheasant encounters an obstacle like tall vegetation, it does not immediately gain height. Flying up to the tree, it continues to maintain forward movement, lowers its tail and then flaps its wings rapidly as in takeoff. Having risen above the obstacle, it again takes to horizontal flight in the same direction.

Apart from the characteristic rapid fluttering flight, Common Pheasants exhibit great maneuverability (Shtegman, 1950). Sensing danger, a flying bird can turn suddenly almost at 180° around its own axis and fly along the same course. During takeoff in the open, it rises at an angle and, after gaining height changes to horizontal flight. Sudden ascent exhausts the birds. A flushed pheasant which has risen several times in succession becomes so exhausted that it can be picked up by hand. Common Pheasants, which prefer to walk rather than fly, do not take to wings willingly. Flights of 200–300 m are resorted to as a means of escaping enemies, after which birds either land in thickets and crouches or run away. However, during courtship and in winter, regular migratory flights have been recorded. In the lower reaches of the Ili River, pheasants undertake flights from places of roosting to feeding sites and back, often distances of 0.5–1 km (Shtegman, 1945). In Hungarian game sancturies, pheasants fly up to 3 km to reach feeding trays put out regularly. Evidently, it is not the duration flight that tires the bird, but the "rocketlike" takeoff which burns considerable muscular energy.

In behavior, pheasants come somewhere between tetraonids and phasianids. Feeding on berries and sometimes on buds, and utilizing snow cover for shelter (e.g. Manchurian Pheasant), they are reminiscent somewhat of tetraonids, but in other aspects they are typical phasianids. These differences emphasize the special adaptability of pheasants and their great endurance.

Feeding

83 A decisive factor in the evolution and dispersal of a species is its ability to utilize a given assortment of food. Thus, characteristic feeding patterns are produced for each individual species or a group of species evolving under similar conditions.

The food of galliform birds of the Soviet Union can be broadly classified into eight groups: *twigs*—needles and fresh cones of conifers ("winter crop"), twigs of shrubs, buds, and male catkins; *green food*—leaves of herbaceous plants, evergreen undershrubs, shrubs, and trees; *flowers* — flowers of herbaceous plants, evergreen undershrubs and shrubs; *seeds*—seeds of wild and cultivated plants, pods, bolls, and other fruits of herbaceous plants, nuts and acorns; *berries*—berries of herbaceous, shrubby and woody plants (including the false berries of juniper); "*roots*"—bulbs, tubers, roots, and rhizomes; *lower plants*—mosses, algae, lichens, and mushrooms; and *animals*—invertebrates and vertebrates.

Each group has a certain nutritive value and generally calls for a special effort to obtain it (e.g., twigs of trees, "roots"). Isolation and the successful dispersal of tetraonids in regions with prolonged snowy winters became possible due to an extensive use of twigs protruding above the snow as food.

Underground plant parts totally absent in the diet of tetraonids are an important constituent in the diet of many species of phasianids.

84 In reviewing the food of galliforms, special attention has been paid to the seasonal changes in food and changes in the biochemical composition of their diet during different periods of their life.

Family TETRAONIDAE

Genus *Lagopus*—Ptarmigans

In foraging habits, the species of this genus fall between forest tetraonids and phasianids. These birds have a wholly terrestrial mode of life. In winter they run on the snow from one bush to another, feeding on the twigs and buds or select herbaceous plants, digging them out of the snow as the Rock Ptarmigan does. Foraging in trees is not common among these birds, but has been observed at times in some regions with extremely heavy snowfall. Chicks early in life feed mainly on invertebrates, but small seeds and green grasses are also found in their diet from the first day.

The food of Willow Grouse has been studied in the various geographic zones—from the tundra to the steppes: Kola Peninsula (Folitarek, 1939; Novikov, 1952; Semenov-Tyan-Shanskii, 1959); Karelia (Ivanter, 1974); Timansk tundra (Mikheev, 1948); southern Yamal (Dunaeva and Kucheruk, 1941); Taimyr (Pavlov, 1971); Kotel'nyi Island (Gorodkov and Korotkevich, 1957); central and eastern regions of the European RSFSR* (Rusakov, 1963; Oliger, 1973); northern Kazakhstan (Ul'yanin, 1949; Kuz'mina, 1968); western Siberia (Yurlov, 1956, 1960); and the Trans-Baikal and the Dusse-Alin' mountain range (Afanas'ev, 1934). Year-round information on the food of this species is available for Kola Peninsula, Karelia, and northern Kazakhstan, while data for only some seasons are available for the rest of its range.

The diet of Willow Grouse in the Soviet Union includes 216 plant species belonging to 45 families (based on an analysis of the gizzard contents of 1,054 birds and isolated references in faunistic records). Food composition is more varied in the southern part of the range. Animal food includes spiders, mollusks, and insects. The latter are represented by seven orders and 18 families. Beetles, moths, and their larvae are commonly eaten by these birds.

In North America the food composition of Willow Grouse is perhaps more uniform. Food of 540 birds was examined in Alaska (West and Meng, 1966); this study revealed only 25 species of plants—far less than in some
85 regions of Eurasia. In winter, willow shoots and buds constitute 94% of its diet. In the snow-free period, green matter and berries predominate while food from trees becomes secondary. Animal food is practically absent.

Not much is known about the food of Rock Ptarmigans. A more detailed study was made in Taimyr (Sdobnikov, 1957; Tikhomirov, 1959; Pavlov, 1971) where 570 birds were dissected at different times of the year; in the Kola Peninsula 77 birds were studied from March through October (Semenov-Tyan-Shanskii, 1938, 1959). Additional information for this region is available in other works (Folitarek, 1939; Novikov, 1952). Data on the intake of the individual diet components are available for Spitsbergen (Birulya, 1910), the northern Cis-Urals—the Pechora-Ilych Preserve (Teplova, 1957), the lower reaches of the Lena (Kapitonov, 1962), Indigirka (Uspenskii, et al., 1962), Anabar tundra (Uspenskii, 1965), upper reaches of Kolyma (Dement'ev and Shokhin, 1939), Kolyma mountains

*The Federal Russian Republic encompasses most of the northern part of the Soviet Union—Editors.

(Kishchinskii, 1968), Kamchatka (Averin, 1948), and the Novosibirsk Islands (Kishchinskii, 1974).

In the Soviet Union, the diet of Rock Ptarmigans in all includes 139 plant species belonging to 29 families, of which mosses and lichens account for 26 species. Animal food is extremely rare and is confined to summer, and includes spiders, ground beetles, ants, flies, mosquitoes, and earthworms. It has been noticed that in the Kola Peninsula the food of Rock Ptarmigans is more diverse than that of Willow Grouse.

In winter, buds and terminal shoot of various scrub willows and birches are the main food of Willow Grouse. The close association of this bird with these food plants also determines its distribution. Of the 32 plant species found during winter in the different parts of its range, 13 were of various species of willow and birch. In regions where the snow is not deep, in addition to twigs, berries, and green shoots of the cranberry family (Vacciniaceae), various seeds usually constituted a small fraction of the gizzard contents. In Finland 19% of the crops examined contained heath (Selskari, 1957). At the southern boundary of its range, a part of the winter diet consists of seeds and vegetative parts of herbaceous plants. The build-up of snow cover permits the birds to reach the top branches of bushes; however, when the bushes are buried under snow, the birds migrate to other foraging sites or take to feeding on trees, as observed in the upper reaches of Kolyma (Dement'ev and Shokhin, 1939; Kishchinskii, 1968), Koryaksk uplands (Kishchinskii, 1962), upper reaches of Lena (Kapitonov, 1962), and the Kola Peninsula (Semenov-Tyan-Shanskii, 1959).

86 The availability and density of shrubbery undergrowth largely control the population of Willow Grouse. According to estimates, in the Kalinin District, 4-9 ha of shrubs are required to feed a single bird (Mikheev, 1948) and 21.4 ha in Syktyvkar (Voronin, 1971). The high mortality of ptarmigans observed during seasonal migrations (Gagina, 1958) is evidently caused by food scarcity when these birds congregate in large numbers.

It was established experimentally (Volkov, 1970) that utilization of shoots of shrubs as food greatly depends on temperature. In severe frosts the shoots become brittle and friable, while in mild frosts and during thaws their resilience makes it difficult for Willow Grouse to feed on them. This affects the bird's biotopic distribution, since changes in the resilience of willow and birch shoots set in at different temperatures.

The uniform winter food (species composition and percentage of twigs in food) of Willow Grouse differs markedly in the different areas of its range. In the Pechora-Ilych Preserve, twigs exceed 90% of its diet, but in Kola Peninsula they constitute only 72%. Additional food components in these regions are the still-green parts of undergrowth, bilberries, and crowberries.

In northern Kazakhstan, in the Barabinsk and Kulundinsk steppes, the main winter diet consists of willow and birch twigs supplemented with dog rose berries and wheat and oat grains collected by the birds from threshing floors and elsewhere. At times they take seeds of various species of Compositae and other plants lying above the snow.

On Novosibirsk Islands, where shrubs are absent, the food of Willow Grouse differs markedly. Judging from gizzard contents, green shoots and fresh roots of dwarf willow, green rosettes and winter buds of tufted saxifrage and pitted cinquefoil, and grains and bolls of polar poppy are consumed in large quantities (Birulya, 1907; Kishchinskii, 1974).

Rock Ptarmigans show a distinct preference for food buried under snow, such as the green parts of herbaceous plants and undershrubs, berries, and seeds; only in areas with abundant snow does it feed on the twigs of shrubs.

The preferential use of herbaceous plants has been recorded in the Komandor Islands, Kola Peninsula, and the less snowy areas of Kamchatka. In Kolyma and the very snowy regions of Kamchatka, Rock Ptarmigan feed only on twigs; in Taimyr, Manchurian alder catkins constitute over 90% of its food.

87 The following are the basic differences in the winter food of Rock Ptarmigans and Willow Grouse. Throughout the vast range of Willow Grouse, not only in Eurasia but also in North America (West and Meng, 1966), their basic food consists of twigs and catkins of willow and birch, which are supplemented with berries; green undergrowth, grains of agricultural crops, and various other seeds from regions with a temporary snow cover. Depending on local conditions, the diet of Rock Ptarmigan and Willow Grouse living in the same territory hardly may differ (regions of northeastern Asia) or, contrarily, may vary significantly (Kola Peninsula and Taimyr). In both cases, the diet depends on their biotopic distribution.

In the spring, as soon as thawed patches appear, the food of both species changes: the volume of twigs decreases while that of the green mass and berries increases. Up to 45 plant species belonging to 12 families have been identified in the diet of Willow Grouse in spring. As before, the main diet continues to be willow and birch; only shoots are replaced by tender leaves and inflorescences. The green grasses and evergreen undershrubs are consumed more avidly. In the northern part of the range, the basic food components in spring are shoots and leaves of bilberry, cowberry, and winter berries of marshwort and crowberry. In the forest-steppe regions, the composition of green fodder differs altogether, consisting of green legumes (meadow pea and tufted vetch), Rosaceae (pitted cinquefoil, spiraea and meadow sweet), and Compositae. Flowers form only a small part of the gizzard and crop contents and include the inflorescences of willow, cottongrasses, and spiraea.

The spring food of Rock Ptarmigans includes 23 plant species; the food composition varies markedly in different parts of its range.

In the Kola Peninsula, Rock Ptarmigans feed on winter berries; black crowberry, alpine bearberry, cowberry, bilberry, and also the shoots and leaves of these plants; shoots of dwarf birch are consumed in very small quantities. In northern Taimyr, the chief food components are the sprouts of punctate dryas and buds and fronds of polar and other willows; in May these constitute 94% of the crop contents (Sdobnikov, 1957).

In the northern regions of eastern Siberia, Rock Ptarmigan feed on the green parts and inflorescences of willow and birch, supplemented with twigs of shrubs—the main winter food in this region (Kapitonov, 1962; Uspenskii, et al., 1962).

88 Differences in the food composition of Rock Ptarmigans and Willow Grouse are manifested distinctly in winter, but become obliterated in spring. Twigs are of great importance in the diet of the Willow Ptarmigan even at this time. In summer the food of these two species is most varied.

Willow Grouse feed on 128 plant species in summer. In the southern part of its range, this species additionally consumes animal food in large quantities. A good portion of the food consists of the green parts not only of herbaceous plants and undergrowth, but also of scrub willow and birch, berries, and seeds. In some areas the ratio of the food groups varies.

In Kotel'nii Island and in Spitsbergen, neither animal food nor berries were detected in the crops and gizzards of Willow Grouse. The main food mass consisted of the green mass, flowers, and seeds of polar plants (Birulya, 1907; Gorodkov and Korotkevich, 1957). On the Kola Peninsula, in the southern part of Yamal Peninsula, and in Timansk tundra, insects constitute a small portion of their diet. The proportion of different groups of food in these areas differs rather sharply. In Kola Peninsula, Willow Grouse feed more on berries in late summer; the total amount of seeds does not exceed 10%.

In the Timansk tundra, on the other hand, seeds (sedges, haircap moss, and bolls of buttercups) constitute a considerable part of the food at this time of year (Mikheev, 1948).

The food composition of Willow Grouse is totally different in the southern part of its range in the forest-steppe and steppe regions of western Siberia and northern Kazakhstan. Here, green plants are important only during the first half of summer when birds readily feed on the leaves of legumes, cereals, and willows. Seeds are their chief food, especially in midsummer: the most important among them are the seeds of wild and cultivated cereals (millet, wheat, cultivated and oats, wavy hair-grass, etc.), knotgrass, legumes, sedges, and various Compositae. Berries (dog rose, stone bramble, strawberry, common strawberry, Russian almond, and steppe cherry) are consumed in smaller quantities than in the north. Animal food was found during June in 30% of birds, in July in 5.7%, and August in 10% of birds (Ul'yanin, 1939; Yurlov, 1960).

The food of chicks in various parts of the range also differs markedly. In the Timansk tundra, fledglings feed on a mixed diet of insects, seeds, bolls of haircap moss, and catkins of dwarf birch right from their first day (Mikheev, 1948). In western Siberia their food consists only of insects; in July the latter were found in 20% of juveniles and in 24.2%
89 in August.

In summer, 47 plant species were identified in the food of the Rock Ptarmigan. This species feeds on berries, flowers, seeds, and the green vegetation but, unlike the Willow Ptarmigan, leaves of willow and birch were rarer than grass and plants of the rocky tundra (rhododendron, saxifrage, etc.) not found in the diet of the Willow Grouse. Berries of the cranberry family are the main food at this time. Among the grasses, particularly abundant are saxifrages (11 species) and Caryophyllaceae, while sedges are also not uncommon. In late summer, berries, flowers, and green vegetation are supplemented with fruits and seeds of herbaceous plants. At this time, mosses and lichens are also found in small quantities in the food of the Rock Ptarmigan. Animal food is found seldomly but earthworms, spiders, ground beetles, and ants enter its food more by chance. The species composition of the food differs markedly in various parts of the range. In Taimyr, during hatching period these birds survive mainly on leaves and flowers; the mosses and lichens found in spring are absent. In Kola Peninsula, the summer food of Rock Ptarmigan contains less green vegetation and more flowers, berries, and seeds, compared to that of Willow Grouse. Animal food forms an important constituent of the diet of the fledglings. In July-August, slugs, spiders, mites, beetles, ants, flies, mosquitoes, and caterpillars were found in their crops.

In autumn, berries, seeds, green vegetation, and insects in the diet of Willow Grouse are replaced by twigs; by late October, buds and twigs of willow become the principal food in many regions. In western Siberia, animal food ceased to be encountered in early September and the main diet at this time was berries and particularly seeds of various kinds. In the forest-steppes of northern Kazakhstan, ptarmigans land on fields in September-October to pick fallen grain. The crops and gizzards of birds I examined contained mainly grains of wheat and oats, seeds of dog rose, and green grass. In the southern part of Finland these birds feed in autumn almost exclusively on seeds (including cultivated cereals—50% by weight); green mass constituted only 1% and animal food 0.03% of the diet. There is a distinct preference for hempnettle containing up to 40% fat (Pulliainen, 1966).

Rock Ptarmigans at this time feed on berries, seeds, fruits, grasses, and green parts of plants. Twigs are constantly found in its diet, but invariably in small quantities. The composition of seeds and fruits of herbaceous plants is fairly varied in autumn; some 20 species have been detected. Among these, seeds of rusk, poppy, sedge, and buttercup families are constant components. The species composition of berries is the same as in summer. Green vegetation is represented by leaves of saxifrages, vetches, willows, and whitlow grasses.

90 A comparison of the food consumed by Willow Grouse in various parts of its vast range confirms that seasonal changes in its food plants are fairly universal (part of the Novosibirsk population wintering on the island is an exception). In any given season, the food composition of birds inhabiting various natural biotopes may vary more markedly than in birds widespread in the same region. This is explained by the extreme floral differences in the natural zones (tundra, forest, and forest-steppe) inhabited by this species.

A comparison of the diet of these two proximate species, Rock Ptarmigans and Willow Grouse, reveals a general similarity while, at the same time, underscores essential differences.

Both species of ptarmigans feed on twigs and buds of shrubs in winter, but the ratio of such coarse food to other dietary items is far from identical. For Willow Grouse, twigs are the main food, and at places almost the only food. Rock Ptarmigans thrive on coarse food only during heavy snowfall with a preference for green vegetation and berries buried under the snow. Thus, the basic difference in the winter diet of these two species lies in the different proportions of "twig" food.

Rock Ptarmigans prefer willow and birch twigs, thin shoots of shrubs, and soft catkins of alder. Hence, its beak and intestinal ceca are more poorly developed than in Willow Grouse (Dobrinskii, 1972; Pavlov, 1974).

During the snowfree period, differences in the diet of these two species are less marked and expressed more in the species composition of plant food than in the proportion of the various components, which is explained by differences in their habitats. Green plants and seeds in the Rock Ptarmigan diet are represented by typical mountain tundra vegetation such as saxifrages, poppies, buttercups, Caryophyllaceae, and a variety of mosses and lichens. Cereal grasses are rare but sedges more frequent. Especially diverse in the diet of Willow Grouse are composites and legumes. The composition of the animal food of the adult birds of these two species is almost similar.

On the whole, despite a general similarity in the types of food, species differences in the food composition of these ptarmigans are clearly evident: of the 139 plant species found in Rock Ptarmigan food, 72 species or 52% are not found in Willow Grouse diet, which points to a fairly distinct selectivity of the latter.

Genus *Tetrao*—Capercaillies

91 The terrestrial-arboreal mode of life of capercaillies is the result of seasonal changes in their food composition. In summer Black-billed Capercaillies are mostly ground birds, but in winter they feed on trees, sometimes breaking the twigs of young trees while moving on snow. The bent tip of the beak helps break shoots; needles are often sheared off by the unique "scissors" present along the sides of the beak.

The food of capercaillies has been studied, almost exhaustively, during all seasons of the year in the European part of the USSR. But in western Siberia only fragmentary data are available for the Altai and none for the eastern parts of the range. Crop and gizzard contents of 1,470 birds have been analyzed in the following regions: Kola Peninsula (Semenov-Tyan-Shanskii, 1938, 1959; Novikov, 1952); Karelia (Ivanter, 1974); Belorussia (Gavrin, 1956; Dolbik, 1967); central and eastern regions of the European part of the RSFSR (Lobachev and Shcherbakov, 1936; Teplov, 1947; Rodionov, 1963; Oliger, 1973); southern Urals (Kirikov, 1952; Kuz'mina, 1968); and southern Altai (Kuz'mina, 1968). In addition to the published results of these analyses, published descriptions of plant and animal food have been utilized to gain a better understanding of the species compositions of diet (Sabaneev, 1876; Menzbir, 1900; Folitarek and Dement'ev, 1938; Snigirevskii, 1946; Rodionov, 1960; Romanov, 1960; Bakeev and Osmolovskaya, 1964). In all, 166 species of plants belonging to 44 families have been identified in the food of capercaillies in the western parts of its range. Animal food is consumed only in summer, but even at this time adult birds prefer plant to animal food both in quantity and frequency of intake.

Food habits of Black-billed Capercaillies have been studied only in Yakutia where crop and gizzard contents of 143 birds were analyzed: 25 in spring, 34 in summer, and 84 in autumn and winter (Egorov, Labutin and Mezhennyi, 1959; Larionov, 1965). Information from a number of faunistic works supplements these data (Kozlova, 1930; Dement'ev and Shokhin, 1939; Pavlov, 1948; Averin, 1948; Andreev, 1953; Mishin, 1960; Filnov, 1961; Vorob'ev, 1963; Barancheev, 1965). In all, 35 plant species belonging to 17 families have been identified; insects, arachnids, mollusks, and frogs have been found among animal food.

92 Seasonal food variations of both species of capercaillies are distinct, especially in the plant food consumed throughout the year. In winter both species feed almost exclusively on twig food. Sprouts, berries, and seeds

have been found in their food in early and late winter or in regions with a temporary snow cover.

The food of Western Capercaillies at this time of the year comprised only 12 plant species, chiefly needles and twigs of pine or cedar, which are mutually interchangeable. Birds living in cedar forests thrive on cedar nuts; twigs, buds, and catkins of deciduous trees are consumed less often. At the southern extremity of the Urals they also feed on oak twigs (Kirikov, 1952); in the former Ryazan Province, after large-scale fellings of coniferous forests, a part of the Western Capercaillie population remained in deciduous forests and had to survive on the buds of birch, linden, elm, and oak (Menzbir, 1900).

Pine needles are the chief winter food of Western Capercaillies throughout the European part of the USSR and in the southern Urals. In the Kola Peninsula, during the snowy half of the year (November–April), pine needles constitute 90% of its diet, juniper 8%, and vegetation buried under snow only 2%. In the Pechora-Ilych Preserve, gizzards of Western Capercaillies contained needles of both species, but those inhabiting regions of boreal taiga contained only needles and nuts of cedar, and those inhabiting pine forests contained only needles and twigs of pine. Cedar needles and nuts are the chief winter diet of Western Capercaillies in the Altai.

The main diet components of Black-billed Capercaillies during this season are a wide variety of twigs, significantly differing in the various parts of the range. Berries completely disappear from its food immediately after heavy snowfall. In eastern Siberia in winter, it feeds almost exclusively on shoot tips and cones of larch, sometimes supplemented with needles of pine and fir. The quantity of food consumed in winter and the duration of foraging have been assessed from the crop contents of a Black-billed Capercaillie killed in Yakutia; it contained 13,350 fragments of Daurian larch shoots weighing 428 g (Larionov, 1965).

In the Amur region the main food consists of larch buds and pine and fir needles. In Sakhalin, buds, cones, and twigs of larch and berries of dog rose are consumed. In Kamchatka, where Black-billed Capercaillies live not only in coniferous forests, but also in birch groves of parks, their main dietary items are catkins and twigs of Erman's birch, twigs and buds of willow, and sometimes buds and twigs of Manchurian alder (Averin, 1948). The intake of twigs not only of coniferous, but also of deciduous
93 plants indicates the wider food preferences of Black-billed Capercaillies during winter.

In spring the diet of both species of capercaillie shows much diversity. The diet of Western Capercaillie includes 39 plant species. In early spring twigs are its main food, but needles of pine, cedar, and spruce are supplemented with buds of aspen, willow, and birch. Toward the end of spring, twigs disappear but young shoots continue to be seen in a small quantity. Usually, toward the end of spring and in early summer, its chief food is green parts of herbaceous plants, flowers, and small quantities of winter berries. The species composition of green plant food may differ not only from region to region, but also in various biotopes in the same region.

In the European part of the USSR at the start of spring, Western Capercaillies begin feeding on shoots of bilberry, cowberry, and marshwort berries, in addition to conifer needles. During the breeding season inflorescences of cotton grass are regularly found in their food and sometimes even fill the entire crop. In the southern Urals, Western Capercaillies inhabiting pine forests subsist mainly on needles and shoots of pine until the end of May. They additionally feed on buds, shoots, and young needles of larch, needles and berries of juniper, birch catkins, aspen leaves, and green herbaceous plants. Shoots and buds of oak predominate in the food of Western Capercaillies of the forest-steppe oak grove; willow and birch are minimal. Shoots are replaced later by green parts and flowers. Among flowers, inflorescences of oak, aspen, willow, and spreading pasque flower are particularly important (Kirikov, 1952).

Significant dietary differences of capercaillies are manifested in spring even in the different regions of the Kazakhstan part of the Altai. Near Lake Markakol' during breeding (first half of May), Western Capercaillies feed predominantly on spruce needles, while in the Uba basin at this time they feed mainly on the flowers and green mass of anemones. All the crops and gizzards examined in late May contained mostly green vegetation, the species composition of which differed markedly in different regions.

During the first half of spring, Black-billed Capercaillies feed on buds and shoots of larch and at some places also on birch buds. During this time the birds begin to nibble on winter berries and sprouts which gradually gain importance. Berries of cowberry, marshwort, low juniper, black crowberry, and dog rose gradually become the main food of this species. However, in different parts of the range, the berry composition differs markedly. Among sprouts and flowers, inflorescences of cotton grass and
94 sedges predominate. In Kamchatka, the crops and gizzards of males in April–May contained birch buds, juniper berries, and year-old berries of black crowberry and cowberry, stems of horsetail, leaves of bog whortleberry and honeysuckle, and foliage (Averin, 1948).

In spring Black-billed Capercaillies change over early to grasses and, instead of greens, consume large quantities of winter berries, while Western Capercaillies prefer sprouts.

In summer, 100 plant species belonging to 33 families, covering all groups, have been found in the food of capercaillies. Animal food is usually of secondary importance (insects, arachnids, and mollusks). Green parts and flowers are found in larger quantities than in other tetraonids, but mainly in the first half of summer. Berries are the main food in the second half, their composition varying in different parts of the range. Larch needles are found in small quantities throughout summer, but by the end of the season constitute a major part of the crop content. Seeds are always less than the green matter and berries by weight but, on the whole, constitute an essential dietary component throughout summer. Western Capercaillies feed everywhere on various species of sedge (15 species), knotgrass, and seeds of May lily. Lower plants have been found in small quantities toward the end of summer.

In the northern part of the range (Lapland and the Pechora-Ilych Preserves), green vegetation acquires extreme importance; only when berries ripen does the quantity of green vegetation decrease. Of the berries found here, capercaillies mostly feed on bilberry, crowberry, cowberry, and cloudberry. Seeds and bolls of herbaceous plants are found regularly but in small quantities. Needles of pine and spruce form only a very small part of the food. The animal food component is always much less than the plant components by volume.

In the southern part of its range (southern Urals) the main summer food is berries: different species of strawberry, raspberry, stone bramble, and bilberry. Among the green foods are leaves of clover, meadow pea, and hawkweed. The bird also feeds avidly on sedge seeds. From the middle of August, Western Capercaillies consume large quantities of larch needles. Insects constitute an insignificant part of the diet of adult birds.

Invertebrates constitute the main part of the food of fledlings. In the Lapland Preserve, they were four to six times more abundant in the food of fledglings than of older birds. In the Pechora-Ilych Preserve in July, animal foods were found in the gizzard contents of 100% of the fledglings, grass seeds in 80%, and green matter in more than 70%. Further, fledglings often feed on twigs of shrubs and conifer needles.

Differences in the summer food of Western Capercaillies in the northern (Lapland and the Pechora-Ilych preserves) and southern (south-
95 ern Urals) parts of their range are fairly distinct. In the north, where berries ripen later, green vegetation and insects constitute the major food items, especially in the early summer months, not only among fledgelings, but also in adults. In the south, where berries contribute more to the diet, fledglings also take to berries early in the season. In the summer diet of Black-billed Capercaillies, twigs are totally absent. This conclusion is based perhaps on a limited study (food of 34 birds analyzed). In the first

half of summer, this species feeds on the seeds and green parts of herbaceous plants, supplemented with insects. In the second half, it prefers berries (bog whortleberry, marshwort, black crowberry, cloudberry, etc.), but also feeds on sprouts and seeds. Insects constitute a larger percentage of the food for the young than for adult birds.

The autumn food of Western Capercaillies contains about 40 plant species; in other words, in variety the autumn and spring seasons do not differ very much but the food groups are different. The importance of conifer needles increases gradually, displacing berries, green matter and seeds. In the northern parts of the range in early autumn, cranberries still predominate, while the green matter and seeds dwindle. By October pine needles become predominant.

There are interesting differences in the composition of the autumn food of capercaillies in the central and northern Urals. In the sparsely inhabited northern parts of the range, Western Capercaillies feed on the food available in the natural biocenoses. In the central regions where cereal and leguminous crops are cultivated inside the forest, they regularly visit fields of oats, rye, wheat, and peas. Some dissected birds contained more than 150 g of oats in their crops. Here capercaillies hunting is even practiced in fields (Bakeev and Osmolovskaya, 1964). In the southern Urals berries are absent in the autumn diet. From early September, (judging from dissected birds) larch needles predominate, with seeds, green matter, and insects found in small quantities. The main difference in the autumn food of Western Capercaillies in the southern Urals is an early changeover to larch needles; however, they begin to consume pine needles at almost the same time as in the north. Thus larch needles replace berries in the south at a time when they are still consumed by capercaillies in other parts of the range.

The food of Black-billed Capercaillies in autumn is determined mostly by dynamics of the snow cover. In less snowy regions this species feeds on berries for longer; on the other hand, in places where the snow cover is deep it starts feeding on larch shoots early. Based on the frequency and quantity, berries in the food of Black-billed Capercaillies of Yakutia
96 can be ranked as follows: cowberry, bog whortleberry, dog rose, red currant, bearberry, marshwort and common strawberry (Larionov, 1965).

In spite of the general similarity in the food of the two capercaillie species, certain differences arising from the present distribution of the species and, possibly, the conditions under which they evolved, deserve special attention. Western Capercaillies, largely feeding on coniferous needles, gradually change to an arboreal mode of life even by the end of summer, while Black-billed Capercaillies feed on the ground for the greater part of the year.

Genus *Lyrurus*—Black Grouse

Black Grouse are ground birds for most of the year. In winter they feed on trees by shearing off the "winter crop" of pine with the sharp edges of their beaks or breaking off catkins and buds of deciduous species. At the southern boundary of their range in the steppe regions, some scattered populations apparently lead a terrestrial mode of life even during winter. While feeding on the ground, black grouse do not dig the soil but simply peck at whatever lies on the ground, in the grass, and among bushes.

Basically herbivorous, Northern Black Grouse feed on insects in summer and early autumn; but even during this period insect food is always less than plant food both in quantity and frequency of intake. The food composition of Northern Black Grouse has been studied in more than 2,400 birds from many regions of its vast geographic range; the Kola Peninsula (Semenov-Tyan-Shanskii, 1959); Karelia (Neifel'd, 1958; Ivanter, 1963); Belorussia (Gavrin, 1956); central and eastern regions of the European part of the USSR (Dubrovskii, 1930; Lobachev and Shcherbakov, 1933; Kuznetsov and Korenberg, 1963; Rusakov, 1963; Nikul'tsev, 1968; Oliger, 1973); southern Urals (Kirikov, 1952; Kuz'mina, 1968); northern Kazakhstan (Ul'yanin, 1949; Kuz'mina, 1968); central Kazakhstan (Kuz'mina, 1968); western Siberia (Yurlov, 1960); southeastern Kazakhstan (Kuz'mina, 1968); southeastern Trans-Baikal (Fetisov, 1934); and central Yakutia (Larionov, 1962). The most complete data characterizing the food components of Northern Black Grouse throughout the year are available for the Kola Peninsula, Karelia, and the Belovezhskaya Pushcha Preserve, the central regions of the European part of the USSR, northern Kazakhstan, and western Siberia; in other regions only food consumed in some seasons of the year has been studied. Supplementary information on feeding habits has been gathered from other ornithological works (Snigirevskii, 1946; Teplov, 1947a; Pavlov, 1948; Shnitnikov, 1949; Bogdanov, 1868). In all, 304 plant species belonging to 58 families and various invertebrates have been detected in the food of Black Grouse. Insects
97 constitute its main animal food component; other invertebrates include arachnids and mollusks. Lizards have been detected as an exception.

The limited range of the Caucasian Black Grouse and its relatively small population are reasons for the insufficient study of its biology and particularly its feeding. Food composition has been analyzed in 54 adult and 17 young birds (Averin, 1938; Khanmamedov, 1965; Tkachenko, 1966; Sirakhulidze, 1974). Individual references to a number of plant and animal foods are available in the works of various authors (Mlokosevich, 1879; Bogdanov, 1879; Radde, 1884; Noska, et al., 1895; Kutubidze, 1961; Chunikhin, 1964). In the food of the Caucasian Black Grouse, 62 plant species belonging to 30 families, insects, and other animals have been detected.

In the winter food of Northern Black Grouse, 27 plant species belonging to seven families have been recorded, but in some regions the crops and gizzards of these birds contained only 3–6 species. In winter these birds feed mainly on catkins and buds of birch and alder, buds and twigs of willow, juniper needles, "winter crops" and needles of pine and spruce, and as an exception, larch twigs. Sometimes the diet is supplemented with berries of juniper, dog rose, cereal grains, seeds of various grasses, and the green matter of undershrubs and grasses. Depending on the latitude of the range, twigs predominate for 5 to 6.5 months, but are common for eight months of the year.

In the forest zones of the European part of the USSR these birds feed mainly on birch buds and catkins, supplemented with "winter crops" and sometimes also on pine needles and juniper berries. The quantity of pine "winter crops" in the food of birds depends on availability while juniper berries are not found in all regions but could be the chief food at places. These berries are consumed in particularly large quantities in the central parts of the European part of the USSR.

In central Yakutia during November–December, Northern Black Grouse feed on catkins and buds of birch and, to a lesser extent, buds and "winter crops" of pine. In addition to twig food, large quantities of cultivated cereal grains were found in 38.1% of the birds dissected. Some collective farms in this region mow cereal crops to feed cattle; haystacks attract Northern Black Grouse, which visit such places regularly (Larionov, 1962).

Food varies in Trans-Baikal. In addition to birch catkins and twigs, juniper berries, and pine "winter crops," Northern Black Grouse feed on buds and twigs of aspen and alder catkins, supplemented with bilberry shoots (Fetisov, 1934).

In the forest-steppes of northern Kazakhstan and western Siberia the chief winter foods are birch catkins and twigs, supplemented with
98 berries of dog rose, buds and twigs of various willows, and seeds of any unharvested cereal left behind in the fields (Ul'yanin, 1949; Yurlov, 1960).

The winter food of Tien Shan Black Grouse reveals maximal variation. On the Trans-Ili Alatau this species feeds on needles of the Schrenk spruce, supplemented with small quantities of aspen buds (Zverev, 1962). Juniper berries are the main autumn food, but could be consumed during early winter too (Shul'pin, 1939). Judging from the position of roosting places and the composition of excreta, these birds spend the entire winter in spruce forests feeding on their needles.

The main difference in the food of the Caucasian Black Grouse is that a considerable part of it consists of plants endemic to the Caucasus. In winter the crop and gizzard contents revealed 15 plant species, mostly buds and catkins of white and common birch, needles and berries of juniper, berries of dog rose, and shoots and buds of willow; leaves of rhododendron, fir needles, and leaves and fruits of bilberry, cowberry, and barberry are consumed in small quantities. When thawing sets in and water puddles form, both species start feeding on the green matter and flowers of ephemerals and winter berries.

Until the middle of May the Caucasian Black Grouse continues to feed mainly on birch and juniper twigs and dog rose berries, but by the end of the month, these are replaced by clover, anemone, bilberry, cowberry, and rhododendron leaves. The spring food consists of only 11 species of plants—less than in winter. The spring food of the Northern Black Grouse is rather diverse, varying markedly in the different parts of its vast range. In all, 73 plant species belonging to 23 families have been recorded; in eastern Siberia alone 20 species are known as its food.

In the northern parts of its range, twigs are consumed for a longer period. In Kola Peninsula these constitute the main food component until the middle of May; the berries, leaves, and flowers of butterburs become the chief food only in late May. Food is more diverse in Karelia. In addition to berries and shoots of the cranberry family, green masses of buttercups, legumes, cereals and horsetails, and cotton grass inflorescences are common here.

At the southern boundary of the range, in the forest-steppe regions of northern Kazakhstan, western Siberia, and the Trans-Baikal green vegetation and flowers are the chief spring food of these birds; berries are found very rarely or not at all. Green fodder is varied and includes not only herbaceous, but also woody plants (leaves of aspen and inflorescences of
99 willow); often seeds are consumed as well. In these regions Northern Black Grouse sometimes lives close to agricultural crops and cereal grains have been found in its food during spring.

In the Trans-Baikal, green fodder constitute the basic diet of Northern Black Grouse, but seeds of field crops have also been found. In central Yakutia, with the appearance of thawed patches, winter berries of cowberry, bearberry, and bog whortleberry come to occupy an important place in the diet, but during the breeding season vitamin-rich food such as the inflorescences of pasque flower and sedge green vegetation predominate.

The summer food of Northern Black Grouse is quite varied. An analysis of crop and gizzard contents revealed 178 plant species belonging to 43 families, including all the fodder groups consumed year round. In fact, twigs are very rarely found in summer and always in small quantities.

Animal food is found in the gizzards of adult birds throughout the summer and early autumn. These are found in larger quantities in the southern parts of the range than in the northern. Animal food mainly comprises insects (32 families from 9 orders), of which the most important are Orthoptera, Hymenoptera (ants), Coleoptera (particularly leaf beetles, weevils, and ladybirds), caterpillars of moths, bugs, etc. Northern Black Grouse also feeds on mollusks, arachnids, and worms. Viviparous lizards have been found as an exception. These foods are of secondary importance, or even incidental.

During the early half of summer, sprouts and flowers are the main food components and seeds and invertebrates are rare. Later, berries predominate, and seeds and insects constitute a significant part, but green fodder is only supplementary. In various parts of the range the proportions of different food groups may differ markedly.

Green vegetation consumed by Northern Black Grouse are exceptionally varied. These birds are particularly fond of legumes (about 20 species), Rosaceae (14 species), buttercups (9 species), and Compositae (27 species); in the northern part of the range cranberries and Ericaceae are also among the preferred foods. From the moment of ripening, berries become the chief food in most regions. In the forest zones the most sought after are cranberries, and Rosaceae in the forest-steppes and steppes.

Seeds and fruits of herbaceous plants are represented by a very large number of species. Northern Black Grouse avidly feed on the seeds of cereals such as oats, rye, wheat, barley, millet (16 species) from cultivated fields and wild, often bitter weeds such as couch grass, common wild oats, and others. Knotgrasses (21 species) are almost invariably found in its
100 food; seeds of legumes and sedges, often, and in some regions, also seeds and fruits of buttercups, the Rosaceae, Compositae, and crucifers.

In Karelia, the green parts of herbaceous plants constitute 64% (frequency of occurrence), seeds 46%, berries in the latter half of summer up to 50%, and insects 28%. In the central regions of the European part of the USSR, in the latter half of summer, berries predominate, especially those of the cranberry family, but insects and green parts of plants are also of considerable importance. At the southern boundary of the range, in northern Kazakhstan and western Siberia, berries are not as important a dietary item as in the forests of the European part of the USSR, and are limited to the Rose family (dog rose, steppe cherry, Russian almond, cotoneaster, stone bramble, and strawberry). Seeds are represented by a wide variety and constitute an essential part of the diet. By the end of summer, Northern Black Grouse regularly fly to fields to feed there on fallen cereal grain. Animal food was found in 72.7% of birds dissected in June, 50% in July, and 53.8% in August (Yurlov, 1960).

In the Trans-Baikal, Northern Black Grouse feeds on green fodder, seeds, berries, and invertebrates. Here the last constitutes 10% of the total intake, while 50% of plant food consists of berries (Fetisov, 1934).

Unlike adult birds, the food of two-week-old fledgelings is up to 95% insect components (caterpillars, mosquitoes, midges, gnats, and spiders), and in one-month-old fledgelings up to 70% while the remainder consists of green fodder, seeds, and berries. The difference between the food composition of adults and young birds disappears gradually, but even in August young birds depend more on insect food, especially in the southern parts of the range.

The food of the Caucasian Black Grouse contains more than 20 species of plants and insects only rarely and in small quantities. During the first half of summer this species feeds on a variety of alpine plants—leaves of dandelion, lungwort, red rattle, campanula, celestial fescue, anemone, bilberry, and cowberry, and seeds of vetches, willow grass, campanula, yellow rattle, etc. From early August, Caucasian Black Grouse feed on the ripe seeds of alpine grasses, but by mid-August the berries of spurge, bilberry, and cowberry predominate. In June–July the food of chicks of this species differs greatly from that of adults. During the first 10–15 days, fledgelings feed almost exclusively on invertebrates—arachnids, mollusks, and insects; the latter constitute the main food item. Insects of 14 families from six orders have been identified in the food of young birds, with
101 spiders, beetles, and especially leaf beetles and weevils predominating. In August the young birds changeover to berries and from then on their food is not different from that of adults.

In early autumn (September) the food of this species is even more varied than in summer. In the southern part of its range its diet during the first half of September contains the maximum amount of animal food, especially grasshoppers and beetles, but by October these disappear almost completely. Berries, too, gradually decrease even though still found until snowfall. Simultaneously, the quantity of different species of seeds increases as does the quantity of twigs. By the end of October in some parts of the range, birch catkins become the main food item, and juniper berries in other areas. In the Tien-Shan mountains, juniper berries constitute the main autumn food.

The autumn food of Caucasian Black Grouse directly depends on altitude. In the subalpine belt it feeds on the seeds of alpine grasses and fruits of bilberry and cowberry; in tall forests, berries of wood sorrel and oriental beech and fir needles are the primary food items.

Seasonal changes in the diet of Black Grouse are as clearly expressed as in the other members of the family. Nonetheless, the winter food of Northern Black Grouse is rather varied (twigs of deciduous and coniferous trees, berries of shrubs, and at places seeds and sprouts). Particularly rich is the species composition of plants in the snowfree period. The fairly diverse combination of foods of the Caucasian Black Grouse, considering its localized distribution, is worth noting. In spite of the distinctive characteristics of the nutrition of Northern Black Grouse, a comparison of the winter food clearly reveals differences typical of geographic forms. The North European Black Grouse in the central regions of the Soviet Union consumes large quantities of juniper berries. For the Tien-Shan Black Grouse conifer needles are primary. The diet of Black Grouse is more diverse; in addition to buds and catkins of birch, this species feeds on the seeds of wild and cultivated cereals and dog rose berries.

In diet, Caucasian Black Grouse differs little from Northern Black Grouse and reveals the same sequence of changes in the annual cycle. The composition of plants consumed by the Caucasian Black Grouse is sufficiently distinct; of the 62 plant species found in its diet, about a third are missing in the diet of the Northern Black Grouse. These differences, however, are wholly explained by the typical floras of the Caucasus. In fact, the difference between the diet of the two species is no greater than that between the different geographic subspecies of Northern Black Grouse.

Genus *Falcipennis*—Spruce Grouse

102 Spruce grouse are one of the most avid consumers of conifer needle among the Paaearctic tetraonids. Their method of feeding is unique. Birds do not break off the tiny needles, nor do they break the terminal shoots along with the needle; instead they pluck the needle leaving the base behind with a relatively uniform edge on the branch. While biting off the needles the bird rotates its head (Yudakov, 1972). This peculiar method of cutting the needles is reflected in the beak structure, which is weaker than in other tetraonids and lacks the characteristic "scissors" along the sides. During the snowfree period, the birds feed on the ground in grass and among bushes, but even then conifer needles remain their chief food.

The mode of life of Siberian Spruce Grouse, the only species of the genus in the avifauna of the USSR and which has a limited range and sporadic distribution, has not been adequately studied and information on its diet is derived from the dissection of 70 birds killed mostly in summer.

In the Dusse-Alin' Range, 14 birds were caught in summer, 11 in March, August, and September on Sakhalin (Mishin, 1960), and several in Yakutia (Vorob'ev, 1963); the remaining data pertain to birds from the Amur and Primor'e regions (Shul'pin, 1936; Kaplanov, 1938; Vorob'ev, 1954, 1966; Abramov, 1962; Yudakov, 1972). In the diet of the Siberian Spruce Grouse, 28 plant species belonging to 12 families have been identified. In some birds killed in winter in the Amur and other places of Khabarovsk Krai, the crop and gizzard contents revealed only needles of Manchurian fir. In the food of all specimens caught in July-August, needles of Siberian larch, Daurian larch, and Manchurian fir predominated; at some places Yeddo spruce needles were also found.

Various berries occupy second place in the diet. These are: bog whortleberry, bilberry, cranberry, raspberry, stone bramble, and cloudberry, together with spikes of sedge, shoots of horsetail, stems of lichens, moss, and reindeer moss. Feeding on birch leaves has also been recorded. The role of animal food in summer is not clear. Vorob'ev (1954) states that insects are generally absent in the food of adult Siberian Spruce Grouse, while according to Afanas'ev, there is no marked difference between the food of young and adult birds. Most of the animal food consists of insects: lepidopterans (caterpillars), orthopterans (locusts), coteopterans (leaf beetles, weevils, and long-horned beetles), and hymenopterans (ants). Spiders are rarely found.

Even from such scanty data it is clear that needles of various woody species constitute the staple food of Siberian Spruce Grouse throughout the year. In summer it often feeds on berries, seeds, and invertebrates, but
103 the chief food during this period, too, remains needles not only of fir and spruce but also of larch.

The predominance of pine needles throughout the year in the food of these birds leads one to assume that Siberian Spruce Grouse leads a rather more arboreal mode of life than its kin, and is the most typical taiga bird among all the galliformes in the avifauna of the USSR.

Genus *Tetrastes*—Hazel Grouse

For much of the year (7–8 months) Hazel Grouse feed in the crown of deciduous trees. Only in regions with temporary winter snow cover are foraging birds found sometimes on the ground. When moving among branches the grouse sometimes leans downward, sometimes stretches upward, and when breaking catkins and buds vigorously shakes its head. Extraordinary skill is required of these birds, which feeds in windy weather along thin ice-covered branches. In summer, while feeding on the ground, Hazel Grouse are extremely mobile: they run about considerably, hop onto felled trees, fly across thickets of brushwood, and feed on the ground among grass and bushes. They never scrape the soil for food.

Food habits of Hazel Grouse have been studied in detail. Data from more than 2,900 birds from various parts of the range have been analyzed: Kola Peninsula (Semenov-Tyan-Shanskii, 1938, 1959; Novikov, 1952); Karelia (Neifel'dt, 1958; Ivanter, 1962); Belorussia (Gavrin, 1956; Dolbik,

1967); central and eastern regions of the European part of the USSR (Formozov, 1934; Donaurov, 1947; Rusakov, 1963; Rodionov, 1963; Popov and Shkylyaev, 1967); southern Urals (Kirikov, 1952; Krasovskii and Troitskii, 1958; Kuz'mina, 1968); Altai (Tsvelen'ev, 1938; Kuz'mina, 1968); Enisei (Folitarek, 1939); western Khamar-Daban (Tarasov, 1960); Trans-Baikal (Afanas'ev, 1934; Rakhilin, 1968); central Yakutia (Larionov, 1965); upper reaches of Lena (Gaidar, 1969); southern Primor'e (Luchnik and Nedetskii, 1938; Nechaev, 1968; Polivanov and Polivanova, 1971); and Sakhalin (Gizenko, 1955).

In most regions of the European part of the USSR, Altai, and southern Primor'e, the food of Hazel Grouse has been studied year round; data from other parts of the range are based only on seasonal observations. Unfortunately, a number of works mention only the main food, ignoring other components. On the whole, more than 2,900 birds were examined.

The vast range and the basic differences between the habitats of Hazel Grouse in the various parts of the range account for the marked diversity in the food of this species. However, it must be conceded that each
104 population consumes only a limited assortment of food due to their exclusively sedentary life.

Hazel Grouse are essentially herbivorous birds; animal food is found in their diet only in summer and early autumn. During this period in most cases the animal food is less in quantity than the plant food, but nonetheless constitutes an important dietary item. Fledgelings in early life are almost exclusively insectivorous, but very soon switch over to a mixed diet, and by October feed on the same food as adults.

The plant food of Hazel Grouse is especially varied: 279 plant species belonging to 64 families have been identified in their diet. Among animal food, chiefly insects, but also mollusks and arthropods, have been found. These birds rather often feed on arachnids; myriapods have also been found on occasion. The quantity of animal food may vary greatly from year to year. In years of population explosions among insects, Hazel Grouse prefer them to berries; the species composition of insects is apparently of no particular significance.

In the winter food of these birds, 55 plant species belonging to 19 families have been identified, but in some regions only 8–10 plant species constitute their winter diet; of these 4–5 species constitute the basic items. The food of Hazel Grouse in Altai is more diverse; 17 species of plants have been found in its winter food. The food composition is even more rich in the Far East. The primary food plants during the winter are various species of Betulaceae, among which species of birch and alder are primary. In most regions where these woody plants grow, they constitute the main winter food of Hazel Grouse. Almost throughout its range the grouse supplements its diet with twigs of various kinds of willow and, at some places, also European mountain ash, aspen, chokecherry, and other deciduous species.

In addition to twigs in early and late winter, the food of Hazel Grouse includes berries of deciduous species and shrubs and green parts of herbaceous plants. In years of abundant berry yield these fruits may replace twigs and buds of woody plants in early winter (November) food, but berries are rare in midwinter. During severe frosts the birds avoid succulent foods, preferring more concentrated fodder such as catkins and buds. Green vegetation constitute only a very small part of the winter diet, but in early and late winter in many regions shoots of bilberry, leaves of wood sorrel, common strawberry, and others have been found in their diet.

In the European part of the USSR, Hazel Grouse feed on twigs and catkins of birch and alder, supplemented with buds of various species of willow, berries of mountain ash and dog rose, and leaves of bilberry.

105 During the first half of winter in Altai, berries are fairly important, especially those of mountain ash, chokecherry, and red currant. During the first half of winter in years of abundant yield, berries almost completely replace twigs. Hazel Grouse here also feed on cedar nuts (Tsvelen'ev, 1938). In Yakutia the principal winter foods are buds and catkins of birch; alder is of secondary importance. In the beginning of winter, berries of bog whortleberry are also consumed (Larionov, 1965).

The winter food of Hazel Grouse in Primor'e varies significantly. Twigs predominate for four months; here, too, as in other parts of the range, chiefly plants of the family Betulaceae are consumed, but the species composition is different (heartleaf, hornbeam, filbert, hazelnut, costate birch, various species of maple, Manchurian elm, Euonymous, Thunberg's willow, and Amur chokecherry). In early winter the food consists largely of berries of Actinidia, Amur grapevine, Maximovich hawthorn, and Maack's honeysuckle. More than 20 plant species from eight families have been recorded (Luchnik and Padetskii, 1938; Nechaev, 1968). However, in each group of birds (small flocks) the food is fairly uniform but may differ notably from that of other groups. This points to food specialization among individuals consequent to their exceptionally sedentary mode of life (Polivanov and Polivanova, 1971).

In April-May (in the north from May to early June) twigs, buds, and catkins are gradually replaced by sprouts, flowers, winter berries, and seeds. During spring the food of Hazel Grouse includes 57 plant species belonging to 20 families. Throughout the range during this season the food of Hazel Grouse consists chiefly of green fodder (sprouts of winter herbs, ephemerals, and tender leaves of trees and bushes), but their species composition differs markedly from area to area.

In the north (Lapland and the Pechora-Ilych preserves). Hazel Grouse feed chiefly on bilberry shoots and sometimes on wood sorrel leaves during spring. Willow inflorescences are also an important constituent of their food. In the central belt of the European part of the USSR, Altai, and Primor'e, a good percentage of the food consists of ephemerals, especially anemones (in Altai up to 50% of food). In the southern Urals, where anemones are not available, the spreading pasque flower replaces them. In the spring, Hazel Grouse readily take to flowers and green parts of corn buttercups, mercury, and Jungarian globeflowers, and in the Primor'e to the green parts and flowers of Adonis, birthwort, moschatels, gagea, and other ephemerals. In addition to ephemerals, it also feeds on the green mass of winter herbs (leaves of wood sorrel, common strawberry, alter-
106 nate-leaved golden saxifrage, etc.). Tender leaves of shrubs and trees are of secondary importance.

During this period, winter berries (bilberry, cowberry, and dog rose) are found only in some regions and are not of much importance. Seeds constitute a small part of the diet, but at the close of spring, early ripening fruits and seeds of ephemerals and, in some regions, a few empty seeds of pine and spruce are constantly found in the crop. Occasionally Hazel Grouse consume some insects in spring, perhaps accidentally.

In summer the food of these birds includes all the food groups found throughout the year. They feed on a wide variety of plant food, comprising 115 species (39 families), although in some regions only up to 50 species have been recorded. Insects make up a large part of their diet in many areas.

The proportion of individual food groups changes significantly from region to region, depending on the phenology of the plant canopy.

Green fodder and flowers are found in the food of Hazel Grouse in early and even midsummer, but are gradually replaced by berries and animal components, so that by the end of the season they are only of secondary importance. During the first half of summer, green plants are consumed in larger quantities in the north than in the south of the range. Berries are the chief summer food, constituting the bulk of the food consumed, and are found in almost every bird examined at this time. In all, 28 species of berries have been identified during this season but these birds seem to prefer only some species, in particular those from the cranberry and rose families, other berries are of local importance only.

Seeds and fruits of herbaceous plants (52 species) are second only to berries in quantity, but constitute a very important dietary component because of their high calorie content. Over much of their range Hazel Grouse have been observed feeding on seeds of various species of sedges, knotgrasses, buttercups, cereals, and violets; seeds of other families are only of local importance.

Animal food constitutes a significant part of the summer diet in many regions, though its quantity may differ intensely from year to year in the same locality. The most abundant animal foods are beetles (leaf beetles, weevils and ladybirds). Hymenopterans (ants and sawflies) are common, followed by Lepidopteran caterpillars (inchworms, owlet moths, hawk moths, and clearwing moths). Hemipterans are usually found, but only in years of population explosions are they found in large numbers, at times
107 even completely filling the crops and gizzard of birds and totally displacing berries (Tsvelen'ev, 1938). Grasshoppers are a regular find but other insects are occasional.

In the northern part of its range (Lapland and the Pechora-Ilych preserves) in the first half of summer, Hazel Grouse feed on the sprouts of herbaceous plants, seeds, and juniper berries. But as soon as other berries ripen (bilberry, cowberry, and crowberry) these become the staple diet. A similar dietary change has been observed in Karelia. At the beginning of summer these birds feed on the green parts and seeds of plants, and berries play a secondary role. From July on, berries begin to predominate (bilberry, cowberry, bog whortleberry, stone bramble, raspberry, and dog rose). By the end of summer, the birds feed on berries and seeds to the total exclusion of sprouts. Throughout the year they consume ants in large numbers, while other insects serve only as a supplement.

In the southern Urals, first place is held by berries (common strawberry, strawberry, stone bramble, and raspberry) and second by seeds (cereals, sedges, cow-wheat, and wonder violets). Animal food is consumed in smaller quantities, but nonetheless constitutes an important part of the diet, especially larvae of butterflies and moths, beetles, and grasshoppers.

The food composition of Hazel Grouse in Altai is unusual. Green plants (clover, meadow sweet, and buttercups) are regularly consumed until autumn. Berries constitute the basic diet, with common strawberries predominating. Seeds have been found throughout the summer and in larger quantities than in the European part of the USSR. Animal food is consumed from May through September in large numbers and its frequency is more significant than in other parts of the range.

In the southern Primor'e in summer this species consumes berries of vine, shrub, and woody plants (chokecherry, hawthorn, raspberry, and Amur grapevine). Its food composition suggests that Hazel Grouse here should be feeding on trees during summer, whereas in other regions they forage only on the ground.

Animal food predominates in young birds, unlike adults. In the Pechora-Ilych Preserve, during the first ten days fledgelings feed only on tiny invertebrates (spiders, leafhoppers, sawfly larvae, mosquitoes, and

ants). For the next ten days they feed additionally on tiny seeds (wood sorrel, sedges, cow-wheat, wonder violet, and buttercups). Their diet range gradually widens with the addition of more varied animal food, supplemented with berries. In August animal food and seeds occupy first
108 place, followed by berries; in September the young birds feed more substantially on animal food than older birds (Donaurov, 1947). During the relatively short autumn, considerable changes occur in the diet of Hazel Grouse. Throughout the range, animal matter is more abundantly consumed in early September than in summer, only to disappear altogether after a month. The basic components of the autumn food are various berries, predominantly of shrubs and trees. Seeds constitute a larger part of the food during this period than in summer. By the end of autumn berries have gradually been replaced by twigs, and seeds disappear. There comes a time when these birds spend almost all their time in trees and feed and roost on trees. The changeover to feeding on twigs and catkins of woody plants takes place earlier in the northern part of the range.

In the Kola Peninsula, twigs predominate during the second half of October. In the central belt of the European part of the USSR, these birds feed on berries in autumn, especially bilberry and cowberry and, to a lesser extent on dog rose, guelder rose, and chokecherry. In October, and sometimes in November too, they consume berries of shrubs and trees in addition to twig food.

In Altai, where berries are especially abundant, they constitute the main ingredient of the autumn food (bilberry, guelder rose, raspberry, redberried alder, chokecherry, mountain ash, and hawthorn). Twigs begin to predominate only in November (Tsvelen'ev, 1938). Until the snow becomes deep, the birds also feed on cedar nuts, plucking them from fallen cones (Dul'keit, 1960).

In Trans-Baikal (Selenga valley) berries of sallow-thorn represent the main food. The high fat content of these berries, not found in other parts of the range, perceptibly fattens the birds (Pil'man, 1961; Smirnov, 1967).

In Primor'e the autumn food consists almost exclusively of berries (chokecherry, raspberry, grapevine, Actinidia, lemon tree, and hawthorn). Twigs are consumed only after November.

During the snowfree period, the food of Hazel Grouse differs little from that of other tetraonids except for a slightly larger intake of invertebrates (not in species composition, but in volume and duration of consumption). In winter Hazel Grouse feed almost exclusively on deciduous trees, consuming not only twigs, but also berries of woody and shrub plants. The greater diversity of the winter food with a distinct preference for leafy food should be considered a consequence of the speciation of

109 these birds in deciduous forests which are diverse in their species composition.

Family PHASIANIDAE

Genus *Coturnix*—Quail

Quail feed on the ground, foraging on the surface and in the grass. They do not dig the soil for food, a fact corroborated by the absence of scraping organs and more so by the absence of underground plant parts in their diet. They feed during maximum diurnal activity, in the evenings, and at daybreak. Nothing is known about the diet of the Japanese Quail. The food of Common Quail has been studied mainly in the southern regions of the Soviet Union; the Crimea (Grekov, 1965); the Caucasus (Formozov, 1964; Khanmamedov, 1965); Sal'sk steppes (Budnichenko, 1965); around Tashkent (Meklenburtsev, 1959); Zeravshan Range (Abdusalyamov, 1964); Kirgizia (Pek and Fedyanina, 1961); Kazakhstan (Ul'yanin, 1941; Kuz'mina, 1968); Barabinsk steppe (Shilo, 1974); and Ob' region (Kuchin, 1963; Shinkin, 1967).

In spite of the fact that the food content has been analyzed in more than 600 quail, information on their diet is still tentative. Much of it covers a short period, the end of summer-autumn, while information for spring is scanty and for the winter stations totally absent. Moreover, most of the publications listed above mention only basic food items. The widely prevailing view that animal food is of secondary importance to these birds is reliable only for the short period of August–October. According to some observers, quail feed chiefly on invertebrates in spring and during the first half of summer. In Odessa District insects and their larvae constituted 80.5% of the food intake in April and May, with 100% frequency of occurrence (Grekov, 1965). In the Sal'sk steppes, the gizzards of five quail mostly contained insects (weevils, Austrian bug, powder-post beetle and darkling beetles) in June–July; weeds were found in small numbers (Budnichenko, 1965). In two birds caught in June in northern Kazakhstan, the crops and gizzards contained only invertebrates—beetles and their grubs, grasshoppers, bugs, and mollusks; in another bird the food consisted of 60% wheat and 40% beetles and grasshoppers. In Naurzumsk Preserve the most preferred food in summer is willow grass followed by insects, but in years of peak grasshopper populations, insects predominate (Ul'yanin, 1941).

110 In 12 quail caught in June–July in Tadzhikistan, the basic food consisted of leaves and shoots of alfalfa and seeds of legumes and crucifers; grasshoppers and tiny beetles were negligible (Abdusalyamov, 1964).

In the Barabinsk forest-steppe, animal foods predominate during spring-summer, especially in young birds (click beetles, leaf beetles, weevils, ground beetles, grasshoppers, bush crickets, larvae of sawflies, horseflies, and tiny mollusks). In autumn the main food components are various grains (wheat, wild hemp, willow grass, corn gromwell, and so on) (Shilo, 1974). This information on the spring and summer feeding of quail, though scanty, nevertheless indicates that in these seasons insects are the primary food, gradually replaced by ripening seeds. The chief plant foods are seeds, while green plants are found only in spring and early summer. Of the berries, only nightshade has been detected in the gizzard contents, with underground plant parts totally absent.

Food selectivity has been observed not only with regard to particular food groups (various parts of herbaceous plants), but also their species composition.

In the Soviet Union 37 plant species belonging to 17 families have been identified in the food of Common Quail. This small assortment of plants comprises only the basic ingredients. In frequency of occurrence and quantity, wild cereals and cultivated plants, knotgrasses, and borages rank first.

For comparison, an analysis of the food of Common Quail in Bulgaria (165 birds) revealed 70 species of herbaceous plants belonging to 17 families. Among them, 10 families and 10 genera are common to those identified in the Soviet Union. Species composition varies even more, with only eight species common to both territories. Insects are represented by 20 families of seven orders, consisting largely of coleopterans (click beetles, leaf beetles, ground beetles, and weevils), grasshoppers, lepidopterans and their larvae, and hemipterans. Apart from insects, other invertebrates, especially mollusks, were found occasionally (Petrov and Donchev, 1969).

In the Soviet Union, seasonal changes in the food composition of quail from the moment of the bird's arrival to its departure have been studied only in the Crimea (Grekov, 1965). Here, in April–May, animal and plant components comprise 80.5% and 19.5%, respectively. From July weeds predominate (99% frequency of occurrence), with 25% cultivated plants and 63% animals (but in young birds up to 80%). Seeds at this time mostly include amaranth, bindweed, green bristle-grass, and black-berried night-
111 shade. In autumn the bulk of the crop and gizzard contents consist of weeds (99.3%), cultivated plants (0.2%), and animal food (0.5%). Greatest preference goes to amaranth (92.7%) followed by bindweed, bristle-grass, and black-berried nightshade.

In the northern parts of Kazakhstan where quail congregate near cultivated crops at the turn of summer, in August their food consists largely of cereals, especially millet (Ul'yanin, 1941). In the foothill region near Alma-Ata, the main food of quail (52 birds) caught at the end of summer and in autumn consisted of the seeds of about 20 plant species (chiefly cereals). Weeds such as yellow bristlegrass formed much of food volume in 84% of the birds and green bristle-grass in 57%. Knotgrass seeds were found in 18% and wheat in only two birds (3.8%). The rarity of cultivated grains may be due to the fact that the birds were shot far from cultivated fields. Sedge seeds (species not identified), observed in three birds, comprised 85% of the entire gizzard content of one. Other seeds such as wild hemp, green amaranth, and tufted vetch are supplementary. Green plant parts were totally absent. Animal food was found quite often, but only in small quantities. In August, insects and mollusks were noticed in 26% of the birds but comprised nearly half the intake in only two (9%) and did not exceed 2 to 3% of the total intake in the remaining birds. In September, animal food has also been found in two quail (9%) and mollusks and beetles constituted 15% of the total food in one bird's fully filled crop. The gizzard of a bird killed on October 25 contained only chitin. Insects in the food of quail from Alma-Ata were represented by grasshoppers, beetles, and bugs.

In the Upper Ob' region, compared to more southern regions, quail begin feeding on insects and young shoots late in spring. In May they still feed on grain from fields of wheat, oats, and willow grass, as well as green shoots; in June their food consists of green shoots and insects—predominantly bush crickets and beetles; in July they feed on insects and weed seeds (tufted vetch, guinea pepper, sheep's sorrel, corn gromwell, and hemp-nettle); and in August the basic food is barley and oats (Kuchin, 1963).

Although the above information is rather sketchy, a general conclusion could be drawn regarding the nature of the diet of this bird at its nesting sites. During the spring-summer period, insects definitely predominate in the food of quail but the period of feeding on animal matter may vary sharply, depending on latitude.

112 In the south, quail begin to feed on insects from the moment of their arrival, while in the more northern regions, even in May, they feed on grains of cultivated crops. Green plants and insects occupy an important place in the northern populations at the end of summer. The consumption of seeds of cultivated crops depends on bird distribution during autumn. At places where quail are close to fields, fallen grain constitutes a significant part of their intake. On the other hand, in less cultivated areas they are satisfied with weed seeds, of which the most important are the yellow and green bristle-grasses.

Food composition is quite uniform, with 4-6 plant species and animals found in the same crop, although one may assume that, in general, the food of widespread, vagrant quail would include a fairly rich assortment of components, varying in different parts of the wide range and in different wintering sites. This assumption is supported by differences in the species composition of food material identified during a study of quail of the Soviet Union and Bulgaria.

Genus *Perdix*—Partridges

Both species of this genus feed only on the ground, either nibbling at the lower branches of bushes or scrabbling in the soil or litter dumps. In winter Gray Partridges can also dig into shallow snow in search of food. It is not known whether Daurian Partridges gather food from underneath the snow cover.

The feeding of Gray Partridges has been studied in Lithuania (Logminas, 1962); southern Ukraine (Budnichenko, 1965); Lower Volga region (Osmolovskaya, 1961); Crimea (Vishnyakov, 1962); Sal'sk steppes and regions of western Kazakhstan (Poyarkov, 1955); northern Kazakhstan (Ul'yanin, 1949); the Caucasus (Khanmamedov, 1969); various regions of Kazakhstan (Kuz'mina, 1968); and western Siberia (Shinkin, 1967). Crop and gizzard contents of about 600 birds have been studied. Additional data on the food composition of Gray Partridges are given in works which regard them as one of the most useful birds to the national economy (Formozov, 1940; Osmolovskaya and Formozov, 1950; Osmolovskaya, 1950, 1961; and Gladkov, 1952).

It must be recognized that our knowledge of the diet of Gray Partridges is still inadequate. Published data are very general and many parts of the range have not been covered.

113 The plant food of Gray Partridges so far identified comprises 109 species belonging to 30 families; animal components are represented by 27 families from nine orders and include insects, mollusks, spiders, and myriapods. A comparison of these data with the vast material published in Czechoslovakia is interesting. There, analyses of 2,038 Gray Partridges dissected in different seasons revealed 81 species of plants belonging to 20 families, more than 130 species of insects belonging to 18 families from 8 orders, and 13 species of mollusks and myriapods (Janda, 1956). Plants common to the Soviet Union and Czechoslovakia comprise only 20 species belonging to 8 families, of which plants of cereal, beet, and Compositae families predominate in both countries.

Gray Partridges feed on green parts, flowers, seeds, berries and subsoil parts of plants. Of these, seeds and sprouts form the staple diet. Berries have been found in small quantities only in northern and central Kazakhstan (totally absent in Czechoslovakia), while roots and tubers are generally rare. From late spring to the middle of autumn, animal matter constitutes the largest part of the diet and different kinds of beetles and grasshoppers are especially favored. In Czechoslovakia animal foods are more diverse (beetles, grasshoppers, and bugs), but their volume in total intake is less.

In Poland, food of Gray Partridges was studied in 420 birds caught in all months of the year. The ratio of animal to plant components in the annual diet was 1:14, with 37.5% (by weight) consisting of seeds of weeds and grasses, 33.4% green parts, 22.4% grains of cultivated crops, and only 6.7% animal components. In winter and spring vegetative parts of plants predominated (90.2 and 50.7%, respectively), while in summer and autumn seeds of weeds and cereals were the main constituents. Animal matter contributes 13.9% of summer food and 0.4% of autumn (Oko, 1963).

Data on the diet of Daurian Partridges are scanty, only covering a total of 200 birds analyzed from the following regions: Primor'e—Prikhankaisk Plains (Nechaev and Nikolaev, 1970); Buryatia (Borovitskaya, 1965; Khankhasaev, 1969); southern regions of Kazakhstan (Kuz'mina, 1968); and Kirgizia (Pek and Fedyanina, 1961).

In all, 29 plant species of 15 families were identified in the food of the Daurian Partridge. The species composition is surely much broader, as only the primary components of the gizzard contents in the few birds analyzed are mentioned. Furthermore, about 1/3 of the plant foods found
114 in the gizzards I examined were not identified. The composition of animal food is almost unknown, since there is practically no information on the summer diet and only some stray grasshoppers and weevils have been recorded. Judging from the scanty data available, Daurian Partridges feed for the greater part of the year on seeds, green parts of herbaceous plants, and occasionally on berries. Animal food is important, if at all, only in summer.

The food composition of Gray and Daurian Partridges differs somewhat. In winter the food of Gray Partridges consists of various seeds, green matter and sometimes hibernating insects. The basic food consists of cereals including cultivated crops. In addition this bird feeds on seeds of orache, cow parsnip, and sedge. Only ten species have been recorded in the winter food of Common Partridge in the Soviet Union. Green plants (Kazakhstan, 13 birds) constituted 26% of the total food volume and seeds of wild and cultivated plants, 74%.

In central Kazakhstan around Lake Kurgal'dzhino, cereal grains constitute the staple food of Gray Partridges. Due to the abundance of food on roadsides and near settlements, the winter birds of this region are fat (Krivitskii, 1962). The food of Gray Partridges in Czechoslovakia is rather uniform: of the 126 birds dissected in winter only six species of plants were identified, 70% of the volume consisting of grass blades and only 30% of seeds of wild and cultivated cereals. A predominance of sprouts in the winter food of these partridges has been observed in the Baltic region and southern Finland (Logminas, 1962; Pulliainen, 1966).

It is interesting that Gray Partridges acclimatized in Canada and the northern states of the USA* (introduced from western Europe) reveal sharp changes in the proportion of different food groups in their new, more severe habitat conditions, with a changeover to a basic diet of cultivated cereal grains. The high-calorie food available greatly fattens birds in winter months (Westerskov, 1966).

The food of Daurian Partridges essentially differs from that of Gray Partridges mainly in the more varied assortment of plant food and the marked predominance of seeds over green fodder. In the winter food of Daurian Partridges, about 20 species of plants have been identified with about ten species (my specimens) unidentifiable. The basic food consists of cultivated cereal grains (barley, wheat, oats, and rye), seeds of wild plants (common wild oats, green bristle-grass, and yellow bristle-grass), and seeds of beets, legumes, knotgrasses, lilies, and members of the Compositae family. Among the greens, leaves of cereals often have been found to be supplemented with leaves of lucerne and wormwood.

Only rice grains were found in three Daurian partridges killed in the Khankaisk plain in February. During the Autumn—winter period in Buryatia the food of the bird consists of seeds of wild plants and cereal grains.
115 According to Khankhasaev (1969), the basic food in autumn comes from grains of cultivated crops (wheat, willow grass, oats, barley, and so on) and in winter, seeds of grasses (black bindweed, paniculate amaranth, and caraway). Borovitskaya (1965) has shown that the proportion of wild grass seeds and cultivated cereal grain in the winter food of Daurian Partridges is almost equal. For the sake of comparison, it would be interesting to note the winter food of Daurian Partridge in Kentei, where 16 birds were dissected (Bannikov, 1957). Cultivated cereal grains predominated by weight, followed by seeds of legumes (vetches of *Vicia* and *Astragalus* genera, and lucerne). Most of the birds (ten) also had an admixture of green shoots of cereals, onions, and mixed grass. Mollusks were found in two birds and insect remnants in three. According to Borovitskaya, Daurian Partridges consume any seeds available in the region in winter.

In Tien Shan (the Trans-Ili Alatau) the proportion of greens and seeds in food perceptibly differs from bird to bird. In some Daurian Partridges caught in December, sprouts and seeds of cereals were more or less equal (by volume), while the stomachs of some birds contained only greens and still others only seeds.

*Introduced species are known as Hungarian Partridges—Editors.

No information is available on the spring food of Daurian Partridges. The food of Gray Partridges is more diverse during this period than in winter. Green food includes, in addition to cereals, leaves of wormwood, tulip, hawkweed, and horsetail stems; statice and tulip flowers have also been found. Cultivated cereal grains were not found in their food during this period. As an exception, in the crop of a bird killed in May in northern Kazakhstan tulip bulbs constituted up to 60% of the total intake. Insects start appearing in the food of the Common Partridge as early as May.

The food of Gray Partridges varies most in summer. In the greater part of its range it feeds on both plants and animals during this period, but their ratio may vary greatly; at some places the partridges are exclusively insectivorous. Of the plants consumed in summer 26 species have been identified. Sprouts and flowers eaten in small quantities include the green parts of bulbous meadow grass, leaves of spiraea, stems of horsetail, and flowers of hawkweed. The composition of seeds is more diverse (11 species); millet, wheat, false acacia, Spanish liquorice, cornflower, goats beard, lettuce, and so on constitute the basic part of the summer plant food. Berries—common strawberry, stone bramble, cotoneaster, honeysuckle, etc.—have been found only in northern Kazakhstan where, along with insects, they constitute the greater part of the diet.

116 The following data illustrate the significance of animal matter in summer food. Fourteen birds studied in June–July in southern Ukraine contained only insects (Budnichenko, 1965). In the Sal'sk steppes, according to Poyarkov (1955), insects comprised a very important food component of adult birds, especially females. At this time 70% of the stomachs of males were filled with seeds and insects and 30% with insects alone. In females, 50% of the stomachs contained mixed food, 25% seeds alone, and 25% exclusively insects. The stomachs of females contained an average of 147 insects, whereas the male average was only 18.

In the Lower Volga region, in addition to insects, Gray Partridges always feed on various plant parts—flowers, leaves, seeds and in some cases, bulbs. During the first half of summer, plant food is considerably less in quantity than animal matter, but it predominates during the second half. Among insects, the most important (in frequency of occurrence and number of specimens) are grasshoppers, followed by ants, beetles, and bugs (Osmolovskaya, 1961). In the Caucasus (Azerbaidzhan) from May through July, this Partridge feeds chiefly on insects and in March–April and August–September on a mixed diet of plants and animals.

Insects, chiefly beetles and grasshoppers, constitute about 60% by volume of the food of partridges in northern and central Kazakhstan. In Kirgizia, summer food has been found to contain barley, wheat, and seeds of various grasses, as well as insects—acridoid grasshoppers and beetles.

Chicks feed in the early days only on insects, with plant material appearing only when they begin to fly (Poyarkov, 1955).

Differences in the food of young and adult Gray Partridges have been studied in a large number of birds in Czechoslovakia. Insects were found in 86.7 to 96.7% of one to four weeks old chicks, in 70.1 to 77.1% five to seven weeks old, and only in 57.1 to 60% eight to nine weeks old. Plant food has been found in 50% of chicks one to two weeks old, in 63.3 to 76.4% three to four weeks old, and 97.1 to 100% in older juveniles. The volume of animal food varied from 94.7 to 91.7% in birds one to two weeks old to 36.2 to 12.5% in birds five to nine weeks old, while the proportion of plant food varied in a reverse order. Among animal components, insects predominate—ants, leaf-hoppers, weevils, and grasshoppers (Janda, 1959).

In early autumn, September, the diet still contains insects, but in October plants predominate (24 species identified). The crop and gizzard
117 contents of some birds contain only the green parts of cereals, but usually the basic foods are grains of wheat, oats, barley, wild hemp, common wild oats, orache, lettuce, three-lobed bur marigold, stickseed, and so on; occasionally berries of honeysuckle and chokecherry are also found.

The range of food found in a given bird is not very large, only one to five plant species. Birds with only wheat in their crops have been found quite often.

The basic foods of Daurian Partridges during this period are the seeds of grasses and cultivated plants. However, in some birds animal food predominates even during this period. Thus, in one of the 16 birds brought down in the Trans-Ilian Alatau in September–October, 80% of the food (by volume) consisted of mollusks, 10% grasshoppers, and 10% dog rose berries; 95% of the food in the other 15 birds consisted of tubers.

There is much in common in the diets of the Common and Daurian Partridges. Seasonal food changes as well as the species composition of plants comprising the basic diets differ little. Gray Partridges in the greater part of their range are typical members of cultivated lands; Daurian Partridges are also attracted to agricultural areas, but since there is less land under cultivation in eastern Siberia than in Europe and western Siberia, cultivated lands are less significant in their life.

The uniformity of species composition of the diet of partridges is apparently due to the nature of the cultivated lands where "vast areas are under a single crop." Under such conditions only a limited assortment of high-calorie plant food is available to the birds. Differences in the winter food of partridges depend on the kind of crop in a given region. Depending on climatic conditions, partridges may use different parts of the same plant. In regions with severe winters, the birds prefer particularly high-calorie grains and, in regions with mild climatic conditions, they feed chiefly on succulent sprouts. They also feed on insects at their wintering grounds.

In a number of regions in the European part of the USSR the benefits that accrue from the partridges destroying insects such as leaf beetles, weevils, and Austrian bugs are far greater than the birds' game value (Gladkov, 1952). In the Lower Volga region they massively destroy the Austrian bug (Nefedov, 1943), pupae of elm springtail, brindled beauty moth, pine beauty moth, and plumed prominent (Osmolovskaya, 1961). It has been stated that the destruction of a large number of insects by the partridges promotes a higher sugar beet yield (Logminas, 1968). It has
118 been estimated in Czechoslovakia that partridges save up to 50% of the sugar beet crop by destroying the sugar-beet weevil.

Genus *Ammoperdix* —Sand Partridges

Sand Partridges feed exclusively on the ground. The detection of underground parts of plants in its food suggests its ability to scrabble the soil for food. However, whether it uses its beak or only its claws for this purpose is not known.

Information on the food of Sand Partridges is very scanty due to its limited distribution and small population. Food analysis was done on 44 birds in central Kopet-Dag (Rustamov and Kogan, 1954) and on only four birds in the mountains of Kopet Dag (Kolesnikov, 1956). In southern Uzbekistan 43 crops and gizzards were analyzed (Salikhbaev and Ostapenko, 1964), but this published work only briefly lists the groups of foods encountered. Some information on the food of this partridge is available for southwestern Tadzhikistan (Abdusalyamov, 1971). This information, though not elaborate, reflects the seasonal food changes and emphasizes the characteristic feeding of See-see Partridges due largely to the xerophytic conditions of the habitats. The species composition of the plants consumed is made up of 48 species of 19 families but the food groups are limited.

Sprouts have been found in the winter food of partridges in Turkmenia and from early spring to the end of April in Tadzhikistan. Insects have been found in the food in Tadzhikistan over a very short period, i.e., when the ephemerals disappear and the late plants have just begun bearing fruit. During this period, acridoid, grasshoppers, various beetles, and other invertebrates were noticed in the food of this partridge. During the greater part of the year its food consists basically of the seeds and fruits of herbaceous plants.

The absence of green plants in its summer food is explained by the early vegetative growth of plants and the fast disappearance of green mass.

Seeds and fruits of legumes, cereals, borages, crucifers, as well as tubers, form food items during this season.

In summer the food of Sand Partridges in the hilly regions of Turkmenia includes seeds and fruits of more than 30 species; the food of some individuals may contain up to ten species. The great diversity of plant composition indicates that these birds are not fastidious in their choice of
119 food and consume whatever is available under the extreme conditions of barren mountains. Seeds and fruits of corn gromwell, storksbill, smooth gold bachia, and spurge occupy first place in summer. Feeding on dry food necessitates frequent visits to watering places.

In Turkmenia the gizzard and crop contents of Sand Partridges weighed more in summer than in any other season. On the basis of this finding, Rustamov and Kogan (1954) concluded that Sand Partridges require more food in summer than in winter.

In the autumn 18 plant species have been found in the diet of this bird, chiefly seeds, and more often prostrate gypsophila, barley, corn gromwell, and small-flowered sainfoin. Onion bulbs were found in 25% of the birds and formed the single largest (by weight) food item.

In winter 13 plant species, or more than a quarter of the known species composition of its food, have been recorded in the diet of Sand Partridges. Considering that as few as nine birds have been dissected during this season, the winter diet of this partridge could be regarded as rather varied. Only during this period were bulbs, tubers, and sprouts found in the crops and stomachs along with seeds and fruits of herbaceous plants. The birds feed avidly on sprouts and bulbs of bulbous meadow grass, found in 44% of the specimens and outweighing any other component in its food.

Thus the limited data available at present on the food of Sand Partridges reveal the specificity of their diet. The short duration of feeding on animal matter, the absence of berries and green plant parts in summer, and a maximum diversity of winter food—all these reflect the characteristic flora of barren lands.

Genus *Alectoris*—Rock Partridges

Chukar Partridges feed on the ground. Only stray cases of the bird subsisting on sallow thorn berries and mulberries are known. On the ground this bird prunes off green leaves and berries, catches insects, and picks up fallen seeds. A considerable part of its food consists of bulbs, tubers, and rhizomes dug up with beak and claws.

Food in the upper layer of the soil is scooped out by its strong claws. Deep-rooted bulbs are dug out vigorously with its beak by making a
120 narrow, deep (6 to 8 cm) pit to the bulb, scooping it out and eating bits of it, leaving behind only the shell in the soil.

The food of Chukar Partridges has been studied more or less exhaustively. The crop and gizzard contents of more than 1,300 birds have been analyzed from the following regions: Caucasus (Khanmamedov, 1955; and Gazan'yan, 1964); Kopet-Dag, Turkmenia (Rustamov, 1945; Fedorov, 1949; Kogan, 1950; and Kolesnikov, 1956); Gissar range, Tadzhikistan (Popov, 1959); Turkestansk and Chatkal'sk ranges, Uzbekistan (Minin, 1939; and Ostapenko, 1958, 1965); Chu-Ili mountains, Trans-Ili, and Dzhungarsk Alatau, Kazakhstan (Kuz'mina, 1955, 1968); and Kirgizia (Pek and Fedyanina, 1961).

Chukar Partridges are predominantly herbivorous birds. Adult birds prefer plant to animal food in all seasons. However, in years of insect population explosions, making them readily available, they often constitute the chief food.

The species composition of plant food is diverse and 252 species of 48 families have been identified; however, in my collection about 40 seed samples were not identifiable. All kinds of fodder groups, except twigs, have been found in the diet of Chukars.

The food of Chukar Partridges differs markedly in different parts of its range, both in species composition of the plants consumed and in the ratio of the different food groups. Bulbs, tubers, and roots are very important dietary items throughout the year.

The food of adult birds contains animal matter during summer, but its importance varies from region to region. In Turkmenia in summer, for example, adult birds rarely feed on invertebrates, while in the hilly regions of Kazakhstan, invertebrates constitute the basic food and, in years of population explosions, they sometimes (autumn and winter) subsist on a single species of insect in wintering sites. Chukar Partridges feed on mollusks, spiders, and insects; the last are represented by 23 families of seven orders and constitute the principal animal food. These insects are relatively less mobile, predominantly phytophagous, and live in the open on low plants. This species exhibits no preference whatsoever and catches any and all that can easily be got at.

In winter, Chukar are confined mainly to snowfree mountain slopes with a southern exposure where food is available. Food at this time is fairly diverse but when variety in winter food is lacking, the bird is least
121 discriminatory. Altogether 72 plant species of 27 families, of which 54 are found in other seasons also, have been identified in their winter food.

The basic food during this period consists of grass shoots, roots, bulbs, shoots of bulbous meadow grass, bulbs of Tatar ixiolirion, tubers of cranesbill, seeds of brome grass, schrenkia, amaranth, stickweed, ebelek, onion, and berries of ephedra.

In the hilly regions of southeastern Kazakhstan, during this period, the underground parts of plants (bulbs and tubers) constitute 42.5% of food intake, seeds 24.9%, shoots 19.6%, berries 12.7%, and insects 0.3% (by weight of dry food). In Kopet-Dag (where many plants begin growing in autumn and thrive through winter) Chukar feed on young shoots supplemented with seeds. Animal matter has been found in winter food but usually is only incidental, except in years of population explosions of the Austrian bug. In such years it feeds voraciously on these hibernating insects; this was noticed also in the southern regions of Kazakhstan in some years.

In early spring, as soon as ephemerals appear and the surface soil layer thaws, the food composition of the Chukars undergoes a major change. During this season (March–April, in southern regions from February) it feeds on green matter, flowers, and subsoil parts of plants. Thirty species of plants have been identified, which is less than the number consumed during other seasons of the year. In the hilly regions of southeastern Kazakhstan, the main food in March comprises bulbs (39.5%), green plants, and flowers (38.1%); seeds are less significant (22.4%) and animal matter an incidental element. In April the quantity of green plants increases sharply, animal matter increases slightly, subsoil parts of plants constitute only 12.3% of the diet, and seeds almost disappear. The basic spring food includes leaves and stems of bulbous meadow grass, Tatar ixiolirion, scaligeria, onion, gagea, and cereals; bulbs of Tatar ixiolirion, gagea, and scaligeria; seeds of Caryophyllaceae, and Chenopodiaceae families; and so on.

In Kopet-Dag the principal spring foods are the flowers and leaves of ephemerals, which are consumed as early as at the end of February.

In summer (May–August) the food of Chukars is highly varied, the greater part comprising plant material (135 species of 41 families); in some regions not more than 50 species, but covering every fodder group are consumed. During this period animal food is also quite significant (but not in all parts of the range). On occasion the crops of felled birds were filled with insects alone.

In the hilly regions of southeastern Kazakhstan, much of the summer food consists of bulbs and tubers. In May these constitute 64% of the total
122 food, and in subsequent months nearly half of the dry food weight. Sprouts constitute 13% of the May and August intake, but are totally absent in June and July. Seeds are found invariably, but usually in small quantities; only in August does their quantum increase markedly (33.3%). Berries have been found in June–July but are only of secondary importance. In summer, 72 plant species have been recorded in the diet, half of which (35) are represented by seeds. The basic food in this season includes a very small group of plants: Tatar ixiolirion (bulbs, berries, seeds, and leaves), wooly

garlic (seeds and leaves), Tien-Shan cherry (fruits), bulbous meadow grass (bulbs), tuberous cranesbill (tubers), and gagea (bulbs). Animal matter is present in the food of Chukar throughout summer, becoming maximal in June (29.7% of the weight of dry food).

In the Turkestan range the summer food of Chukar Partridges comprises 38 plant species; bulbs and tubers predominate in May to early June and essentially sprouts in the latter half of June and in July. The chief food plants during this period are dandelion, oxytropis, tuberous cranesbill, gagea, and ungernia. Animal matter is minimal and consists of beetles, butterflies (caterpillars), ants, flies, and leaf-hoppers (Minin, 1939).

In the Nuratau Range in summer, Chukars regularly feed on mulberries (berries) (Salikhbaev, Kashkarov, and Sharipov, 1970).

In Kopet-Dag, 45 plant species have been found in the summer food of this bird. Succulent foods are preferred—small-fruited cherries, cotoneaster, Persian honeysuckle, madder, and corms of Turkmenian gladiolus. It also feeds readily on the berries of dog rose and ground burnet, and frequently on seeds of crupina and cornflower.

These birds prefer cereals when living close to cereal farms, taking readily to wheat and barley. Animal matter, berries, and seeds are more important, however, in the food of fledgelings than of adults. Subsoil parts of plants appear in large quantities in the food of young birds only at the end of summer, but are still less significant than in the food of adults.

More than 90 plant species have been identified in the autumn food of Chukars. Animal matter is found fairly often even in September, but only in small quantities. In the hilly regions of southeastern Kazakhstan, subsoil parts of plants constitute over 60% of the total food of Chukars. Especially popular are the bulbs of Tatar ixiolirion, onion, bulbous meadow grass, tulips, and tubers of scaligeria. These are followed by a wide assortment
123 of seeds (more than 26% of intake), among which the important are those of the families Gramineae (at places also cultivated), Chenopodiaceae, Boraginaceae, Cannibidaceae, Compositae, and so on. During this period the birds feed readily on berries (about 7% of intake), including ephedra, dog rose, fallen fruits of iron-tree, black-berried nightshade, and so on. Animal matter constitutes just a little over 3% of the intake and greens even less.

In southern Kazakhstan during this season, Chukar Partridges feed throughout most of the day. Crops of birds caught at different hours of the day contained a large quantity of food, up to 40 to 45 g in some. Increased food consumption makes the birds grow quite plump with the formation of large fat deposits in the body cavity (16 to 20.5 g) and a thick hypodermal layer of fat reserve for the long winter.

The autumn food composition of this bird is completely different in

Kopet-Dag. The basic food consists of seeds, including various species of legumes (several species of vetch and small-flowered sainfoin), seeds of desert candle, dog rose, cornflower, corn gromwell, and smooth goldbachia. Fruits of cotoneaster and dog rose constitute the major part of the diet. Chukar avidly feed on young sprouts, especially the tender leaves of sedge and Tagar ixidirion bulbs. Underground parts of plants are not so important as in Kazakhstan and among them have been found corms of Turkmenian gladiolus and onion bulbs. Here intense feeding has not been observed in this bird. The maximum weight of the crop and gizzard contents was 26 g in summer, did not exceed 15 g in autumn, and was barely 9 g in winter (Kogan, 1950). In this snowfree region Chukar enjoy an adequate food supply in winter and hence do not accumulate body fat.

The food of Chukar Partridges is particularly diverse in regions with prolonged and fairly snowy winters; during this difficult period it feeds on almost everything available except twigs. On the other hand, during the snowfree period it is more discriminatory in choosing its food and selects only particular types from the abundant food assortment available. Indiscriminate feeding during the difficult part of the year could be considered a special adaptation of this species.

Genus *Francolinus*—Francolins

Francolins forage for food in thick undergrowth as well as in open areas. In the latter case they are alert, stop often, and if suspicious either run away quickly or fly to nearby bushes. They pick up food from the ground and more rarely scrabble for subsoil parts of plants. While feeding
124 on berries they occasionally perch on bushes, but have never been sighted on trees. Francolin feeding has been studied in the Trans-Caucasus and Turkmenia. In the Caucasus the crop and gizzard contents of about 400 birds collected in all seasons have been analyzed. Of these, only 10% covered the summer season (Kravtsov, 1936; and Khanmamedov, 1956). In Turkmenia the diet of 140 birds has been studied (Vorob'ev, 1946), mostly in winter.

Compared to the other galliform birds of the Soviet Union, Francolins eat more animal food. In the Trans-Caucasus their food contains 210 insect species of 45 families of eight orders (beetles, grasshoppers, true bugs, dragonflies, bees, butterflies, flies, and earwigs); in addition to insects, wood louse, spiders, mollusks, and even small snakes are consumed.

In May-July Francolins feed chiefly on invertebrates; in April, August, and September mixed food is more characteristic, and from October through March, chiefly plant food. However, during the winter period,

Khanmamedov (1956) observed remnants of beetles in 84% of the crops and gizzards examined. Plant food is less diverse among francolins—109 species of 30 families—compared to other phasianid species. This is logical since animal matter constitutes the main food for the greater part of the year; moreover, the data available reflect the food habits of francolins in only a limited part of their range. The most important plant species are wild and cultivated grains. In the Trans-Caucasus 17 species of this family have been recorded in 47% of the birds dissected. Legumes were represented by 15 species found in 40% of the birds, roses and crucifers by eight species each, and borages by seven species. Other families were represented by a few species, but on the whole were still significant in the food composition of the francolins. The chief foods include slender knotgrass, sorrel dock (knotgrasses), greater bindweeds, corn bedstraw (beets), black-berried nightshade, and others.

The main part of the plant food consists of seeds. However, Francolins also feed on the green parts of plants and small quantities of berries and roots. Seasonal changes in the species composition of food are less marked in this genus than in other galliform birds.

In winter in the Trans-Caucasus, Francolins feed on 60 plant species of 19 families. Maximum diversity has been observed in cereals, followed by legumes (clover, vetches, and French honeysuckle). Among roses, dog
125 rose berries are more common. At times the birds eat insects also. In Turkmenia the chief foods during this season are tender sprouts and barley grains, the latter found on the pathways and in dung; these birds also frequent threshing floors. Supplementary foods include reedmace rhizomes and ants (Vorob'ev, 1946, 1967).

In spring, plant food is gradually replaced by animal food. In the Trans-Caucasus in March, 21 species of invertebrates and 29 plant species were identified, 32 and 16 in April, and 91 and 19 in May. By the end of spring, Francolins feed mainly on invertebrates. At this time all the orders of insects found in their food throughout the year were detected, except for dragonflies. The birds also feed on spiders and wood louse.

In summer in this part of the range, plants (28 species) simply supplement the diet of invertebrates and may sometimes even be considered incidental components. Cereals are found more often than others while legumes are completely absent. Occasionally these birds consume grapes and black-berried nightshade fruits but, on the whole, these are not significant. Animal food composition is more varied, but beetles and grasshoppers predominate. In Turkmenia, in addition to insects, saltwort berries form an important food component.

In autumn the crops and gizzards of Francolins are largely filled with a variety of plants (cereals, knotgrasses, borages, beets, crucifers, and roses). In their breeding places animals are a constant though minor component of food (dragonflies, grasshoppers, and beetles) and are consumed more readily than plants.

Of all the galliform birds of the Soviet Union, animal food is highest in the diet of francolins. Invertebrates have been found in varying proportions in their food throughout the year. In winter they are supplementary or incidental elements. During the long snowfree period, insects are the chief food and are represented by an exceptionally diverse variety of species. This type of food is possible only in mild and rather humid climates when animal foods are accessible throughout the year.

Genus *Tetraogallus*—Snowcocks

The chief food of snowcocks is the vegetative parts of herbaceous plants. While feeding, birds rise upward, seize the grass stems on the way and, with a sharp jerk of the head, break them along with flowers or seeds.
126 Bulbs and roots are dug up in all seasons of the year, not only from soil but even from under snow of moderate depth. Digging is done by raking the soil alternately with the claws and the powerful bent beak.

The feeding habits of snowcocks have not been studied thoroughly. Even the fragmentary data available on different species underscore the diet characteristics of snowcocks.

The food composition of Caucasian Snowcocks is somewhat better known than that of other species (Chunikhin, 1960; Ivanov and Baziev, 1961; Baziev, 1965; Tkachenko, 1966). The food of 101 birds has been studied and 100 plant species of 26 families identified.

Published reports on the food of Caspian Snowcocks *per se* are not available but Baziev (1965) mentions that the crops of 28 birds dissected by him contained 41 species of plants of ten families, mostly similar to those of Caucasian Snowcocks.

The food of Himalayan Snowcocks has been studied in various parts of its geographic range and a total of 77 birds examined; Tersk Alatau in winter (Zimina, 1956); Gissar and Karategin ranges (Popov, 1959); and Trans-Ili, Dzhungarsk Alatau, and Ketmen' range (Kuz'mina, 1960, 1968). Moreover, individual references to its food are made in faunistic works (Shnitnikov, 1949; Korelov, 1956; Stepanyan, 1959; Abdusalyamov, 1964; Potapov, 1966). In all 121 plant species of 35 families have been identified in the food of Himalayan Snowcocks and insects were found occasionally. Such a variety is explained by the wide geographic range, broad vertical distribution, and great ecological adaptability of this species.

Reliable data on the food of Altai Snowcocks are almost negligible. The only positive report is the finding of shoots and bulbs of wild onions in the crops (Folitarek and Dement'ev, 1938). The reference in *Ptitsy Sovetskogo Soyuza* (*Birds of the Soviet Union,* vol. IV, 1952) that it occasionally catches and eats small rodents and Hartert's (1905) statement that it feeds on the shoots and buds of dwarf birch and berries of barberries are rather unreliable.

Data on the food of Tibetan Snowcocks are likewise fragmentary. It is known to feed on over ten species of plants (Maklenburtsev, 1946; Abdusalyamov, 1957, 1971; Potapov, 1966). In the Pamir during summer, sprouts and flowers have been found in its food (pitted oxytropis and
127 saussurea). In October it feeds on the lily bulbs and sedge rhizomes. Winter food is quite varied. Shoots and larvae of wood stitchwort, litvinov's meadow grass, celestial fescue, catmint, marsh saxifrage, and other plants have been observed here in the food of Tibetan Snowcocks. Most of these have also been found in the food of Himalayan Snowcocks. In Kansu the main summer food of Tibetan Snowcocks is bulbs of the yellow onion [*Priroda i zhivotnyya Tsentral'noi Azii*, 1875].

The food of all snowcocks is similar not only in food groups but also in composition. Some plant genera are common to all snowcocks but represented by local species. Such genera are oxytropis, meadow grasses, sheep's fescue, saxifrages, vetches, and many others.

The food composition of different species of snowcocks whose geographic ranges overlap or are slightly contiguous is even more similar. Thus, Baziev (1965) emphasizes that the food of Caspian and Caucasian snowcocks does not differ basically, while according to Potapov (1966) the food of Himalayan and Tibetan snowcocks in the Pamir is essentially identical.

One of the main characteristics of the food of all snowcocks is the extremely insignificant quantity of animal food, and more importantly, the complete absence of invertebrates in the diet of chicks during their early life. The report by Nasimovich (1947) that insects predominated in the crops of two chicks of Caucasian Snowcocks caught in mid-June is not confirmed by later data.

Observations on Himalayan Snowcock chicks hatched in captivity but raised freely near their nesting site (in the Trans-Ilian Alatau) showed that their first basic food was the tender leaves of white clover and vetch. Only at two weeks of age did they occasionally peck at weevils; before eating an insect the chick trampled the ground for quite some time as though attempting to catch it.

Animal matter in the diet of snowcocks can rightly be considered incidental. Stray bush crickets, butterflies, and snails (Chunikhin, 1960) have been found in Caucasian Snowcocks, and one saranchuk (*Gomphomastax clavata*), three butterflies, and wireworms of the family Elateridae in a Himalayan Snowcock. However when grasshoppers appear *en masse* in the high mountains, Himalayan Snowcocks avidly consume them. In the Trans-Ili Alatau (Tien-Shan) for about two weeks in August, grasshoppers form a significant food item for snowcocks; their excrement during this period contains much chitin.

The basic food of snowcocks in all seasons of the year consists of plants of the Gramineae, Leguminaceae, Ranunculaceae, Liliaceae, Rosaceae,
128 Compositae, and Cyperaceae families. Seasonal changes are almost negligible. The same plants are consumed but, depending on vegetative growth and stage of desiccation, some plant parts are exchanged for others. In spring the basic food is green plants. As the alpine grasses begin flowering, flowers, leaves, and stems become the chief food. When ripe, seeds gradually appear in the food but only in relatively small quantities. The quantity of tubers, bulbs, and roots consumed differs markedly from species to species.

A study of the feeding of Himalayan Snowcocks provides some idea of the seasonal changes in the diet of snowcocks. In winter Himalayan Snowcocks feed on a variety of plants. In the crops and gizzards of 45 dissected birds 76 plant species belonging to 23 families were identified. These comprised mainly the leaves, stems, dry flowers, seeds, and berries of herbaceous plants. The underground parts of plants however, are a constant and very important element. During this period the snowcock crops almost always contained pebbles and soil. Zimina (1956), who found that food was always mixed with soil, points out that soil contains salts necessary for their nutrition. This assumption is fairly convincing since adult birds regularly visit places with alkaline soils (solonetzes), indicating that salts are a physiological necessity. Independent chicks, three to four weeks old, keenly dug the soil near a salt bag and pulled out tiny salt crystals through the sack cloth.

More complete data on the winter feeding of Himalayan Snowcocks (37 birds) were collected in Tersk Alatau; 67 plant species were identified in its food. In terms of weight, 51% comprised leaves, stems, and seeds of thermopsis and 29.2% shoots of various cereals (Tien-Shan sheep's fescue, Kentucky blue grass, annual blue grass, meadow grass, and spiked oat-grass) and cobresia (Zimina, 1956).

In the Ketmen' range, I studied the food of seven birds and identified 24 plant species. Of these, stems and leaves of potentilla, oxytropis, vetch, and honeysuckle were the most common and formed the bulk of the food. The assortment of food differed from bird to bird (6–10 plant species); the food of a male caught on November 16, 1954, contained leaves and seeds of 15 plant species, roots of four kinds, and some soil. Such a variety of food indicates that snowcocks are not very selective in winter and eat everything available in their limited territory, primarily and naturally, the abundant plants.

129 During the snowfree period the plant food of Himalayan Snowcocks varies very little. When food is abundant, it exhibits definite preferences. In spring the green matter and flowers of ephemerals such as anemones and pasque flower, and onion stems predominate. In summer the chief foods are green mass of cereals (couch-grass, sheep's fescue, and meadow grasses), legumes (vetch, oxytropis, and clover), buttercups (Jungarian globeflower, Alberts buttercup, and callianthemum), and onion bulbs. During the second half of summer Himalayan Snowcocks, in addition to shoots and flowers, feed on the seeds of cereals, sedge, legumes, knotgrasses, etc., swallowing them along with leaves and stems. Almost throughout the year bulbs of onions and crocus and roots of various plants constitute a considerable part of the diet. They become secondary elements only in spring and early summer, when shoots rich in vitamins are abundant.

The species composition of green food consumed by adult birds and chicks differs markedly. In Caucasian Snowcock fledgelings, plants of the family Leguminaceae constitute 51.3% of the intake and cereals 16.7%; in adults these percentages are 39.5 and 42.4, respectively (Baziev, 1965).

I found that legumes (leaves of clover and vetch) also predominated in the food of Himalayan Snowcock fledgelings. Apparently these differences between the food of adult birds and chicks are of special significance. In the food of fledgelings of other galliform birds in the USSR fauna, various invertebrates predominated and provided the proteins necessary for vigorous growth. In alpine conditions very few insects are found during the hatching period and the necessary proteins are obtained only from plants, primarily various legumes.

Similarities in food for all species of snowcocks, the slight seasonal changes in their food, and the absence of animal matter in the food of fledgelings can be regarded as a distinct adaptation of this entire genus which evolved under severe alpine conditions.

Genus *Phasianus*—Pheasants

These birds forage in a variety of ways. They feed on the ground, collecting food from the soil, grass, and bushes, scrape the soil with their claws like hens, or dig with their beaks. In winter they even dig out food from a thin layer of snow. They also feed on bushes and trees, peak at berries, and in some regions also consume twigs in winter (buds and
130 catkins), although this is exceptional. However, their movement from branch to branch is clumsy and foraging on trees is apparently difficult for them.

Several reports on the food habits of pheasants are available from Tadzhikistan, where 576 birds have been dissected (Stroganov, 1957). Syr-Darya valley (Vereshchagin, 1931; Shaposhnikov, 1936); and Ili valley (Shtegman, 1946). However, these works do not present a complete analysis and only list the basic food components. The complete food composition of 800 birds is known from these regions: the Caucasus (Krasavtsev, 1940; Bannikov, 1955); Volga delta (Shatas, 1963); Syr-Darya (Grachev and Kuz'mina, 1947; Tyurekhodzhaev, 1967); Balkhash region (Kuz'mina, 1968); Trans-Ilian Alatau (Zhirov and Vinokurov, 1959); Zerevshan valley (Mekhlenburtsev, 1940); and Kirgizia (Pek and Fedyanina, 1961).

Animal food plays an important role in the diet of pheasants, which is probably the reason why plant food is represented by a limited number of species. To date 160 plant species belonging to 49 families have been recorded. It should be noted that in some regions the list of food plants is considerably shorter than in the case of many other galliform birds. Animal food has been found almost throughout the year and in fairly large quantities for nine months, while being the chief food in mid-summer. In winter, I found a few insects in the pheasant food in Syr-Darya while Shatas found them in the birds from the Volga delta. The important animal foods are invertebrates, mainly insects (nine orders, 32 families). Most of all pheasants feed on grasshoppers, especially acridoid grasshoppers and beetles, and additionally mollusks, spiders (particularly karakurt spiders) and argasine and ixodid ticks. In addition to invertebrates, mice (specially harvest mice), voles, lizards, and frogs are often found. According to a verbal communication from A.F. Kovshar', at the end of October, seven of 28 pheasants from the southern Kazakhstan and the Lake Biilikol'sk region contained tiny frogs in their crops; one bird held 12 of them. Murines have been found in the food of pheasants from Ili, Issyk Kul', and Primor'e.

The plant food of pheasants is varied and differs markedly from region to region not only in species composition, but also in assortment. Seeds form the greater bulk of the food in autumn and winter. Their diverse composition (72 species) includes not only wild, but also cultivated plants. Berries are eaten throughout. The use of subsoil parts is a local phenomenon; bulbs, roots, and tubers are consumed in those regions where winter berries are not available (Russian olive, sallow thorn, and saltwort),
131 especially in the Volga delta (Shatas, 1963). Sprouts have been found throughout the year and twigs (buds of willow, Asiatic poplar and birch catkins) were found as an exception during severe winters in the Ili valley and in Ussuriisk Krai in spring.

Winter is the most difficult period for pheasants, particularly in those regions where they feed exclusively on the ground. During this period 72 plant species of 32 families have been identified in their food. In the Volga delta, 25 species of plants and small quantities of beetles and grasshoppers have been detected. Pheasants in these regions in winter confine themselves to ilmens*, kultuks**, and sand bars, where some aquatic plants survive the winter as green rosettes under the shelter of dry rushes and reedmaces. They feed on the seeds and green parts of plants, tubers, roots, and berries. Roots of various species of sorrel dock, common watercress, and farinaceous nodules of subsoil suckers of sea scirpus and arrowhead are the main subsoil plant food. Rosettes of fine-leaved water dropwort and ragwort, and tender leaves of northern bedstraw, buttercups, and bent grass comprise green food. Often the birds feed on the seeds of water plantain, water lily, and occasionally cockspur grass and other plants. Only bittersweet and dewberry were the fruits found in winter (Shatas, 1963).

In the Syr-Darya valley during winter, the birds haunt vegetation-covered bottomland and salt tree bushes. Their food comprises 18 plant species at this time. Insects are consumed very rarely and incidentally. Green fodder is negligible during this period with the main food consisting of berries of oleaster and seeds of various plants, of which salt tree, wooly saltwort, and Siberian petrosimonia are very important. The yield of Russian olive berries varies greatly from year to year and, in those winters when berries are few, pheasants face acute food shortages (Tyurekhodzhaev, 1967). Underground plant parts are not found, but pheasants feed on algae and hornwort discarded by fishermen from their nets.

In the Ili valley, where thickets of Russian olive and sallow thorn prevail, their berries constitute the main diet of pheasants during the first half of winter. In the second half various seeds, especially of the families Chenopodiaceae (summer cypress, herbaceous sea blite, European winterfat, winter saltwort, and so on), Leguminaceae (salt tree, Spanish liquorice, lucerne, and dentate melilot), and Gramineae (timothy grass, couch grass, and so on) predominate. At places where Russian olive and sallow thorn are absent, pheasants largely feed on seeds throughout the winter. Underground plant parts are rarely found. In particularly snowy winters, when the food shortage is extremely acute, twigs and buds of Asiatic poplar have been found in addition to seeds. In Issyk-Kul' the main winter food is sallow-thorn berries and Russian olive berries in the tugais
132 in Vakhsha. In central Tien-Shan in the upper reaches of the Tekes River, the diet comprises berries of sallow-thorn, barberries, dog rose, and leaves and seeds of oriental clematis, and a variety of grass seeds and occasionally

*Ilmen—a delta lake—Editors.

**Kultuk—a deeply indented bay—Editors.

willow bugs; mollusks and insect pupae were also rare finds of animal matter (Zhirnov and Vinokurov, 1959).

In winter in Ussuriisk Krai, pheasants concentrate near fields and feed on a variety of cultivated plants (rice, wheat, oats, millet, willow grass, soybean, maize, and potato tubers). During severe, snowy winters they remain close to farms and gather food remnants from dung (Sleptsov and Gorchakovskaya, 1952).

With the onset of spring, plant seedlings, shoots of ephemerals, and also various insects appear in the food of pheasants; the proportion of insect matter increases steadily. In the Volga delta the quantity of animal food (spiders and ladybird beetles) increases slightly as early as March. In April the content of animal matter markedly increases and that of seeds decreases, while green fodder continues to be significant. During spring floods, more often than at any other time, the birds feed on mollusks. In the crop and gizzard contents vegetative parts of plants constitute 23%, seeds 59%, and animal matter 18% (Shatas, 1963).

In the Syr-Darya valley animal matter weighed slightly less than half the entire spring intake. Ground beetles and ants predominated in the crops of five birds collected in the Chu valley on April 28th. A large number of animals have been found in the birds of the Ili valley. In the Zeravshan valley pheasants feed in spring on sallow-thorn berries and partly on shoots. In early spring their chief foods in the Vakhsha valley are year-old berries and seeds and to a lesser extent shoots and other green plant parts. During the vegetative period the role of green fodder increases and that of seeds decreases; at the same time the proportion of animal matter increases. The food composition of Ussurian Pheasants during spring is typical. This species often pecks at sprouting Mongolian oak acorns and consumes willow buds and birch catkins. From mid-April to early May, Pheasant crops contained 20–30 acorns (Sleptsov and Gorchakovskaya, 1952).

In summer pheasants feed chiefly on invertebrates. Plant food comprises 39 species, the composition differing perceptibly in various parts of the range; moreover, in a given place, depending on the phenology of the vegetative cover, some groups of plants might be substituted by others. Animal food also varies widely from region to region. It has been observed
133 that pheasants are not fastidious in their choice of animal food and consume whatever is readily and abundantly available.

In the Caucasus in mid-June, land mollusks become the basic food of pheasants. In August the bird feeds on the fruits of dewberry, Turkmenian box-thorn, and Russian olive, supplemented with acridoid grasshoppers. In the Volga delta the chief summer food is insects, which constitute 70% (by weight) of the food. Green matter is found in very small quantities. The importance of seeds and berries increases by the end of summer. Commonly found insects are beetles (especially weevils, jewel beetles, and ground beetles), grasshoppers, and ants. Fruits of dewberry and bittersweet are the most important of the 19 plant species identified.

In the Syr-Darya valley in summer animal food constitutes 57.9% (by weight) of the food (beetles—27.4%, grasshoppers—22.8%, and bugs—7.7%) while plant food (42.1%) is made up of seeds and berries 35.5% and leaves and flowers 6.6%. Among seeds, asparagus and wild cereals are common and Turkmenian box-thorn among berries (Tyurekhodzhaev, 1967). In the hilly regions of the Zeravshan valley and in the upper reaches of the Tekes, berries of sallow-thorn are the chief summer food while animal food is less important.

In the Ili valley the summer food of pheasants consists of insects, green parts of plants and, in August, also the berries of saltwort and sallow-thorn. Juveniles feed mainly on insects, supplemented with small quantities of seeds and berries; unlike adults, there is no green food in their diet.

In autumn the food of pheasants includes over 50 species of plants, seeds, berries, and sprouts; animals constitute the larger part of the food during the first half of autumn and comprise both invertebrates and vertebrates (frogs and rodents). Later seeds, berries, and sprouts become the chief food. Where pheasants live among cultivated areas, they readily take to cereal grains.

In the Volga delta during this period only 13% of the food consists of animals. The intake of seeds (cockspur grass, willow grass, lilies, water plantains, pondweed, bindweed, and so on) and berries (dewberry, black-berried nightshade, and bittersweet) rises sharply while the quantity of green fodder increases (common watercress, sheep's sorrel, fine-leaved water dropwort, ragwort, couch grass, and so on).

In the Syr-Darya valley in autumn, plant food constitutes (by weight) about 80% of the diet. The food of those living close to fields largely
134 consists of rice left behind in the field after the harvest. At the end of autumn, pheasants inhabiting vegetation-covered bottomlands feed exclusively on berries of Russian olive, but in years when the yield of fruits is inadequate, they largely depend on the seeds of various plants, especially salt tree, asparagus, and bindweed.

In the bottomland forests of Vakhsha, the chief autumn food of pheasants is fruits of the Russian olive. In the upper reaches of the Tekes the birds feed on berries of sallow-thorn, barberries, and a variety of seeds. In the Ili valley berries of Russian olive, saltwort, and various seeds predominate, mostly of saltwort. The autumn food of the Ussurian Pheasant in cultivated lands consists of rice, wheat, oats, millet, willow grass, soya bean, maize, pea, and potatoes. In September–October nourishment is abundant and feeding maximal by the time the snow cover forms.

The pheasant's indifference to food is distinctly expressed in winter (even twigs and washed-up algae are consumed). The bird's ability to adjust its basic food easily to what is available abundantly underscores its high adaptability (Brull, 1960). This is also reflected in the advancing northern boundary of its range (Primor'e), the colonization of cultivated terrain with the irrigation of desert areas, and its particularly successful naturalization in west Europe and the USA.

SEASONAL CHANGE OF THE COMPOSITION OF FOOD AND ITS NUTRITIVE VALUE

The food of galliform birds in the USSR avifauna is known to include 1,282 species of plants (of which, 35 are lower plants—masses, lichens, and occasionally mushrooms and algae). Higher plants are represented by 105 families. Tetraonids consume 735 plant species belonging to 80 families and phasianids, 645 species belonging to 79 families. The greater diversity of the food of tetraonids is explained by a whole series of factors. The food of tetraonids has been better studied (in 8,400 individuals) covering most parts of its wide geographic range, in various terrains and geographic zones, from arctic deserts to steppes, and from the European part of the USSR to the Far East. The food of pheasants, on the other hand, has been studied much less (in a little over 4,000 individuals) and in very limited territories. Consequently, data for a number of species of pheasants are
135 fragmentary. Finally, in some phasianids (Francolin, Pheasant, and Gray Partridge), the assortment of plant foods is rather limited, either because of significant intake of animal food or the limited assortment of cultivated cereal crops available to them.

The differences between the plant food composition of tetraonids and phasianids are reflected in the fact that only 7.6% of the species are common to both groups. This may be explained in two ways. First, even at places where the ranges of some species overlap, the members of the two families are usually biotopically isolated. Second, even though 51.9% of the plant families are common in the food of these two families, and often represented by the same genera, species selectivity and availability differ.

The basic plant foods of most galliform birds are widespread polymorphic families such as the Gramineae, Cyperaceae, Polygonaceae, Leguminaceae, Chenopodiaceae, Cruciferae, Rosaceae, and Compositae. However, local plants in the geographic zones, often represented by a limited number of species, are equally important to individual species of birds. For many tetraonids, therefore, the Pinaceae, Betulaceae, Salicaceae, Vacciniaceae, and Ericaceae are especially important, while the Liliaceae, Boraginaceae, and Elaeagnaceae are significant for pheasants. Danilov (1958) has also pointed out that these same plant families provide food to several game animals and birds. This may be due to their high nutritive value, wide distribution, and availability of woody-shrub fodders in winter. The greater part of these plants is not consumed at all by the birds, thus indicating their definite food selectivity.

Invertebrates predominate in animal food (worms, mollusks, and arthropods). Worms and myriapods have been found in some regions, but very rarely. Mollusks have been found in small quantities in the food of all species. The major part of the animal matter consists of arthropods, particularly insects (93 families from 15 orders). The birds readily feed on beetles (27 families), grasshoppers, true bugs, and larvae of butterflies and lacewings.

There is no essential difference between the animal food of tetraonids and phasianids. In both groups there are species which feed on animal matter to a limited extent (Willow Grouse, Rock Ptarmigans, and snowcocks) and those for which animal food is important only during certain snowfree months (Gray Partridges, Hazel Grouse, and Northern Black Grouse).

136 It must be emphasized that the rapid digestion of animal food and varying degrees of accuracy in its identification make it difficult to compare reports obtained from various sources. Also, when analyzing the food of different species, the importance of animal food has been somewhat underrated.

In studying feeding habits, not only the food composition but also its nutritive value should be considered. The biochemical characteristics of many plant foods of farm animals have been detailed in a number of works (Ivankin, 1935; Arkharov, 1939, Aleksandrova, 1940; Sokolov, 1949; Larin, 1950; Yunnatov, 1954; Fedorova, 1958; Tikhomirov, 1959; Znamenskaya, 1961; and Tomme, 1964). Analyses of plants in the food of galliform birds help trace the seasonal changes in the nutritive value of their diet in relation to the bird's requirements during different periods of its life (breeding, molting, and low activity in winter).

Winter is the most difficult period for Russian galliform birds. Among various adaptations of tetraonids to the unfavorable conditions of this period, a radical change in diet, i.e., using large quantities of readily available twigs of limited assortment, is without doubt the most important.

Forest tetraonids enjoy ample food in winter. Pine needles, the chief food of capercaillies, are available in large quantities in forests. There is no significant change in the fruiting of birch and alder and Hazel Grouse are never short of food. The diet of Northern Black Grouse is more varied. It readily changes from one food to another and utilizes the twigs of conifers and deciduous species and even the seeds in some localities. The availability of twigs for forest Tetraonidae is fully ensured; only when the thin branches of deciduous trees are covered heavily with ice is it difficult for the Hazel Grouse and Northern Black Grouse to forage. Ptarmigans experience even greater difficulty since their winter food is strictly controlled by the snow cover. Willow Grouse feed on the shoots and buds of willow and birch shurbs and is confined to areas with very deep snow. If the bushes are covered with snow, it either migrates to less snowy regions or flies onto trees for feeding. Rock Ptarmigan, which feed on herbaceous plants and catkins of willow and alder shrubs, prefers less snowy localities.

Twigs are low in moisture, but their nutritive value is lowered by the high cellulose content which in needles ranges from 8.0 (cedar) to 13.9% (common juniper), in birch catkins, 13.5%, and in twigs of birch and willow, from 21.4 to 22.8%. Fat is found in all twig fodders, with the maximum (15.05%) in birch catkins, 5.3% in pine needles, 4.4% in cedar needles, 4.6% in common juniper needles, and 1.2 to 1.5% in twigs of birch and willow. Cedar nuts, found in the food of Western Capercaillies and Hazel Grouse in some regions, are particularly rich in fat—30.25%.

Nitrogen-free extractable matter (NEM) is found in considerable quantities in twigs, but protein compounds are low, especially in conifer needles.

Of all the "twig" foods, the most nutritious are birch catkins which contain the least moisture, sufficient proteins, and much fat. Their NEM content is notably less but the cellulose content is the same as in needles. Twigs represent a poor winter diet as they contain large quantities of cellulose, NEM, varying quantities of fat, and are poor in proteins. All the same, these are adequate to maintain an energy balance because of their abundance and the low activity of birds when they remain immobile in the characteristic microclimate of their "snow homes" for the greater part of the day during frosts.

The food of phasianids is totally different. If the narrow food specialization of tetraonids has led to the use of large quantities of one or another species of food found above the snow cover, phasianids, contrarily, feed on whatever can be obtained in the snowfree or less snowy localities, such as the seeds, sprouts, tubers, bulbs, and berries of shrubs, but not twigs. Only the Common Pheasant is somewhat unique, feeding sometimes not only on ground food, but also on the berries of sallow-thorn and Russian olive and, occasionally, even the buds of Asiatic poplar, willow, and birch catkins. Different species of phasianids, living in the same locality, often subsist on the same kind of food; contrarily, the food of the same species may differ intensely in different parts of its geographic range. This once again reveals an utter lack of fastidiousness in food selection among these birds in winter. The type of food somewhat differs only for the Common and Daurian partridges, whose food composition is similar in winter and consists mainly of a limited assortment of seeds and sprouts of wild and cultivated plants.

The diversity in the winter food of phasianids can be judged from
138 Table 1. The food of different species during this period includes 60 (Francolins) to 74 (Himalayan Snowcocks) species of plants, including various fodder groups: seeds, green matter (fresh), dry leaves and stems, subsoil plant parts, berries, and in some cases also insects. Among plants, the preferred species are those of the Leguminosae, Chenopodiaceae, Boraginaceae, and Gramineae families.

In addition to seeds and green plants (sometimes also dry parts of plants, subsoil parts of plants (roots, rhizomes, and bulbs) play an important role in the food of Himalayan Snowcocks and Chukar Partridges during winter. This food group is less important in the nutrition of francolins and pheasants. Compared to other members of the family, the food of pheasants is especially varied, if not in species composition at least in the utilization of various fodder groups. In localities where the berries of Russian Olive and sallow-thorn are the main food, pheasants withstand winter satisfactorily even in years of repeated snowfall. However, in regions where they feed on various parts of herbaceous plants, bird mortality due to starvation has been observed, especially in years of severe winter. The assorted winter diet of pheasants varies in its nutritive value. Green plants, used mainly in regions of mild winter, are low in calories and fat due to their high moisture, but fairly high in proteins; the quantity of NEM varies greatly and the cellulose content is low. Seeds are the chief food of pheasants in regions subject to very severe winters. Low moisture and cellulose content, high protein content, NEM, and the presence of fat
139 assure the particularly high calorie content of this food (Danilov, 1938, 1958; and Kendeigh and West, 1965).

However, the seed harvest, i.e., availability of seeds, fluctuates highly. Seed yield is low after a dry summer with adverse consequences for wintering birds. In many regions, berries of Russian olive and sallow-thorn are of great importance in the winter diet of pheasants. Fruits of Russian olive contain 10.5% protein, ample glucose and fructose, and are also rich in potassium and phosphorus salts [*Flora SSSR* (*Flora of the USSR*), vol. 15,1949]. Berries of sallow-thorn contain sugar (glucose and fructose), acids (malic and tartaric), and nitrogenous matter. The fat content in them varies from 2.8 to 7.8% (Devyatkin and Obodovskaya, 1955) and the subsoil parts of the plants are high in nutritive value. Mealy tubers of subsoil shoots of sea scirpus and arrowhead, the favorite food of pheasants in the Volga delta, contain up to 10% protein and 30% starch (Shatas, 1963).

Table 1. Seasonal changes in the composition of plant food among some galliform birds*

Bird species	Number of plants in annual diet	Winter, November to March			Summer, June to August		
		Number of birds	Plants found	Percent of annual diet	Number of birds	Plants found	Percent of annual diet
Willow Grouse	216	430	32	14.8	124	128	59.2
Western Capercaillie	166	332	12	7.2	250	100	60.2
Northern Black Grouse	304	147	27	8.8	600	178	58.5
Hazel Grouse	279	250	55	19.7	550	116	41.5
Chukar Partridge	252	250	72	28.5	400	135	53.6
Black Francolin	109	122	60	55.0	40	28	23.8
Himalayan Snowcock	121	47	74	61.1	26	41	33.8
Common Pheasant	161	170	72	44.7	160	39	24.2

*This table does not include data on the food composition of birds for which the collection date was not available.

A comparison of the winter food of tetraonids and phasianids reveals marked difference in the food groups used, their species composition, and nutritive value. Large quantities of twigs are consumed by tetraonids but they are less nutritious because of their high cellulose content; however, when such food is abundant and the birds are less active, it helps maintain the heat balance in the body.

The more wholesome foods of phasianids are mostly buried in snow and are inaccessible. When food is scarce, the birds become less exacting but still refrain from the available twigs (except for pheasants which, under particularly trying conditions, sometimes feed on the buds and twigs of deciduous plants). Food shortages force the birds to spend considerable time and energy in foraging. In very severe winters with repeated snowfalls, a number of species (Francolins, Common Pheasants, Chukar, and Gray Partridges) suffer high mortality due to starvation.

Spring is the most active period in the life of these birds. Courtship displays by males, the sharply improved nourishment of females, and maturation of gonads (the weight of a full clutch of Gray Partridges or Chukar constitutes about 50 to 55% of the body weight of the female) call for high energy output which can only be replenished by the intake of large quantities of wholesome food.

Data are not available on the food requirements of galliform birds during the snowfree period. Food requirements during winter have been
140 estimated by examining birds caught with full crops. In the spring–summer period there is no specific foraging time; nevertheless, Donaurov (1947), based on the maximum weight of the crop contents of Hazel Grouse and assuming two feeds per day, estimated the daily food requirement of this bird as 58 g in spring and 41 g in winter. Judging from the tremendous energy expended during nesting and again when molting, the quantity of wholesome food consumed during the spring–summer period should be considered maximum not only for Hazel Grouse, but for other species as well.

The basic ingredient of the spring food for all the species studied is green plants, rich in vitamins and proteins but lower in NEM than twigs, and containing almost no fat. During this time the birds avidly feed on young plant sprouts, green leaves, and flowers of various ephemerals, cereal grains, and legumes. In many species (Black Francolin, Common Pheasant, and Gray Partridge), animal matter providing wholesome proteins acquires great importance especially in late spring. In other birds, the protein requirement is met by the green portions of two plant families—legumes and cereals.

It can be assumed that the high vitamin content of green plants is of considerable importance to galliform birds during spring. It has been observed that Capercaillies, feeding in winter on coniferous needles rich in vitamins, changes to green plants later than any other tetraonid. That green plants are required for successful breeding is shown by the fact that in years with a late spring, when green plants cannot be used fully, there are fewer eggs in Hazel Grouse clutches.

Differences in the food of tetraonids and phasianids are not so marked in spring as in winter. The food of phasianids in this season becomes more monotypical and selectivity is apparent; in tetraonids, on the other hand, the species composition of food plants is more varied than in their limited winter diet.

The life styles of tetraonids and phasianids differ less in summer than in winter and there is much in common in their food groups and in their species composition. The breeding period and then molting require much energy. During this time their diet is characterized by a rich species composition and a particular diversity of food groups; sprouts, seeds, berries, and animal matter make up the food composition of most species.
141 Moreover, some phasianid species use underground parts of plants while tetraonids use small quantities of twigs.

The diverse combination of animal and plant components in the diet and gradual replacement of one food group by another varies from species to species depending on the phenology of the plant cover and the time of the mass appearance of different invertebrates. When food is abundant food selectivity comes to the fore. This is more pronounced in phasianids which inhabit diverse terrains.

Among different species of tetraonids in the Soviet Union, the composition of summer food differs little. During this season they feed on sprouts, berries, and seeds, as well as on animal matter to some extent. Twigs are used in small quantities, but not by all species. However, the birds readily feed on the leaves of trees and shrubs. Berries constitute the basic food of tetraonids during the latter half of summer. Among them, the Vacciniaceae are of great importance but berries of the Rosaceae, Ericaceae, Solanaceae, Empetraceae, and others are also eaten. In many localities in the summer distribution of Northern Black Grouse, Western Capercaillie, Hazel Grouse, and Willow Grouse is closely associated with the availability of bilberry, bog whortleberry, and cowberries. In more southern regions where berries of the family Vacciniaceae are not available, the birds feed on common strawberry, strawberry, dog rose, raspberry, dewberry, stone bramble, and so on. According to Raus (1973), the frequency of occurrence of berries in the crops of tetraonids based on long-time data, is about 20% (stone bramble 40%, raspberry 40–45%, and marshwort 3.7%).

Succulant berries have a high water content (80.99% in raspberry and 88.25% in marshwort) and a low cellulose content (1.49% in cowberry and 4.46% in raspberry). An abundance of sugars and acids and their easy digestibility are their chief distinguishing features. The sugar content of berries in percentage) is 6.35 in common strawberry, 6.58% in raspberry, 2.91 in cloudberry, 2.84 in marshwort, 6.91 in cowberry, and 5.59 in bilberry (Sokolov, 1949). The high tannin content, especially in bilberry, imparts medicinal properties. In years of high Vacciniaceae yield, tetraonids enjoy excellent nourishment and their resistance to diseases such as coccidiosis and helminthiasis improves. On the other hand, when the yield is low, Hazel Grouse become undernourished, which leads to languor, reduction in population, and increased susceptibility to coccidiosis (Formozov, 1964).

142 All tetraonids, especially Western Capercaillies, feed on green food in summer. Seeds exhibit a rich species composition and are of great importance to all these birds during the second half of summer. Compared to other species, seeds are the most important food of Hazel Grouse.

Animal matter is found in the food of all tetraonids, but their proportion is always less in adults. The maximum amount of invertebrates is consumed by Hazel Grouse and Black Grouse, especially in the southern parts of their geographic range. It forms only an incidental item in the food of Willow Grouse and Rock Ptarmigans, especially in the far north. The general similarity in the diets of tetraonids is supported by the fact that in summer the food composition of Hazel Grouse, Western Capercaillies, and Northern Black Grouse in different parts of the geographic range of each of these species differs more intensely than their food at places where they coexist.

Differences in the summer food of phasianids are better expressed and particularly noticeable in the ratio of animal and plant components. Invertebrates are the chief food of francolins and pheasants in summer (actually, in summer they are insectivorous), while animal matter is almost always absent in the food of snowcocks. The summer food of Gray, Daurian, and Chukar Partridges is an admixture, with animal or plant matter predominating according to local conditions. Berries are less important to phasianids than to tetraonids; they are almost absent in the food of quails and snowcocks while being only a local supplement for partridges. For Common Pheasants, Black Francolins, and Chukar, berries of shrubs and sometimes of trees are of special importance.

Green plants are the chief food of all snowcocks; for other species of phasianids this food is common only in spring and eaten in small quantities in summer. As for tetraonids, the Leguminosae, Gramineae, and Ranunculaceae are the most important among green foods. Seeds are consumed more readily by See-see, Daurian, Gray, and Chukar Partridges, while bulbs and tubers are generally found in the food of Chukar and snowcocks during this season. Thus the summer food of each species of phasianids has been distinctly traced.

The summer food of different species of tetraonids includes up to 60% of the plants found in their food throughout the course of the year. Hazel Grouse are an exception; the winter food of these birds more diverse and only 41.5% of all the plant food is found in its summer diet.

The composition of the summer diet is different for phasianids; only Chukar consume 53.6% of the species of its food plants. For francolins, pheasants, and snowcocks this value varies from 23.8 to 33.8%. Francolins (and in most regions, pheasants) feed mainly on insects, with plant food
143 simply a supplement. The Snowcock, almost exclusively herbivorous, feeds on a limited assortment of plants.

The food composition of fledglings of most of the species studied differs markedly from that of adults. During the first few weeks the basic food of fledglings consists of various invertebrates. As they grow the animal components are substituted by plants but until autumn, as long as the juveniles are growing vigorously, animal food is found in their diets more often than in that of adults.

The food of the fledglings of snowcocks, Willow Grouse, and Rock Ptarmigans is quite different. Snowcock chicks never feed on animal matter during the first two weeks. Among ptarmigans, the young usually commence life feeding on both invertebrates and tender parts of plants. The unique food composition of chicks of these species may be explained as the result of severe climatic conditions in their habitats, where insects are scarce by the time chicks emerge.

Animal matter, being the chief source of protein, is particularly necessary for the growing young birds and molting adults. Animal components have been found in nearly every bird examined. For those few species that exclude animal matter from their diet (snowcocks, Rock Ptarmigan, and Willow Grouse in the Far North), green food is of great importance, especially sprouts of legumes and cereals rich in plant proteins.

The fat content of summer food (green matter) is less than that of twigs in winter or totally absent (berries). Possibly the birds have low fat requirements in summer although it is the main source of body heat so essential in winter. However, it is quite probable that some fat does enter the bird organism with seeds of various wild, and later, cultivated plants. Berries, the chief summer food, contain considerable NEM and, of course, sugars and acids. For Chukar and snowcocks, the chief sources of these components are bulbs and other underground parts of plants with a high starch content.

It should be mentioned that in spite of the rather marked differences in the food composition of various species of galliform birds, the nutritive value of the diet is by and large similar. During the first half of summer, when most galliform birds feed on sprouts and invertebrates in varying proportions, their diets contain a good amount of plant and animal
144 proteins and vitamins. During the second half of summer, green vegetation is replaced in most species by berries, and in others by subsoil parts of plants and seeds. As a result, the nutritive value changes somewhat; the carbohydrate content (sugar and starch) increases and, possibly, the fat content also. The protein level is maintained with insects and green parts of plants especially rich in protein (legumes and cereals). The composition of the autumn food of all galliform birds is fairly varied and changes fast, reflecting the transition from summer to winter diet. Since differences between summer and winter food are more marked in tetraonids, the change in their autumn food is also more distinct. The reduction in animal food, compared to that consumed in summer, is characteristic of both phasianids and tetraonids, but in the latter this change occurs rapidly and by the end of September most species feed only on plants. For some phasianids animal food is of importance throughout autumn and is consumed even in winter.

Tetraonids change over to twigs at different times. As early as mid-August, larch needles appear in the food of Western Capercaillies (southern Urals), and predominate by mid-September. By the end of September pine needles, the chief winter food, also appear in the diet. In autumn berries predominate in the diet of Hazel Grouse and Northern Black Grouse. Seeds and berries of trees and shrubs constitute the chief components of their diets for quite some time even after the snow cover has formed. At some places the basic food of Northern Black Grouse is the grain of cultivated cereals. Catkins, the chief winter food, are found only after the end of October. Twigs of various species of willow, the chief food of Willow Grouse, begin being consumed only after snowfall; until then it subsists mainly on various berries and seeds.

In phasianids the shift from summer food is far more gradual. Francolins and pheasants continue to consume animal matter long into autumn, but the proportion of seeds, often of both wild and cultivated plants, as well as berries of shrubs, markedly increases. For partridges and quails, various seeds represent the basic food during this period. For Chukar and snowcocks, tubers and bulbs are of primary importance and seeds supplementary.

Nourishment in autumn-winter is directly dependent on the wintering conditions. Tetraonids, which enjoy access to stable feeding stations even during the snowy season, exhibit high nourishment (subcutaneous and

internal fat deposition) only among some species in regions of the far
145 north. On the other hand, phasianids, which face food shortages in winter, feed intensely in autumn and build up fat.

The changeover to twigs in a number of tetraonids, especially capercaillies, commences much before the first snowfall and the arrival of winter. This indicates that during the process of evolution these birds adapted to coarse food to such an extent that even when nature provides more nutritious ground food such as berries, seeds, and green fodder, twigs remain their major dietary component.

Thus the basic difference in the diet of the two families of galliform birds of the Soviet Union lies in that for a considerable part of the year tetraonids feed chiefly on coarse twigs, while phasianids refrain from them even in the face of a food shortage.

The type of food common to tetraonids of the Soviet Union is proof of the unitary direction of the evolution of this group under severe climatic conditions with prolonged cold and snowy winters. The consumption of large quantities of woody-shrub food is reflected not only in their winter mode of life (foraging on trees and a marked decrease in activity in restricted feeding sites), but in the structure of their organs used for food collection (beak and legs) and of their digestive system (structure and function of the cecal processes).

The sharp differences in the winter food of tetraonids and phasianids are distinctly reflected in the beak structure of these birds. The main functions of the beak of phasianids are picking seeds, berries, insects, fruits, flowers, and plucking the leaves and stems of herbaceous plants. Most species pick food even from the soil. In summer, tetraonids feed on the same food groups as phasianids, but never scrabble for the underground parts of plants or soil invertebrates. It may be assumed that the characteristic beak structure of phasianids, serving mainly as a pincer, is more primitive and less specialized, and at one time was probably typical of tetraonids as well.

It has been found that the shape of the beak, zygomatic arches, and jaw muscles of phasianids are proximate to the structure and function ("pincer") of the jaw apparatus of most birds (Dzerzhinskii, 1974). The beak of phasianids is powerful, pointed and somewhat falcate at the tip. The culmen is broader than the gonys and has overhanging margins; there is a
146 wide gap between the two portions of the beak. Such a structure facilitates grasping tiny objects and plucking soft parts of herbaceous plants.

Leaves and stems held between the culmen and gonys are bent sharply and, with vigorous head movements, cut through by the sharp edge of the upper culmen as though with a knife. However, the birds cannot cut twigs from trees and bushes. The sharp beak with its massive horny sheath (sometimes hooked at the tip of the culmen) facilitates digging up bulbs, tubers, and roots. The shape of the beak is determined largely by the structure of the horny sheath and its rate of growth (Ludicke, 1964). By marking the culmen of Chukar with an incision, I could discern the extremely rapid growth of the horny sheath, which can be explained by the quick wearing away of the horny layer due to constant digging in gravel.

The use of twigs has led to important structural changes in the beak of tetraonids. Unlike phasianids, tetraonids characteristically molt their horny sheaths or slough the surface layer. During summer, when food is soft, the horny sheath is weak; in winter, conversely, it is considerably thicker. The beak of tetraonids is broad, bulging, and in most species has a comparatively small horny sheath. The sharp margins of the culmen are intensely lowered and somewhat incurved so that, unlike in phasianids, the culmen and gonys close tightly along the sides and form the fairly sharp "scissors" so useful for plucking twigs. This change in the structure and functions of the beak has resulted in an intense development of the masticatory muscles (Fig. 13).

The characteristic beak structure (scissors), according to Dzerzhinskii (1974, 1974a), who studied in detail the jaw apparatus of Western Capercaillies, Hazel, and Willow Grouse, is typical of the family Tetraonidae. An external examination of the beaks of most tetraonids (Kuz'mina, 1968) led me to conclude that, depending on food, the degree of development of the "scissors" may vary sharply from species to species.

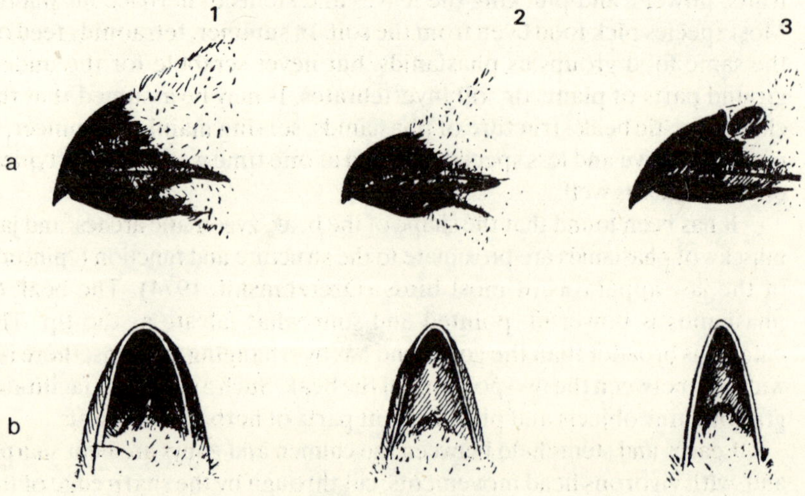

Fig. 13. Beaks: 1—Willow Grouse; 2—Spruce Grouse; 3—Chukar. A—lateral view, B—ventral view.

The capercaillies, which cut whole bunches of needles and fairly thick twigs, have highly developed masticatory muscles and a massive horny sheath; the tip of the culmen is curved in a hook (rhamphos). In Willow, Northern Black, and Hazel Grouse, the mode of feeding requires less effort and their beaks are consequently rather weak. The structure of the beak in Spruce Grouse is slightly different. In these birds the cutting
147 margin of the culmen does not close tightly against the gonys and, hence does not form the "scissors" typical of most tetraonids. The beak in Spruce Grouse is weaker in general than in Willow and Hazel Grouse, even though it weighs almost twice as much as that of Hazel Grouse.

Spruce Grouse do not shear off fir needles, their chief food, but cuts them at the stem base. Thus the strain on the jaw is considerably less than when breaking off twigs or bunches of needles.The degree of development of the "scissors" in the beak of tetraonids is directly dependent on the hardness of the winter food consumed. The weakest beak, that of Spruce Grouse, is closer to that of phasianids than of any others and is probably the most primitive among tetraonids. On the other hand, the beak of the Capercaillie reveals its distinct food specialization.

Potapov's (1974) view on the beak structure of tetraonids is somewhat different. He considers that the type of "scissors" typical of the family is found in the capercaillie and then only in males. On the basis of the fact that the winter food of males and females is the same, he proposes the following unsubstantiated hypothesis: "The sharp sexual dimorphism in size and structure of the beak of capercaillies, even though the food of
148 males and females is essentially identical, suggests that this is associated with the sex of the bird and the reproductive cycle."

In my opinion, the perceptible difference in the beak size of capercaillies (males and females) is simply due to differences in feeding processes. During winter, in severe frosts, the feeding of both the male and female capercaillies is limited to one or two flights to the feeding site extending for 1.5 to 2 hours. During this short time the birds must gather enough food to last for the entire day. The female, much smaller in size, requires considerably less food and, having a weaker beak, fills her crop in the same period with the softer parts of twigs by cutting whole needles and bunches. In addition to stray needles, clusters of needles and twigs are found in the food of the male. Feeding on such coarse food helps the male fill his enormous crop in the same time the female takes to fill hers with only conifer needles.

The legs of galliform birds perform a very important role in procuring food; their structure is dealt with in detail in the next chapter.

The digestive system of galliform birds has not yet been thoroughly studied. Quite recently it has been proved (Baziev, 1965) that the crop of galliform birds not only stores food and keeps it warm in winter, but also performs a digestive function. Glands that secrete digestive juices are found in the walls of the crop and the food is processed with the crop juices before it reaches the gizzard.

The chief function of the muscular gizzard is to grind the food. It has been observed that the gizzard is largest in Western Capercaillies and Willow Grouse, birds which feed on very coarse food in winter. Seasonal changes in the muscular gizzard have been noted in Willow Grouse (limited data). In May–July its relative weight is 40‰ less than in November–December. This is explained by the insignificant work required of the muscles for grinding softer food material (Semenov-Tyan-Shanskii, 1959).

A comparison of the relative sizes (weights) of the muscular gizzard of the two groups of galliform birds reveals no marked differences. According to Semenov-Tyan-Shanskii (1959), the weight of the muscular gizzard in tetraonids varies in relation to the body weight from 26.9‰ (Hazel Grouse) to 39.0‰ (Western Capercaillie). In my data this value varies considerably from species to species of phasianids and in several cases is even higher than in some tetraonids: Himalayan Snowcock
149 22.2‰, Daurian Partridge 27.8‰, Common Quail 34.3‰, and Chukar Partridge 37.8‰. This is explained by the fact that the food of phasianids contains, along with soft food, many hard seeds and berry stones, used not only as an aid for grinding, but as food *per se* (stones of Russian olive); hence the load on their gizzard muscles is quite considerable. The relatively small size of the muscular gizzard of Himalayan Snowcocks, predominantly herbivorous, and of Hazel Grouse, which live on very soft food compared to other tetraonids, is therefore understandable.

The intestines of tetraonids and phasianids differ rather markedly not only in external structure, but also in functional features. It is known that the intestine is relatively longer in herbivorous birds; furthermore, within the same group larger birds have a relatively longer intestine than smaller ones. This is explained by an increase in the absorptive surface of the guts with their increased area and volume; in order to maintain sufficiently effective food digestion, an increase in the absolute and relative lengths of the intestine is necessary.

To establish the relative sizes of the intestine in various species of galliform birds, I calculated the ratio of the length of intestine to the body length, and the ratio of both the caeca to the overall length of the gut (Table 2).

From the data given in Table 2 it is apparent that the caeca are better developed in tetraonids than in phasianids; furthermore, their volume is considerably greater, which notably increases their absorptive capacity. They are largest in Willow Grouse, which consume food with a very high cellulose content (birch twigs, 21.4% and willow twigs, 22.8%). Rock Ptarmigans differ markedly from Willow Grouse whose caeca are very short in spite of a comparatively long gut (Dobrinskii, 1972; and Pavlov, 1974).

Among phasianids, the longest caeca are found in snowcocks and pheasants, and the shortest in quails. Tetraonids have a relatively long intestine which differs little in different species. The intestines of the phasianids are shorter and their relative sizes vary widely in different species. The longest intestine is found in pheasants, snowcocks, and Chukar Partridges and the shortest in See see and Daurian Partridges. The relative size of the intestine varies with age; in young birds, particularly in chicks, it is relatively longer than in adults. The ontogenesis of
150 Caucasian Snowcocks even shows a reduction in the absolute length of the intestine (Baziev, 1965). I also observed this phenomenon in Himalayan Snowcocks. In adult females (2), the length of the intestine was 150.3 cm and of the caeca 43 cm; in adult males (5) these values were 161 cm and 47 cm, respectively, and in 6- to 10-month-old females (3) 169 cm and 51 cm.

Table 2. Relative lengths of intestines and caeca

Species	No. of specimens	Ratio of length of intestine to length of trunk	Ratio of length of caeca to total length of gut
Willow Grouse	9	3.5	0.99
Western Capercaillie	8	3.8	0.80
Black-billed Capercaillie	2	4.5	0.74
Northern Black Grouse	9	3.9	0.76
Spruce Grouse	1	3.6	0.64
Hazel Grouse	16	4.1	0.78
Common Quail	18	2.9	0.32
Gray Partridge	9	3.1	0.49
Daurian Partridge	8	2.6	0.44
See-see Partridge	6	2.0	0.55
Chukar Partridge	13	3.4	0.40
Black Francolin	3	2.8	0.52
Himalayan Snowcock	12	3.4	0.60
Altai Snowcock	2	3.5	0.61
Common Pheasant	19	3.6	0.67

Potapov (1974) conducted a histological analysis of the cecal region of the intestine. However, the physiology of digestion in this part of the system still remains vague. The widespread view that cellulose is digested in the caecal section under the action of the glands and bacterial flora present there was supported by Dement'ev (1940) and has not been disputed to date. It is known that birds consuming food with very high cellulose content also have highly developed caecal sections in the intestine, which change with the seasons and reach maximum size in winter (Willow Grouse). This shows a definite relationship between the function of the caecal region and the cellulose content of the food. In this context, experiments showing that Willow Grouse assimilate up to 27.7% of the cellulose in the shoots of bilberry and cowberry are indeed interesting (Pulliainen, Paloneimo and Syrjala, 1968).

151 Potapov holds a totally different view. He studied the chemical composition of the food in the intestine and of the caecal fluids. In his opinion, not only cellulose but carbohydrates in general are not assimilated by tetraonids in winter. He considers the digestive function of the caeca as an additional process to extract other nutritive substances such as proteins and fats from the intestinal contents, which may have decomposed and been absorbed here. The absence of experimental studies under laboratory conditions casts doubt on these assumptions, especially the almost complete indigestibility of carbohydrates. Twigs (needles of pine, cedar, common juniper, and birch and willow twigs contain 21.8 to 24.1% carbohydrates (NEM), which is 8.0 to 22.8% more than the cellulose content (Tomma, 1964). Albumen, protein, and fat together account for only 6.9 to 10.5%. Such quantities of nutrients are distinctly insufficient for supporting life at low temperatures, even of birds with very low activity. It is strange that Potapov should assign a major role to the caecal section of the intestine in the phylogeny of tetraonids.

There is no doubt that the major trend in the evolution of tetraonids has been to adapt to a new type of feeding and a change in the mode of acquiring food (arboreal mode of life). Naturally this adaptation has influenced the behavior, morphology, and physiology of these birds, which is particularly well reflected in organs that assist in the procurement of food (beak and legs) and its assimilation (digestive system).

Ecomorphological Adaptations to Climatic Conditions

152 Severe climatic conditions in high and temperate latitudes with a prolonged snowy period leave an imprint of fauna. Most birds nesting in such latitudes are migratory; only a small number of species which have evolved a whole range of adaptations to severe climatic conditions are resident. Deep snow cover deprives terrestrial birds of much of their nutritive food and hinders their movement. Maintaining energy balance during very low temperatures and food shortages becomes crucial for such birds. Food and thermoregulation are the basic factors influencing energy balance, and the two may be considered as operating analogously, at least in homoiothermal animals (Dement'ev, 1947).

The evolution of tetraonids is a striking example of adaptations to the changing conditions of food and thermoregulation. The utilization of abundant and readily available twigs resulted in decreased activity (brief periods of foraging and long rest periods in snow holes), which helps reduce heat loss and thus maintain the heat balance.

The phasianids in thc USSR avifauna, even in regions with prolonged winters, permanent snow cover and, in some places, very low temperatures, preserved features of their southern origin. Unlike tetraonids, species of this family spend the greater part of the daylight hours in search of scarce food; the lower the temperature, the more active they become.

Snow cover exerts considerable influence on the winter life of animals.
153 Depending on its condition, the snow cover may exert a favorable or adverse influence. Physical properties of snow (density, moisture, hardness, looseness, and gas permeability) depend on a variety of factors, primarily temperature (Rikhter, 1948). Depending on the topography, vegetation, and soil, the structure of the snow cover may vary significantly even in small contiguous localities. Sharp changes of snow cover from the long-term mean suffice to change not only the populations of the larger animals, but also the range of their geographic distribution (Formozov, 1961).

For pheasants, which are essentially ground feeders, the snow cover has only an adverse effect because it covers their food. Movement on soft snow is difficult for all species and deep snow, like glaze ice, deprives them of food. The influence of snow cover on the life of tetraonids varies depending on its structure. Snow does not deprive them of food. Deep soft snow is necessary for these birds, especially during severe frosts when they spend the greater part of the day in snow, where the temperature is a much higher and helps maintain the heat balance. Forest tetraonids move little even though movement on soft snow is somewhat easier for them than for phasianids. The characteristic foot structure of species of genus *Logopus**, essentially ground feeders, ensures fairly free movement. Crust formation prevents digging in the snow, hence, during protracted severe frosts, birds often die in spite of abundant food.

Rapid changes of snow cover, when soft snow is suddenly covered with a thick crust, are fatal for the birds resting in snow holes. In general, the duration of the snow cover and its ten-day mean depth markedly influence the distribution of phasianids and less so of tetraonids. In localities where members of the two families coexist, phasianids are confined to less snowy areas while tetraonids prefer deeper snow.

There is no doubt that glazes are fatal to phasianids which pick up food from the ground, but they affect tetraonids too. For example, the favorable wintering conditions of Hazel Grouse depend largely on the density of the snow cover. Frequent thaws promote crust formation which not only hampers burrowing but causes icing of thin branches and catkins because of which warming of the food in the crops consumes considerable body
155 heat (Teplov, 1960). The varying relationship of these two families to the nature of the snow cover is better depicted on the map (Fig. 14).

Among the resident species of phasianids, Common Partridges reach the northernmost area of their distribution. Both Daurian and Gray partridges inhabit regions with snow lasting for 140–180 days with a ten-day mean thickness of 10–50 cm. However, Gray Partridges in the European part of the USSR inhabit regions with a maximum ten-day mean snow cover thickness of 50–70 cm. True, here they seek less snowy regions and move to inhabited areas, waysides or migrates to regions where the snow is not deep.

The distribution of Daurian Partridges coincides with less snowy regions where the ten-day mean thickness of the snow cover does not exceed 10–30 cm (Trans-Baikal) or inhabits hills with uneven snow cover. Futile attempts at acclimatizing this species in the more snowy central

*[sic]; should be *Lagopus*—Editors.

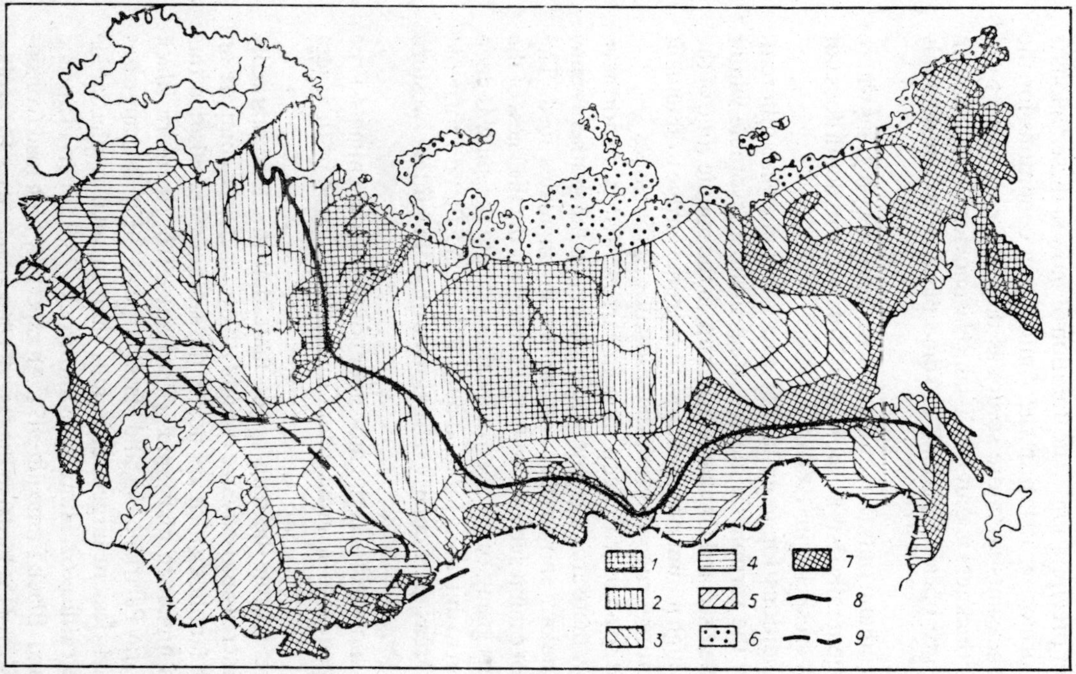

Fig. 14. Snow cover zones in the Soviet Union (Rikhter, 1948) and boundaries of distribution of tetraonids and resident phasianids. Maximum 10-day mean thickness of snow cover: 1—more than 70 cm with prolonged snowy period; 2—50–70 cm with prolonged snowy period; 3—30–50 cm with brief snowy period; 4—10–30 cm in the season; 5—less than 10 cm in the season; 6—20–50 cm with very prolonged snowy period; 7—hilly region; 8—northern boundary of range of resident phasianids; and 9—southern boundary of range of tetraonids.

Urals and Moscow suburbs confirm the great preference this species to the less snowy regions (Skripchinskii, 1962; Sergeeva and Sumina, 1963).

Much of the northern boundary of the range of pheasants coincides with the zone of less snowy plains, foothills, and continental arid plateaus with little winter precipitations. The duration of snow in these regions is 60–100 days (Formozov, 1946). In the Amur Basin conditions for the habitation of pheasants are more severe as the snow cover lasts 5-5½ months, but its thickness is only 10–20 cm (Nasimovich, 1950). Only in Ussurii region are pheasants found at some places with considerably deeper snow.

Francolins inhabit only the southernmost, less snowy regions where the snow cover generally lasts no more than 30 days and its depth does not exceed 20 cm (Vereshchagin, 1960).

Winter conditions for mountain species are somewhat different. Scanty snow in the xerophytic foothill areas assures favorable winter conditions for See-see and Chukar. Higher up, in the middle zone of the mountains and high mountains the snow cover is uneven due to abundant winter precipitations, rugged relief, and insolation of the slopes.

Alongside the areas with deep snow in narrow ravines on northern slopes, invariably there are large, snow-free patches on the southern sunny
156 sides where Chukar, snowcocks, and mountain ungulates feed. The tetraonids do not exhibit such a specific preference for the thickness of the snow cover. This group often lives in localities where the snow is deep enough for them to dig freely for food but, all the same, they do not avoid the less snowy areas, particularly in regions with mild winters (western Europe).

Deep, soft snow restricts the movement of terrestrial animals. Free movement depends to a great extent on the load shared by their feet (Table 3).

It is apparent from the data in the table that this load markedly increases with increasing body size (weight). Nevertheless, a comparison of birds of both families which are nearly the same size reveals that this index is always higher in phasianids (Pheasant—60.6, Northern Black Grouse—40.1, Gray Partridge—35, and Hazel Grouse—30.7). In many cases the load is much less in larger tetraonids than in smaller phasianids (Western Capercaillie—62, Himalayan Snowcock—75.9, Siberian Spruce Grouse—31.9, and Black Francolin—47). An especially high load (consid-
157 ering the small size of the birds) in the case of Chukar and Black Francolins puts them to maximum disadvantage in winters with heavy snow. Without doubt, a lower load facilitates locomotion on snow; however, even Willow Grouse, with the lowest load factor, sinks deep in snow because of the peculiar structure of its feet. Walking on snow is difficult even for forest

groups with the supporting area of their feet enlarged by horny fringes, while it is almost impossible for phasianids with narrow bare feet. It must be noted that the load factor may vary greatly within the same species depending on the plumpness of the bird. This is particularly seen among phasianids which are much fatter during early winter than at the end of the season. This would also explain the differences between my data and those published. For example, the published data are somewhat lower for Willow Grouse—14–16 g/cm^2 (Formozov, 1946; Dul'keit, 1960) and higher for the Common Partridge—40–41 g/cm^2 (Formozov, 1946). For Himalayan Snowcocks the difference is even higher—85–88 g/cm^2. It has also been reported that these birds could feed and move about only on 5–10 cm deep snow (Zimina, 1959).

The area of distribution of tetraonids (as can be seen from the map) is limited neither by the duration nor the depth of the snow cover, although observations conducted in the Pechora-Ilych preserve revealed that forest grouse (for which movement on soft snow is difficult) avoid moving on snow. From November through March (when snow is fairly loose), of the total number of birds seen moving on snow, 83% were Willow Grouse, 9.5% Western Capercaillies, and Black and Hazel Grouse only 1.5% each. Moreover, Hazel Grouse were not sighted at all in two months (January and February) and capercaillies in three (January, February and March). Footprint measurements on loose snow showed that Willow Grouse sink 4–6 cm while the body makes a 2–3 cm deep furrow. Footprint of Hazel Grouse, small birds, were 14 cm deep and the furrow made by the body 6 cm deep; the corresponding measurements for capercaillies were 30 and 17 cm (Teplov, 1947; Donaurov, 1947). The labored movements of capercaillies on fresh snow are also evident from observations made in northeastern Altai, where as many as six temporary, shallow snow holes, in which the birds rested, were found along a track only 50 m long (Dul'keit, 1960).

158 That tetraonids do not venture on the snow during the colder months can be explained not only by difficulty in movement, but also by the low temperature on the snow surface. The reflection from the snow surface cools the near-ground air and this phenomenon is even more marked in calm and clear weather (Rikhter, 1948).

Movement on loose snow is very difficult for phasianid species with a heavy load on their feet. Gray Partridges can negotiate up to 18 cm deep loose snow while foraging, but in the process leave deep trenches (Krivitskii, in lit.). Pheasants usually do not venture onto 10–15 cm deep snow; to them even a very thin layer of snow crust presents a formidable barrier (Nasimovich, 1950).

No data are available on the movement of other species on snow. However, the senseless "hunting"of Chukar, francolins, and pheasants on fresh snow indicates that they are almost incapable of moving on soft snow. During severe winters with heavy snowfall, francolins, pheasants, Chukar and Gray partridges die in large numbers, and their population dwindles at times to 1/10 or even disappears altogether (Formozov, 1935, 1970; Dement'ev and Shimbareva, 1941; Stroganov, 1946, 1957; Nasimovich, 1950; Sludskii, 1965).

However, if the climatic conditions in subsequent years usually are favorable, the populations of Chukar, pheasants and francolins are restored to their former level. In Gray Partridges the process is rather slow. After winters with particularly heavy snows, shrinking of the northern boundary of the pheasant's range is a common phenomenon. According to Sludskii (1965), in the 18th century, pheasants nested in the Ural valley as far as Fort Tatishchevo. These birds later disappeared totally from this area and from the northern coast of the Caspian Sea. Sludskii assumed that the reason for their disappearance was the severe winters of 1826–27 and 1848–49. In the middle of the last century pheasants perished in Tarbagatai and Zaisansk basin; in the foothills of Dzhungarian Alatau and the Alakul'sk basin they appear occasionally, disappearing during severe winters.

It is clear from the examples given above that the snow cover is the
159 main factor limiting the northward distribution of phasianids. Mass mortality of these birds during severe winters accounts for the fluctuating northern boundary of their range sometimes shrinking it. This demonstrates that the life of the few phasianids species inhabiting regions with prolonged snowy winters is precarious, to say the least, and that their survival has been possible only because of their endurance and, in particular, high fecundity. Such severe conditions have no adverse effect on forest grouse. Actually the winter distribution of capercaillies in the forest depends on the shaggy ice-rime on tree branches. Most birds congregate in the forest sections exposed to the wind and where the trees are less covered with snow (Teplova and Teplov, 1947).

The depth of the snow cover greatly influences the life of Willow Grouse and Rock Ptarmigans. Although movement on snow is not very difficult for them, food shortages in especially snowy winters, when bushes are almost buried under snow, cause mass migrations of Willow Grouse, which usually end in their death. If starvation is the reason for migration, then its intensity and extent are apparently determined by population density and severe frosts. If the population density is low, the winter food available even in the tundra is sufficient for ptarmigans; but when it is high, a large number of birds must migrate to the forest-tundra or even more southward. Lower temperatures while accentuating heat loss demand abundant food.

Severe winters with heavy snowfall cause a marked decline in the population of Willow Grouse as recorded in the winter of 1939–40 in western Siberia where the number of birds decreased by 50 to 60%. At places they disappeared altogether, for example in the upper reaches of Bukhtarma in Altai in 1959–60 (Sludskii, 1965).

Delayed spring also has an adverse effect on the population of several species of galliform birds. When thaws are delayed plant growth may be retarded and the much needed green fodder may remain inaccessible for long. Phasianids are especially affected by a shortage of spring food, since by this time their body fat is totally used up and they became highly emaciated. As for tetraonids, late thaws extend their foraging for less nutritious winter food, leading to emaciation prior to molting and reproduction. The population reduction of Willow Grouse after prolonged snowy winters has been attributed to a shortage of green fodder in the
160 spring (Teplov, 1960). Late changeover to green plants reduces the clutch size of Hazel Grouse.

A comparison of the ranges of tetraonids and phasianids based on the monthly mean temperature in January revealed that the former not only inhabit regions with the lowest temperatures, but that the range of temperature in various sections of the range is very wide. Their species composition in the eastern part of the Soviet Union, i.e., northeastern Siberia and the Primor'e where frosts are very severe, is more diverse than elsewhere. The range of Black-billed Capercaillies falls in a territory with average January temperatures of –15° to –45°C and that of Siberian Spruce Grouse is –20° to –35°C. In widespread species the temperature fluctuations are very great within their ranges. Rock Ptarmigans inhabit regions where winters are mild and the average January temperature only –5° to –10°C (Scandinavia and the Kola Peninsula), as well as regions where winters are extremely severe with temperatures of –45°C (northern Yakutia). Northern Black and Hazel Grouse can tolerate a still wider range of mean January temperatures; they inhabit frozen eastern Siberia as well as some regions of western Europe where the mean January temperature dips only to 0 to –5°C without any persistent snow cover.

The distribution of phasianids is confined, in general, to a territory with less frosty winters. Francolins and See-see Partridges do not penetrate north of areas with a mean January temperature of 0 to –5°C. In Soviet Central Asia and Kazakhstan the northern boundary of the range of pheasants does not cross –10° to –15°C in January, but in the Far-East these birds live in much more severe conditions, where the mean January temperature is as low as –20°C. The northern boundary of distribution of

Table 3. Load on supporting area of feet

Bird species		Number of specimens, n	Period of capture	Body weight, g	Supporting surface of both feet, cm^2	load g/cm^2
Willow Grouse		12	April, October, November	734	35.4	20.7
Western Capercaillie	♂♂	7	April, May	4,036	65.5	62.0
Black-billed Capercaillie	♂	1	Winter	3,800	65.6	58.0
Northern Black Grouse	♂♂	6	March, April	1,344	33.5	40.0
	♀♀	3	March, April	992	24.2	40.9
Siberian Spruce Grouse		3	Winter	708	22.2	31.9
Hazel Grouse		11	April, May	375	12.2	30.7
Gray Partridge		8	October, January	368	10.5	35.0
Daurian Partridge		5	May, December	336	10.4	31.8
Chukar		10	November	518	10.6	48.8
Black Francolin		6	February	466	9.9	47.0
Himalayan Snowcock	♂♂	6	November	3,168	41.6	75.9
	♀♀	3	April, November	2,075	29.5	70.3
Common Pheasant	♂♂	3	November	1,425	23.5	60.6
	♀♀	9	October, November	986	20.8	47.4

Gray Partridges lies within the limits of the -15 to -20°C January isotherm, but Daurian Partridges of Trans-Baikal inhabit regions where the mean temperature of the coldest month is -25° to -30°C.

Low ambient temperatures accentuate the loss of body heat and large quantities of food are required to maintain the heat balance. In the less snowy regions of Mongolia and southern Balkhash pheasants are known to endure frosts of up to -40°C in some winters but food must be adequate. However, in winters with heavy snow the birds die en masse of starvation even at higher temperatures. Gray Partridges feed voraciously at low temperatures, their maximum activity being recorded at -20°C (Pulliainen, 1966). On frosty days, Gray Partridges introduced into Canada feed
161 throughout the winter day, consuming 50–75 g seeds, or 12–18% of its body weight (Westerskov, 1966). These examples reveal that ecological factors like food availability and ambient temperature not only exert great influence on maintaining the energy balance, but are inseparably and mutually linked.

The heat balance of an organism is maintained by the regulation of two aspects, namely heat production and heat transfer, of a single process. Heat production (chemical thermoregulation) generated by oxidation in the muscles and internal organs utilizing the energy received from the environment (food, water, heat, and oxygen) is a purely physiological aspect of thermoregulation. Heat transfer (physical thermoregulation) regulates the heat produced by the organism in relation to the environment. The state of the plumage is of great importance in this process. The biological role of the plumage lies in controlling any sharp decrease in the level of heat loss, and consequently, the compensatory expenditure of energy to maintain a constant body temperature (Shilov, 1968). Heat-insulating properties of the plumage depend on the quality of the feathering—its structure, density, sizes of apteria, and color—as well as on the reflex action which controls the movement of feathers, and thus the thickness of the heat-insulating air layer. The subcutaneous fat layer, size of bird, and to a considerable extent its life style and behavior, influence the level of heat loss.

The presence of additional vanes on the contour feathers, which enhances the heat insulation property, is characteristic of the plumage of galliform birds. The most significant difference between the plumage of tetraonids and phasianids is that tetraonids have at least two molts a year—a complete molt in autumn and a partial one in spring—and correspondingly two plumages—winter and summer. As an exception, Sage Grouse (*Centrocercus urophasianus*), which inhabit North America near the southern boundary of distribution of the family, like phasianids have only one complete autumn molt. Poorly developed aftershafts on the

contour feathers and sparser plumage (not all the downy feathers are replaced immediately and hence the plumage thins) are typical of tetraonids. According to Snigirevskii (1950), in both species of capercaillies, Northern Black Grouse, Spruce Grouse, and Hazel Grouse, molt of the head, neck, crop, part of the breast, and the back commences in spring, after (in males) or during (females) breeding. Feathers of the summer plumage are
162 variegated and additional vanes are almost completely absent. Simultaneoulsy feathers are shed on some parts of the body. It is known that some of the feathers of Hazel Grouse are shed in spring, but new ones appear only during the period of complete molt. Western Capercaillies and Northern Black Grouse have two plumages—summer and winter—and three molts since the autumn molt continues with a long pause, simulating an additional prewinter molt. Willow Grouse have only one complete autumn molt, but three partial molts result in the formation of four seasonal plumages. This complex molting pattern and warmer winter plumage of tetraonids with better developed vanes on the feathers indicate that low temperatures were responsible for adaptive changes in the plumage of this group of birds.

Molt has not been fully studied in phasianids. In every work devoted to this problem, attention has mainly been paid to changes in juvenile plumage and the order of molting of wing and tail feathers (Kuz'mina,1955, 1960; Khanmamedov, 1956; Grachev, 1960; Baziev, 1966; Niethammer, 1968).

The absence of supplementary molts leading to seasonal plumages in tetraonids is interesting. In birds of this family there is one complete autumn molt and consequently one plumage is a distinct phenomenon. However, in some species, feathers are known to be shed in spring without being replaced thus thinning the plumage and increasing the heat conductivity of the plumage. To what extent this phenomenon occurs in members of the family is not known.

I have observed the thinning plumage in Himalayan Snowcocks in April–May. During this period the number of feathers on measured areas of the skin of the head, neck, and sides decreased to nearly half. Snigirevskii considered such partial shedding of feathers in spring as another molt. In my opinion, this is not correct since shedding feathers is only the initial phase of molt and does not lead to the formation of a special summer plumage, even though it undoubtedly thins the feather coat.

For characterizing plumage an index is sometimes used, namely the ratio of weight of feathers (P_f) to weight of bird—(P_b) (Kaftanovskii, 1951; and Shilov, 1968). This index cannot be used in comparing birds that differ greatly in size (in my material the weight of a male Western Capercaillie is more than ten times that of a Gray Partridge). While volume rises by a
163 cubic dimension, body surface increases only by square units; thus the

surface area is relatively small in large birds. In my opinion, the ratio of the weight of the feathers (P_f) to the surface area of the body (S_b, cm^2) after it has been plucked provides a realistic picture (Table 4).

For comparison Table 4 also gives the P_f/P_b ratio, but it does not agree with the P_f/S_b ratio. This discrepancy may be due perhaps to the considerable differences in the sizes of birds compared.

The quality of winter plumage (according to such indices as weight of feathers per unit surface of body and degree of development of the vaned part of feathers) exhibit significant differences within the family Phasianidae. These differences are not so marked in the Tetraonidae.

The maximum weight of feathers per 100 cm^2 of body surface has been
164 recorded in winter for Himalayan Snowcocks and Common Pheasants, and the minimum weight for Gray and Daurian partridges. In tetraonids this ratio lies midway between the especially well-feathered snowcocks and pheasants and the other phasianids with a very low weight index of feathers. The well-known principle that as the body size (volume) increases its relative surface area decreases is distinctly demonstrated by the S_b/P_b ratio. It is interesting, however, that the index is low in equally sized Tetraonidae and Phasianidae or even in very small phasianids (Himalayan Snowcock—Western Capercaillie; Chukar — Willow Grouse). On this basis it can be suggested that in phasianids the S_b/P_b ratio is more favorable for maintaining the heat balance at low temperatures.

Heat conductivity of the feather coat largely depends on the structure of the plumage. In tetraonids with distinct summer and winter plumages, the winter feathers have very long and dense vanes and the down on the main feathers is well developed. But it cannot be generalized that the vanes of the winter feathers in tetraonids are better developed than in phasianids. Even though a comparison of the structure of feathers is difficult due to differences in size and shape, it can be stated that the quality of feathers is usually superior in birds which have a high feather weight per unit of body surface. Down feathers are well developed in snowcocks, pheasants, Willow Grouse, Black-billed Capercaillies, Hazel Grouse, and Siberian Spruce Grouse. Supplementary vanes are poorly developed in Gray, Chukar and Daurian partridges (Figs. 15 to 20).

The warmest plumage (judging from weight of feathers and their quality) is found in snowcocks, pheasants, Willow Grouse, and Black-billed Capercaillies, and the least warm plumage in Gray and Daurian partridges*. Due to the tarsometatarsus being feathered, all tetraonids

*In the absence of special equipment, I could not measure the heat conductivity of the skin of birds. Tentatively, I assume that the plumage is the warmest among species with a high feather weight and well-developed down on the vanes—Author.

Table 4. Weight of contour feathers and its ratio to weight and surface area of birds

Bird species		Number of specimens, n	Period of capture	Body weight, P_b, g	Weight of feathers			Surface area of body	
					P_f, g	% of body weight	per 100 cm^2 of body area, g	S_b, cm^2	per g body weight
Willow Grouse		7	Winter, April, October, November	747	40.8	5.6	6.1	665	0.89
Western Capercaillie	♂♂	3	April	3,980	149.8	3.7	5.6	2,666	0.67
	♀♀	3	April, May	1,965	91.6	4.6	6.8	1,332	0.68
Black-billed Capercaillie	♂	1	April	3,800	180.0	4.7	5.6	3,215	0.84
	♀	1	April	2,200	120.7	5.4	7.1	1,683	0.76
Northern Black Grouse	♂♂	8	March, April, May	1,318	63.7	4.7	6.3	999	0.75
	♀♀	2	April	1,037	55.7	5.3	6.8	819	0.79
Siberian Spruce Grouse		2	March	712	35.1	4.9	–	–	–
Hazel Grouse		8	April, May	375	21.5	5.7	4.5	487	1.30
Common Partridge		8	January, October, November	374	18.7	5.0	4.1	461	1.20
Daurian Partridge		4	December	351	16.1	4.5	4.4	361	1.00
Chukar		11	March, December	449	22.4	5.0	6.0	371	0.81

Black Francolin		5	February	465	24.5	5.2	5.1	479	1.00
Himalayan Snowcock	♂♂	6	March, May, November	3,054	181.0	5.8	11.0	1,643	0.53
	♀♀	3	April, November	2,408	131.4	5.4	9.4	1,401	0.58
Common Pheasant	♂♂	3	November	1,425	68.0	4.8	7.1	959	0.65
	♀♀	8	October, November	900	50.8	5.4	6.6	780	0.86

have an advantage over phasianids in snowy, frosty winters. The dependence of the feather coat of the tarsometatarsus on climatic conditions is shown well in Hazel Grouse. The tarsometatarsal feathers are longer and denser in the eastern Siberian forms than in the western forms. In Black-
168 billed Capercaillies, long feathers on the tarsometatarsus cover the toes in such a way that only the claws peep out. Such feathering protects the feet from frosts and also increases the supporting area, which facilitates movement on snow.

Among tetraonids, the most perfect plumage is found in Willow Grouse. The feathers covering the body are very long. Down covers more than half the length of the main feather while the aftershaft is denser. In feathers from other parts of the body the aftershaft is 1/2–3/4 the length of the main feather. According to Snigirevskii (1950), the white color of the winter plumage arose as a supplementary thermoregulatory mechanism in direct response to low temperatures, and became established in birds due to its cryptic importance. It has been observed (Johnsen, 1929) that the white feathers of the Ptarmigans are less conductive than the colored ones since their white coloration is due to the presence of extremely minute air sacs in the horny structure of the vane.

Unlike in other tetraonids, the toes of Willow Grouse are also densely feathered in winter. Feathers not only keep the feet warm, but also considerably enlarge the supporting area and thus facilitate movement on snow.

Fig. 15. Feathers of snowcock: 1—breast feather; 2—back feather; 3—feather from sides of body.

Among tetraonids, the minimum weight of feathers per 100 cm^2 of body surface area is found in the Hazel Grouse. However, the intense development of the down compensates for the sparse feather density. Plumage in Hazel Grouse is sparse and soft. Feathers covering the body are very long. The down extends for 1/2–3/4 the feather length while the aftershaft is very long and dense. On frosty days the bird perches with its feathers ruffled and spread in such a way that a large air pocket forms between them. In the structure of feathers Siberian Spruce Grouse is close to Hazel Grouse. The former has very soft but denser plumage which could be regarded as possessing high heat-insulating properties.

Among the phasianids studied, the plumage of Himalayan Snowcocks is most perfect with the highest weight index of feathers. Feathers covering the body are quite narrow with long, dense additional vanes. In most feathers the greater part of the vane is fluffy. The feathers, especially on the back, overlap one another so that only the vane tips of the lower feathers are visible. Such an arrangement of feathers forms a multi-layered coat like a thick "feather jacket." In addition, the apteria in snowcocks are much smaller than in other galliform birds of the Soviet Union. The feathers of Chukar are rather hard and dense with a poorly developed but fairly dense down on the vane. The feather coat of Gray and
169 Daurian partridges is markedly softer than that of Chukar but is less compact, with large air pockets.

Body fat plays an important role in the thermoregulation of the organism; subcutaneous fat deposits are not only an important source of

Fig.16. Feathers of pheasant: 1—breast feather; 2—back feather; 3—feather from sides of body.

energy during food shortages, but also greatly lower the heat conductivity of the feathers. In tetraonids very plump birds are found only among some species in regions of the Far North and East Siberia. In Verkhoyan, an extremely cold but less snowy region, Black-billed Capercaillies become very plump, since the birds cannot always burrow in the snow to warm themselves and often have to roost in open holes. It is here that Black-billed Capercaillies change late over to twigs and fattens up by feeding on berries. The weight of fat in these birds reaches up to 12% of the body weight and is preserved until spring (Kirpichev, 1960). In other species of tetraonids, formation of fat reserve have been detected only in populations living in regions with especially low temperatures. Thus, the subcutaneous fat layer of Willow Grouse in Novosibirsk Islands is well developed in winter (Kishchinskii, 1974). In Spitsbergen Rock Ptarmigans gain much fat (Birulya, 1910). In Yakutia, in addition to Black-billed Capercaillies, fat deposits have been found in Northern Black, Hazel, and Willow grouse. In the tugais of Khamar-Daban Hazel Grouse have been found to become extremely obese by feeding on berries of sea sallowthorn (Smirnov, 1967).

Among the phasianids of the USSR avifauna seasonal changes in plumpness are very pronounced. After breeding and molting, when the birds are quite emaciated, they begin to feed voraciously, and by late autumn-early winter gain maximum weight. A thick subcutaneous fat layer grows and numerous fat deposits appear in the body cavity.

The degree of plumpness of various phasianid species differs mark-

Fig. 17. Feathers of Chukar Partridge: 1—breast feather; 2—back feather; 3—feather from sides of body.

edly. Pheasants, snowcocks, quail, and Chukar become most obese in autumn. Distinct differences in the plumpness of various populations of Chukar and Gray partridges, have been recorded, depending on wintering conditions. In Turkmenia, where the snow cover is not permanent, Chukar do not fatten as much as those in the hilly regions of southern Kazakhstan, where winters are more prolonged and severe. Gray Partridges in the European part of their range usually do not fatten much by winter, but the European forms naturalized in the northern parts of the USA and Canada have modified their winter diet under severe climatic conditions. Instead
170 of green plants they consume grains of cultivated cereals and fatten considerably with weight gains reportedly to 70 g from September through January (Westerskov, 1965). High fat deposition has also been observed in Gray Partridges in the steppes of central Kazakhstan (Krivitskii, in litt.) where the birds likewise feed on cereal grains.

The fattening of phasianids and tetraonids toward winter is apparently due to several different factors. In phasianids the formation of the fat layer should be regarded primarily as an adaptation of different species (or populations) to food shortages. A rapid decrease in plumpness, after the snow cover has formed, suggests that the main purpose of their fat is to compensate for food shortages. However, there is no doubt that the subcutaneous fat also helps reduce heat loss. Fat deposition in quail may be considered an energy reserve for use during migration. The growth of subcutaneous fat in tetraonids in regions with very frosty winters but sufficient food availability, and its retention for the greater part of winter,

Fig. 18. Feathers of Willow Grouse: 1—breast feather; 2—back feather; 3—feather from sides of body.

suggest that this reserve is more important for controlling body heat. By lowering heat loss it increases the heat-insulating properties of the skin, which is particularly necessary at low temperatures.

The foregoing discussion shows that, on the whole, the thermal insulation of the plumage is about identical in tetraonids and phasianids even though the winter plumage in tetraonids is warmer than the summer plumage. The feather coat of tetraonids is superior to that of phasianids in that their tarsometatarsus is feathered—even the toes are covered among Willow Grouse. Furthermore, the growth of a thick horny layer, a poor conductor of heat, on the underside of toes protects the feet from freezing (Madsen and Wingstrand, 1958).

Thus there is no marked qualitative difference between the feather coats of phasianids and tetraonids despite distinct differences in their life style and behavior. While phasianids lead an active life in winter by foraging almost throughout the day, forest tetraonids are far less active. After the first heavy snowfall they spend most of the time in the snow, emerging briefly to forage. Willow Grouse and Rock Ptarmigans are
171 intermediate: they spend more time foraging and rest less in snow holes. The low winter activity of tetraonids and their long rests in snow holes, where the temperature is higher than on the surface and less subject to fluctuation, help maintain the heat balance and supplements the inadequate heat-insulating properties of their plumage.

Comprehensive studies on the use of snow cover by tetraonids have been made in preserves. In the Pechora-Ilych Preserve, at the commence-

Fig. 19. Feathers of Siberian Spruce Grouse: 1—breast feather; 2—back feather; 3—feather from sides of body.

ment of winter and during thaws, capercaillies generally feed twice a day, usually once in midwinter, while on frosty days they spend no more than 1-1½ hours foraging. During blizzards, strong north winds, and frosts below -30°C, they can stay in snow holes for as long as three to four days (Teplov, 1947).

According to Semenov-Tyan-Shanskii (1959), the low winter activity observed among tetraonids is particularly marked in Western Capercaillies (which spends about 80% of its time in snow holes). Semenov-Tyan-Shanskii considers this method of adaptation to winter conditions as somewhat similar to hibernation among mammals.

In the Pechora-Ilych Preserve, Hazel Grouse spend almost 20 hours a day in snow holes up to 30 cm deep (Donaurov, 1947). In frosty weather their activity is reduced: they spend 21-22 hours in snow holes (Formozov, 1946). According to Formozov, in severe frosts Northern Black Grouse spend more time in the open than do Hazel Grouse but they rest in shallow pits in the snow several times a day to warm. Willow Grouse and Rock Ptarmigans, which feed on top of the snow cover, spend much more time foraging. According to observations in Taimyr, they feed five to six times a day. It was noticed that Rock Ptarmigans keep to less snowy regions and do not dig holes in the snow. They are often carried away by blizzards while crouched on the leeward side of snow banks. In less snowy regions Northern Black Grouse and Black-billed Capercaillies try to bed in open snow holes when there is no possibility of burrowing into the snow.

Donaurov (1947), in the Pechora-Ilych Preserve, was the first to study

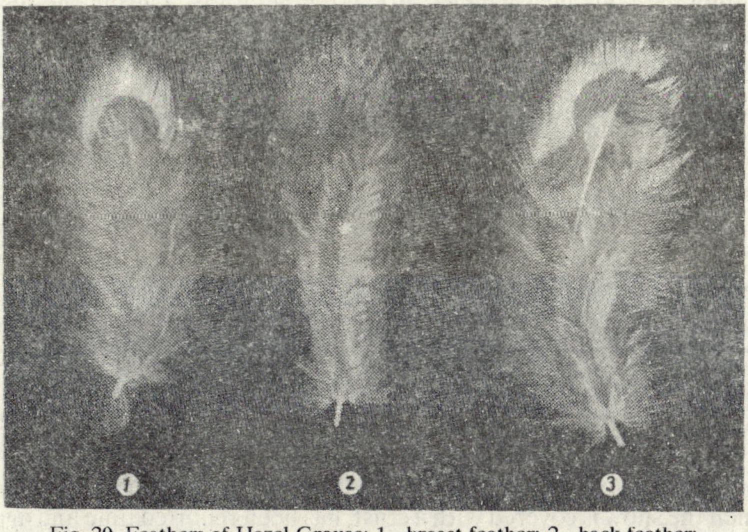

Fig. 20. Feathers of Hazel Grouse: 1—breast feather; 2—back feather; 3—feather from sides of body.

the temperature of the snow in which tetraonids warm themselves. According to his observations in December, 1938–January, 1939, the ambient temperature varied from -16° to -39°C and at a depth of 20 cm from the surface the temperature was -10° to -16°C. Further experimental
172 studies of temperature conditions in artificial holes for Black Grouse and Hazel Grouse were later conducted in the Central Forest Preserve (Volkov, 1968, 1968a).

Experiments showed that the differences between the temperatures of air, snow surface, and in the snow holes were maximum during severe frosts. At low temperatures the free water vapor between the snow particles condenses and the snow becomes dry and less conductive. On the other hand, with increasing temperature, heat conductivity also increases due to vapor saturation. When the temperature of the snow is -3°C, it rises about 0° in holes where the birds rest, and thawing sets in. It has been observed that at this temperature Northern Black Grouse and Hazel Grouse avoid making deep holes but remain instead on the snow surface in shallow pits.

Long rest in the snow protects the birds not only from low temperatures, but also from any sharp fall in daily temperatures. In experiments with Hazel Grouse, the temperature in the holes changed only from -1.8° to -3.9°C, while the ambient temperature dropped to -26°C. According to Volkov, the most favorable temperature in the holes for the birds is -1°C, when the heat loss is minimal. These experiments demonstrate how important is the utilization of the protective properties of snow in maintaining the body heat balance at low temperatures. During the severe frosty and snowy winter of 1939–1940 no mortalities were reported among tetraonids. Deep, loose snow enabled Hazel Grouse and Northern Black Grouse to dig holes right down to the ground surface. Willow Grouse, which usually do not dig so deep, used the deep tracks of Northern Black and Hazel grouse during this particular winter (Dement'ev and Shimbareva, 1941).

A study of the bioenergetics of Northern Black Grouse in the Leningrad District at ambient temperature of -15° showed that the energy needed to sustain life in these conditions (when buried in snow where the temperature is around -4°C) is equal to 184 to 205 Cal/kg weight per day (Potapov and Andreev, 1973). However, the combination of low temperatures and little snow or snow crust, when the birds cannot hide in holes, usually results in their mortality. Frozen Northern Black Grouse (Blokhin, 1929) and Hazel Grouse (Formozov, 1946) have been observed under these circumstances. Large-scale mortality among birds in frosty winters, when the snow is covered with a crust, has been reported time and again. The foregoing examples of the behavior of tetraonids in winter resulting

in decreased activity and the use of snow cover for protection against cold,
173 prove convincingly that such behavior is a very important factor in maintaining heat balance, since winters represent a difficult period for birds of this family (Volkov, 1968).

In this connection an interesting assumption has been made by Volkov (1969) and Osmolovskaya (1970) that the northward recession of the southern boundary of the range of a number of tetraonid species observed in the 1930s–1950s, is due to the warming of the climate. Frequent thaws, causing the formation of a snow crust, followed by severe frosts may lead to freezing conditions. When a snow crust or a thick ice layer forms after a sudden rain, birds perish in the snow bed (Formozov, 1970).

The above examples show that the snow cover plays an important and varied role, depending on its state, in the winter life of tetraonids. The depth and compaction of the snow and the formation of a crust considerably influence the course of their hibernation and the intensity of thermoregulation.

The winter behavior of phasianids in the USSR avifauna is altogether different. Shortage of their food, which generally is more nutritious than that of tetraonids, compels phasianids to be more active for the greater part of the day. Their choice of less snow-covered areas for foraging while moving freely precludes the possibility of utilizing snow cover for protection from low temperatures. On the whole, the phasianid species inhabit southern regions of the country where winters are more mild and the snow cover not deep. However, in the Trans-Baikal Daurian Partridges inhabit areas where the mean January temperature is −30°C, and in some parts of the range pheasants live in localities where the winter temperature falls to −40°C in some winters.

Adaptations by phasianids in the USSR avifauna to low temperatures vary notably. As mentioned earlier, in every species of this family a marked increase in obesity has been observed in autumn, accompanied by the deposition of subcutaneous and internal fat, which not only serve as a reserve in periods of food shortage, but also reduces the body-heat loss. Snowcocks and pheasants additionally possess a dense feather cover. Phasianids roost under various shelters in the most favorable microclimatic conditions. Chukar and snowcocks roost under overhanging rocks and in caverns. Francolins and pheasants utilize shrubbery and reed canopies, while Gray and Daurian partridges rest and roost among brush reeds where they make shallow commonal holes in the snow. Group
174 roosting is characteristic of all phasianids; they roost in packs pressed close to one another which makes for excellent heat conservation.

In regions with extremely heavy snowfall and low temperatures, pheasants and Gray Partridges roost at times in covered snow holes like tetraonids. This phenomenon has been recorded for pheasants of the Far-East (Formozov, 1961, citing Sleptsov). In the Trans-Volga forest belt, where the snow accumulates in sizable drifts, Gray Partridges are known to roost in snow banks. Gray Partridges, naturalized in Canada, reportedly roost in covered snow holes when the depth of the snow exceeds 30 cm (Westerskov, 1965). Observations on the behavior of a covey of Gray Partridges under various climatic conditions showed that in frosty, windy weather an entire flock (18 birds) roosted in a single community hole. In still weather the birds rested in small groups in holes not far from each other (Knysh, 1974).

Some North American tetraonids—Prairie Chicken *Tympanuchus cupido* and *T. pallidicinctus,* as well as Sage Grouse—constitute intermediate ecological types between the species of Tetraonidae and Phasianidae, since they lead a terrestrial mode of life and feed on the ground in winter. Prairie Chickens avoid particularly severe and snowy winters by migrating southward from the northern part of its range, even beyond its breeding grounds. Sage Grouse migrate on foot from regions of deep snow cover. The need for using snowcover to maintain body temperature is not distinctly manifest in these birds. In winter, Prairie Chickens roost in shallow snow pits in meadow grass Bent (1932). Sage Grouse congregate in thick undergrowths of wormwood but, if the snow is deep, it burrows into it during blizzards. Its relation to the state of the snow cover is nearly the same as that of Gray Partridges and pheasant of the Far-east, i.e., they avoid deep snow by migrating to less snowy regions, but may at times burrow into the snow as do other tetraonids.

It should be noted that adaptation to prolonged winters in the temperate and cold zones of the northern hemisphere, which are characterized by low temperatures and deep snow cover, is shown very distinctly in the suppressed activity of forest tetraonids by their continual use of snow holes.

175 Adaptive characteristics of Willow Grouse and Rock Ptarmigans are reflected in their cryptic coloration and plumage on their feet, which is very essential for birds that forage on snow. Since such foraging requires a great deal of energy they remain on the snow for a very short while. Mortality among Willow Grouse in winters with heavy snowfall indicates that this group of birds has only a limited assortment of food available to it. However, their ability to move freely on snow and to take advantage of the microclimate of the snow cover ensures the most northern distribution of this group of Galliformes.

For the heterogeneous group of phasianids, which rely only on ground food, survival during winter is entirely dependent on the depth, state, and duration of the snow cover—decisive factors in good wintering of these birds. The combination of deep snow or glaze ice with low temperatures is especially lethal when thermoregulation is disrupted by food shortages and high losses of body heat.

The effect of severe climatic conditions on behavior varies widely not only in different species but in different populations of the same family. A diet of berries from shrubs and trees reduces mortality among phasianids in years with heavy snow in the same manner as burying in snow in the very snowy, frosty regions of the Ussurii Region. Local differences in the winter behavior of Gray Partridges are distinctly manifested in widespread migrations and even in the use of snow holes for protection against frosts in the snowiest regions.

Energy balance is maintained by birds through their selection of the most favorable microclimatic conditions for roosting places and snuggling close to one another to reduce the loss of body heat. Snowcocks and Chukar roost under overhanging rocks and in niches under stones; Gray and Daurian partridges choose shelters with maximum protection from wind—hollows, overhanging bushes, and undergrowths of weeds—where they burrow into a common snow hole to roost pressed against each other. Pheasants and Black Francolins often roost in bushes and reeds.

Mass mortality among phasianids in winters with heavy snowfall, however, indicates that, for some members of this group, adaptations to continuous snow still do not ensure survival in the temperate zone. The only factor ensuring the life of a few species of phasianids in the temperate latitudes is an exceptionally high fecundity that makes for rapid restoration of populations soon after severe winters.

Ecomorphological Characteristics of Locomotor Organs

176 Some ecomorphological characteristics of the organs of locomotion in tetraonids and phasianids are reviewed in this chapter. The selection of this organ system is not incidental. Describing the behavior of animals Tinbergen (1969) states: "In general, I am inclined to consider the most varied movements and their changes, including total immobility, in short, all external characteristics of movement, as "behavior". In this context the function of locomotor organs distinctly reflects the basic features of behavior and the life styles of animals (mode of foraging, migration, courtship displays, behavior during other seasons, etc.). This is why the ecomorphological adaptations reflected in the structure of the locomotor organs are the result of speciation under specific conditions.

The evolution of the locomotor apparatus is responsible for the morphological integrity of the class Aves. Modification of the forelimbs into an organ of flight marked the main trend in avian morphology. Flight is not the principal mode of locomotion in all birds, however, very often the hind limbs are most important in this respect. This latter case is very distinctly manifested in terrestrial, arboreal, and aquatic birds. Flight is of overwhelming significance in only a very small group of birds such as marine birds, many predators, and highly specialized "hovering birds" of various taxonomic groups.

The fundamental importance of the hind limbs of birds in locomotion was recognized even by Kessler (1840) who remarked, "to classify birds
177 and understand their manner of seeking food, a knowledge of the structure of the hind limbs is no less important than the structure of the beak."

In paleornithological investigations, the hind limbs are of paramount importance especially their distal parts which reflect all trends in narrow specialization and ecological specificity of fossil forms (Kurochkin, 1971). At the same time, the study of the morphology of flight organs should not be ignored even in such birds as Galliformes which have limited flight capabilities. In the structure of the forelimbs (as also of the hind limbs), differences are more marked in the distal part of the flight organs. The structure of the sternum displays typical characteristics of the order (Shtegman, 1958), but the details depend on the nature of flight under given conditions, and especially on the degree of development and structure of the pectoral musculature, which differs sharply in different species.

Before discussing the morphological characteristics of the locomotor organs, it is necessary to note that locomotion in birds, especially flight, depends not only on the structural peculiarities of the limbs, but also on the size and shape of the body. It is known that as the bird size increases, an incompatibility is noticed between the physical and physiological features. While the body size increases in a cubic dimension, the wingspan increases only as a square of the linear measurements. Thus a size increase hampers speed; this is the reason for the restricted size of flying animals (Kokshaiskii, 1971).

The group of birds (Galliformes) under study differs markedly in body size and shape. This must be kept in mind while comparing the morphological peculiarities of the locomotor organs (a male capercaillie weighs up to 6 to 8 kg while a quail is only slightly over 120 g).

In studying the morphology of the locomotor organs I have followed the method for measuring the size of organs adopted by Böker (1927), Yudin (1950), Shtegman (1958), and Yakobi (1960). A uniform method and calculation of similar indices makes it possible to compare diverse results from various systematic groups of birds.

Certain terms and values I adopted to measure and weigh birds are explained below.

I. SKELETON MEASUREMENTS

178 1. Truncus*—length of trunk—distance from the anterior margin of the last cervical vertebra to the dorsal margin of the neural canal on the posterior surface of the synsacrum.

2. Humerus—distance between the proximal end of the caput articulare and the distal end of the epicondylus meadilis.

3. Ulna—distance between the proximal end of the processus olecranon and the distal outer margin of the bone.

4. Manus—distance between the proximal outer margin of the carpometacarpus and the tip of the second phalange.

5. Sternum—distance between the base of the carina externa sterni and the posterior margin of the body of the thorax.

*This method of measuring the trunk was used by Yudin (1950) for the family Falconidae, among which there are species with last cervical vertebra fused with the thoracic. This basic measurement has also been used by other ornithologists (Yakobi, 1960; and Shtegman, 1960).

6. Width of thorax—distance between the points of attachment of the sixth pair of ribs.

7. Crista sterni—distance between the tip of the keel and the median part of the posterior end of the sternum.

8. Length of sternum together with the coracoid—distance from the acrocoracoideum to the medial part of the posterior end of the sternum.

9. Femur—distance between the tip of the ridge of the trochanter major and the distal end of the epicondylus lateralis.

10. Tibiotarsus—distance between its proximal and distal ends.

11. Tarsometatarsus—distance between its proximal end and the trochlea digitale mediale.

12. Length of synsacrum—distance between the proximal and distal ends of the spina neuralis.

13. Dorsal width of synsacrum—greatest distance between the antitrochanters.

14. Ventral width of synsacrum—greatest distance between the ventral margins of the os pubis.

II. EXTERNAL MEASUREMENTS

179 1. Length of wing—distance between the anterior margin of the wrist flexion of the folded wing and its tip.

2. Alula—distance between the anterior margin of the wrist flexion and tip of the alula.

3. Longest primary wing feather—distance between the distal end of the second digit to the tip of the wing.

4. Wing projection—distance between the apex of the wing and the distal end of the first secondary feather.

5. Area of the wing-measured from the contour of the outspread wing.

6. Wingspan—distance between the apices of the maximum wing spread.

7. Length of tail—distance between the point of the emergence of the median tail feathers from the skin and their tips.

The body length and wingspan were measured in killed birds with a centimeter measuring tape and the rest with sliding calipers; accuracy of measurements of skeleton up to 0.1 mm and of feathers up to 1 mm.

III. WEIGHT INDICES

A. Total body weight (together with crop content). Weight of heart.

B. Thoracic musculature (m. pectoralis and m. supracoracoideus from both sides separately).

C. Musculature of pelvic girdle. Weight of entire musculature of hind limbs (from both sides without m. obturator).

D. Weight of separate muscles:

1. Extensors of femur (M. caudatus iliofemoralis, M. pubi ischiofemoralis, M. ischiofemoralis).

2. Extensors of femur and flexors of tarsus (M. iliofibularis, M. ischioflexorius, M. caudatus ilioflexorius).

3. Flexors of femur (M. trochanterius–posterior, anterior, medium).

4. Flexor of femur and extensor of tarsus (M. sartorius).

5. Extensors of tibiotarsus (M. iliotibialis, M. femorotibialis).

6. Flexor of tibiotarsus and extensor of tarsometatarsus (M. gastrocnemius, M. peroneus sublimus).

7. Extensor of phalanges and flexor of tarsometatarsus (M. extensor digitorum).

8. Flexor of phalanges, extensor of tarsometatarsus, and flexor of tarsus (M. flexores perforati).

9. Flexor of tarsometatarsus (M. tibialis).

10. Rotator of tarsometatarsus (M. peroneus profundus).

11. Flexor of femur, extensor of tibiotarsus, and flexor of phalanges (M. ambiens).

Total body weight, weight of the thoracic muscles, and total weight of musculature of the legs are determined with an accuracy of 1 g, and the weight of individual muscles and heart, with an accuracy of 0.05 g.

IV. INDICES CALCULATED FROM LINEAR MEASUREMENTS

1. Ratio of the length of skeletal components to truncus (femur, tibiotarsus, tarsometatarsus, humerus, ulna, manus, length of synsacrum, length of sternum, and total length of the coracoid and sternum).

2. Ratio of the length of external measurements to truncus (maximal length of primary flight feathers, bastard wing, wingspan, and tail).

3. Ratio of the length of the humerus, ulna, and manus to total length of wing.

4. Ratio of the length of the femur, tibiotarsus, and tarsometatarsus to total length of hind limbs.

5. Ratio of the width of the sternum to its length.

6. Ratio of the length of the keel to the length of the sternum.

7. Ratio of the dorsal width of the synsacrum to its length.

8. Ratio of the ventral width of the synsacrum to its dorsal width.

9. Ratio of the length of wing up to alula to the entire length of the wing.

10. Elongation—ratio of the length of outspread wing to its average width (measured along wing contour).

V. WEIGHT INDICES

1. Ratio of the weight of M. pectoralis major of both sides, M. pectoralis minor of both sides, all thoracic musculature, heart, and musculature of the hind limbs of both sides to the total body weight.

2. Weight of individual muscles of the hind limbs to the total weight of musculature of the limb (from one side).

All indices are calculated in percentages; the only exception is the index of the heart which is given per thousand.

LOCOMOTION AND MORPHOLOGICAL FEATURES OF THE HIND LIMBS

Movement of the hind limbs is more important than flight for ground and ground-arboreal species of galliform birds.

The marked differences in the structure of the hind limbs of galliform birds are good systematic characters for distinguishing families. The
181 families Tetraonidae and Phasianidae differ notably in the structure of the synsacrum and hind limbs.

Locomotion by the hind limbs like flight depends on the size and general body shape of birds. Great dexterity, speed, and freedom of movement are characteristic of small birds, while larger birds move far more slowly. Body shape is of great importance in species which inhabit dense undergrowth.

The adaptation of tetraonids to more severe climatic conditions and their change to a ground-arboreal life style have altered the structure of the synsacrum and hind limbs, especially the distal parts or feet, which is an important taxonomic feature of the family.

The structure of the synsacrum does not vary much; nevertheless the synsacra of tetraonids and phasianids (judging from species of the USSR fauna) differ considerably. A broad and low synsacrum formed from vertebrae fused over a considerable distance and wide-set ilia are characteristic of tetraonids. The postacetabularis os ilium and fused vertebrae form a broad, slightly convex surface in the caudal part. The ischia are small and form a small ridge where they join the distal part of the ilia. In the proximal part of the synsacrum the crista transversa is fairly low in tetraonids and hence the pars gluteus of ilia occupies a comparatively small area. In general, the synsacrum in tetraonids is broader than in phasianids, with a larger cavity formed by the pars gluteus of ilia and the large area of the fused vertebrae. The surface for the attachment of the hind limb muscles (proximal part of ilia and ischia) is small. Despite the general similarity in the synsacrum structure in tetraonids, the specieswise differences are rather distinct.

The index of the dorsal width of the synsacrum (distance between the antitrochanters in relation to the truncus) in tetraonids varies in the range of 42.8 to 50.6%, and the ratio of the width of the synsacrum to its length 75.5 to 94.0% (Table 5). The synsacrum is broadest at this part in the Rock Ptarmigan and narrowest in the Black Grouse. However, the dorsal measurement of the width of the synsacrum alone does not provide a complete idea of its width, since the caudal part of the synsacrum is largely determined by the ventral margins of the ischia and the adjacent os pubis. The second measurement of the synsacrum that I adopted, also covers its caudal part from the ventral side.

186 The synsacrum is much wider ventrally (than dorsally) in Spruce Grouse, Western Capercaillies, Northern Black Grouse, and Hazel Grouse, and narrower in Willow Grouse and Rock Ptarmigans (this ratio is characteristic of all Phasianidae). Furthermore, in Willow Grouse the proximal end of the ilium is broadened, which is also characteristic of phasianids. The pelvic girdle of capercaillies resembles most that of Spruce Grouse and differs markedly from that of Hazel Grouse. The synsacrum of Northern Black Grouse is slightly higher and more narrow than in other tetraonids. In these measurements, it is somewhat closer to phasianids, but the general structure (as in Willow Grouse) remains typical of tetraonids. A broad and low synsacrum makes for stable body equilibrium, which is very important for movement from branch to branch. The same structure of the synsacrum in the Willow Grouse can be regarded as proof that it, too, led a groundarboreal life style in the past. Also, a wide synsacrum makes for a larger body cavity, essential for its voluminous intestine. In my opinion, some similarity in the proportions of the synsacrum of Willow Grouse and ground phasianids should be considered as probable subsequent changes arising from the change to a terrestrial life style (Fig. 21).

Compared to tetraonids, the synsacra of phasianids is flattened medially, higher, and narrower. The fused vertebrae form a relatively narrow strip sunk below the margins of the ilia and pushed deeper into the body cavity, but convexities (pars renalis) separated by the fused vertebrae are formed in the caudal part of the synsacrum. Because of the laterally compressed body, the large ischium runs deep under the distal part of the ilia, forming a deep depression between these bones under the linea ilio lateralis, where the powerful muscles are attached. The body cavity in phasianids, delimited by the large lobes of the ischium, is narrower than in tetraonids and divided into two deep cavities by the vertebral column.

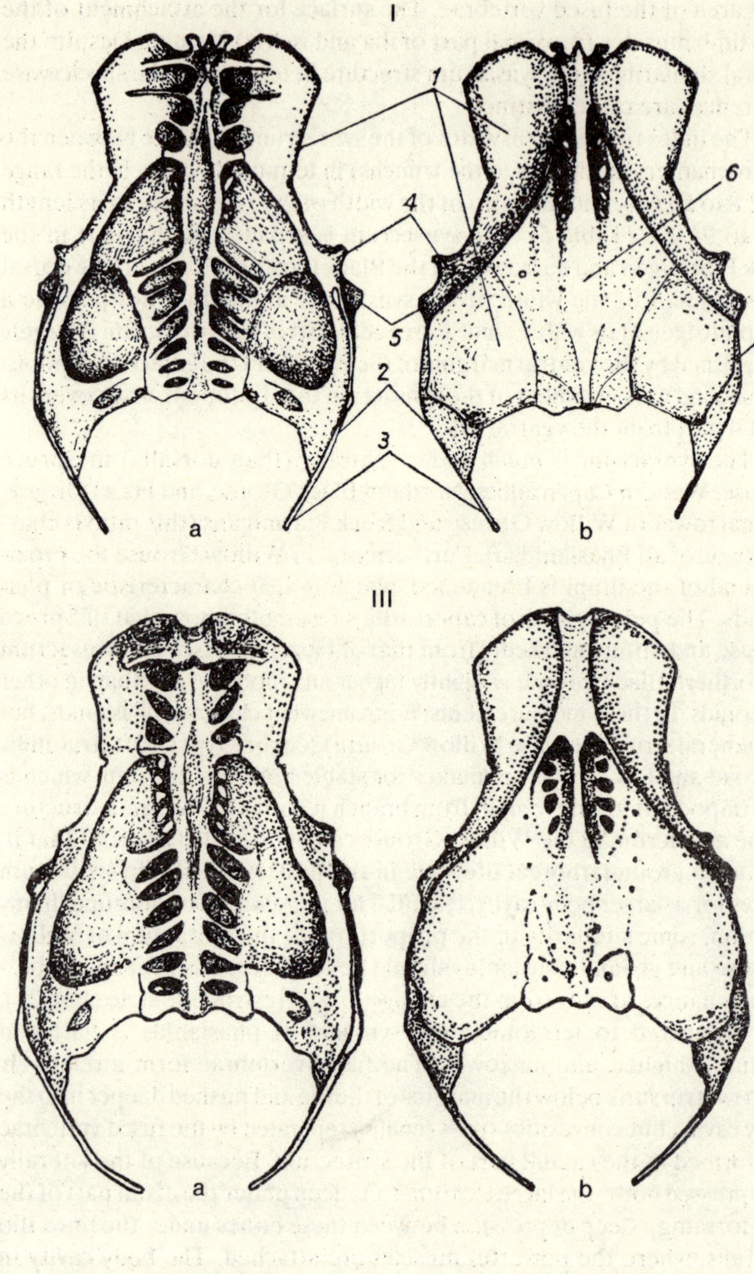
I
1
4
6
5
7
2
3
a
b
III
a
b

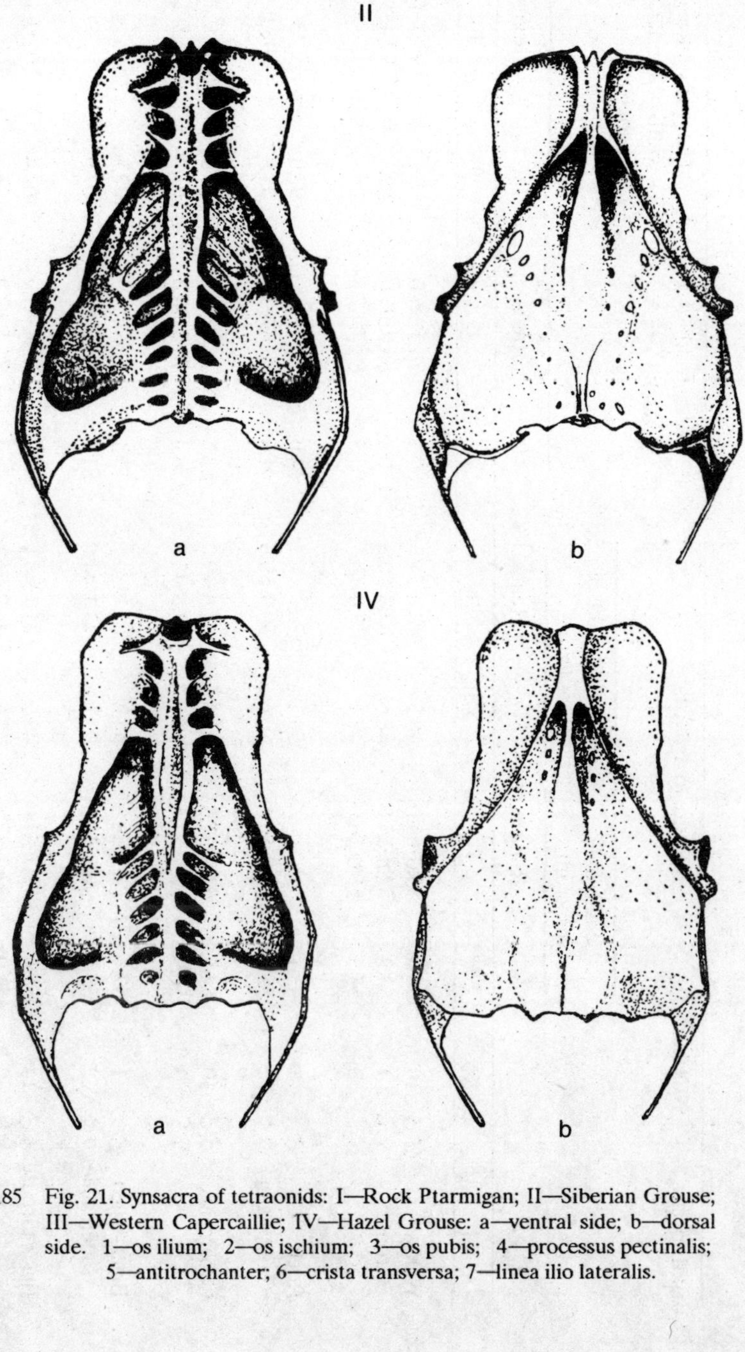

184-185 Fig. 21. Synsacra of tetraonids: I—Rock Ptarmigan; II—Siberian Grouse; III—Western Capercaillie; IV—Hazel Grouse: a—ventral side; b—dorsal side. 1—os ilium; 2—os ischium; 3—os pubis; 4—processus pectinalis; 5—antitrochanter; 6—crista transversa; 7—linea ilio lateralis.

Table 5. Indices of the synsacrum

Bird species		Number	Ratio to truncus, %				Dorsal width to length, %		Ratio of ventral width of synsacrum to dorsal, %	
			Length of synsacrum		Dorsal width of synsacrum					
			M	Range	M	Range	M	Range	M	Range
1		2	3	4	5	6	7	8	9	10
Willow	♂♂	10	56.5	53.3–65.0	48.5	46.0–54.0	87.1	76.5–94.0	97.2	95.7–100.0
Grouse	♀♀	4	53.9	52.8–55.0	48.8	47.3–50.3	90.3	86.0–95.0	99.1	97.4–102.0
	o*	3	55.5	55.0–56.0	47.7	47.0–48.5	85.5	83.5–87.5	—	—
Rock	♂	1	53.1	—	50.1	—	94.0	—	93.0	—
Ptarmigan	♀♀	2	55.1	54.0–56.1	50.6	50.3–50.8	91.6	89.8–93.5	95.8	93.0–98.5
Capercaillie	♂♂	10	54.5	51.3–57.5	45.5	42.8–47.0	82.1	76.0–92.3	110.9	107.5–122.0
	♀♀	3	56.9	56.4–57.5	44.7	43.8–45.3	78.4	76.0–80.3	107.3	106.5–109.0
Black-billed	♂	1	57.4	—	44.3	—	77.1	—	109.5	—
Capercaillie	♀♀	2	53.5	50.5–56.5	43.6	43.1–44.0	81.5	78.0–85.0	107.0	—
Black Grouse	♂♂	14	55.4	53.7–57.0	43.2	41.3–44.4	76.2	72.4–81.4	104.4	102.0–109.0
	♀♀	9	56.0	53.7–57.8	42.8	41.2–44.1	75.5	71.5–83.2	103.1	100.4–107.0

1		2	3	4	5	6	7	8	9	10
Siberian	♂	1	58.0	—	46.4	—	79.8	—	—	—
Spruce	♀♀	2	55.7	55.4–56.0	45.6	44.8–46.4	82.2	81.0–83.3	108.3	105.5–111.0
Grouse	∘	3	54.3	50.8–56.5	46.1	46.0–46.2	83.2	81.4–85.4	107.2	106.0–109.4
Hazel	♂♂	11	55.9	53.8–60.0	47.1	43.9–49.3	84.5	73.7–90.0	102.0	98.8–105.0
Grouse	♀♀	8	55.7	53.8–56.4	48.2	45.0–50.7	86.4	81.2–92.0	102.1	101.0–102.5
Common	♂	15	51.2	48.6–54.3	28.6	25.7–30.7	55.8	49.9–62.7	85.7	81.0–92.5
Quail	♀	1	49.8	—	27.4	—	55.0	—	77.8	—
Common	♂	1	55.0	—	39.3	—	72.3	—	—	—
Partridge	♀♀	6	55.8	53.4–57.4	38.7	36.0–41.0	71.6	65.8–78.3	95.3	93.7–96.8
	∘	8	56.3	53.2–58.2	40.1	37.2–42.8	71.3	65.2–78.6	93.0	84.5–100.0
Daurian	♂♂	2	51.6	51.4–51.7	36.7	36.0–37.4	68.8	67.5–70.2	100.2	—
Partridge	♀♀	4	56.1	54.0–58.5	40.3	37.0–42.3	72.4	67.5–77.2	94.3	93.2–95.3
	∘	2	54.5	52.2–56.8	39.8	39.2–40.3	72.5	70.8–75.2	96.0	—
Chukar	♂♂	21	57.3	50.7–61.7	38.6	35.0–40.8	67.7	60.9–72.0	91.6	85.0–98.0
	♀♀	12	57.5	55.8–59.2	38.2	36.2–40.5	67.8	64.0–73.0	91.4	89.4–93.5
See-see	♂♂	2	52.5	50.5–54.5	41.8	40.8–42.7	79.6	78.8–80.3	96.0	—
Partridge	∘	2	54.3	54.0–54.5	38.0	37.3–38.7	75.9	71.6–80.3	—	—

(continued)

1		2	3	4	5	6	7	8	9	10
Black	♂♂	4	54.7	53.2–56.2	32.5	31.6–33.3	60.4	60.0–60.7	95.0	—
Francolin	♀♀	5	56.3	54.5–58.5	33.9	32.2–35.0	60.1	57.0–63.5	96.2	95.6–96.3
Himalayan	♂♂	8	58.1	55.0–61.5	41.5	39.0–43.5	72.1	66.8–78.4	87.7	83.5–95.0
Snowcock	♀♀	7	56.9	55.6–59.9	42.3	39.6–44.7	75.2	69.5–80.5	88.5	83.7–97.5
Altai	♂♂	2	58.5	58.2–58.7	44.7	43.7–45.7	73.8	70.3–77.3	86.2	83.8–88.6
Snowcock	♀	1	56.1	—	43.2	—	77.5	—	85.2	—
Caucasian	♀	1	58.4	—	44.0	—	75.3	—	—	—
Snowcock	o	1	57.7	—	43.0	—	73.7	—	—	—
Common	♂♂	11	56.0	51.8–60.5	36.1	34.3–39.6	64.6	63.5–67.0	92.3	89.0–96.3
Pheasant	♀♀	12	56.3	51.3–62.5	35.7	33.3–37.7	63.9	58.7–69.3	92.7	86.9–95.6

* Here and in other tables 'o' denotes "sex not known"—Editors.

The structure of the preacetabulum of phasianids differs markedly from that of tetraonids. The crista transversa separates the proximal part of the ilium and is considerably higher in phasianids; the lobes which extend downward from it are very wide, a necessary condition for the attachment of the powerful musculature. Compared to the pelvic girdle of tetraonids, the processus pectinalis is better developed in phasianids; its
187 larger size perhaps restricts the sideward shifting of the femur, a characteristic of the phasianid species.

The index of the dorsal width of the synsacrum (in relation to the truncus) varies much more in phasianids than in tetraonids (27.4-44. 7%), and the ratio of the dorsal width of the synsacrum to its length, 55-79.6%. The narrowest synsacrum (between the antitrochanters) is seen in quail followed by francolins and pheasants, and is broadest in the See-see, Gray and Daurian partridges. In phasianids the ventral width of the synsacrum is almost always less than the dorsal (81-100% of the latter). The pelvic girdle of See-see Partridges is unique. Although it differs little from that of Daurian and Gray partridges in the width of the postacetabulum, in the relative width of the proximal part of the ilium it greatly resembles quail and snowcocks, particularly Altai Snowcocks. Both francolins and pheasants have a narrow synsacrum but in pheasants the preacetabulum is markedly wider. The intense development of the ilia, with elongated and highly developed lobes projecting far beyond the caudal end of the ischia and broader than the caudal part of the pars renalis, in the extremely narrow and high synsacra of quail merits attention (Fig. 22).

It must be emphasized that in general the structure of the pelvic girdle (unlike the pectoral girdle) clearly differentiates the two families. The finer details which distinguish the species throw light on the conditions of their speciation.

Differences in the structure of the skeletal components of the hind limbs in members of the two families are less than those in the synsacrum, although some differences do exist in the proportions of the different bones and in their shapes.

Phasianids have relatively longer limbs and, usually, a long tarsometatarsus (Tables 6 and 7). Snowcocks are an exception. As a consequence their body is set higher, which facilitates movement on land using longer strides. On the other hand, the low-slung body of the tetraonids facilitates free movement from branch to branch, but somewhat restricts movement on the ground.

Compared to other tetraonids, Hazel Grouse have a relatively high-set body and relatively longer legs. These features enable easy movement in the forest, especially in a cluttered forest.

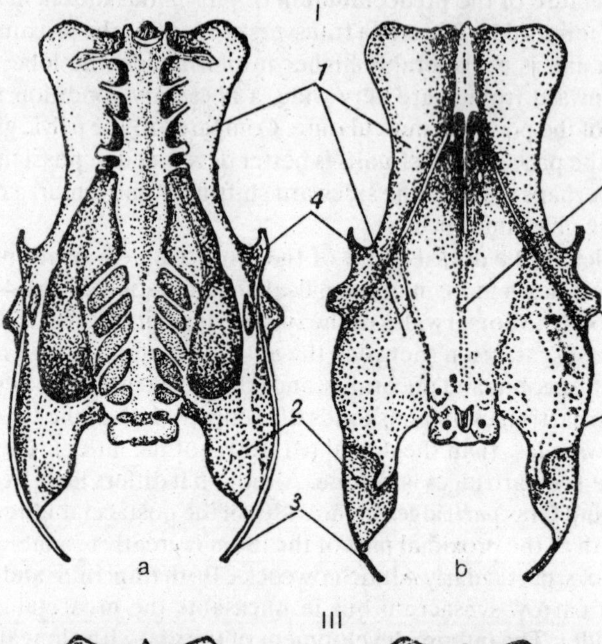
I
1
4
6
5
7
2
3
a
b
III
a
b

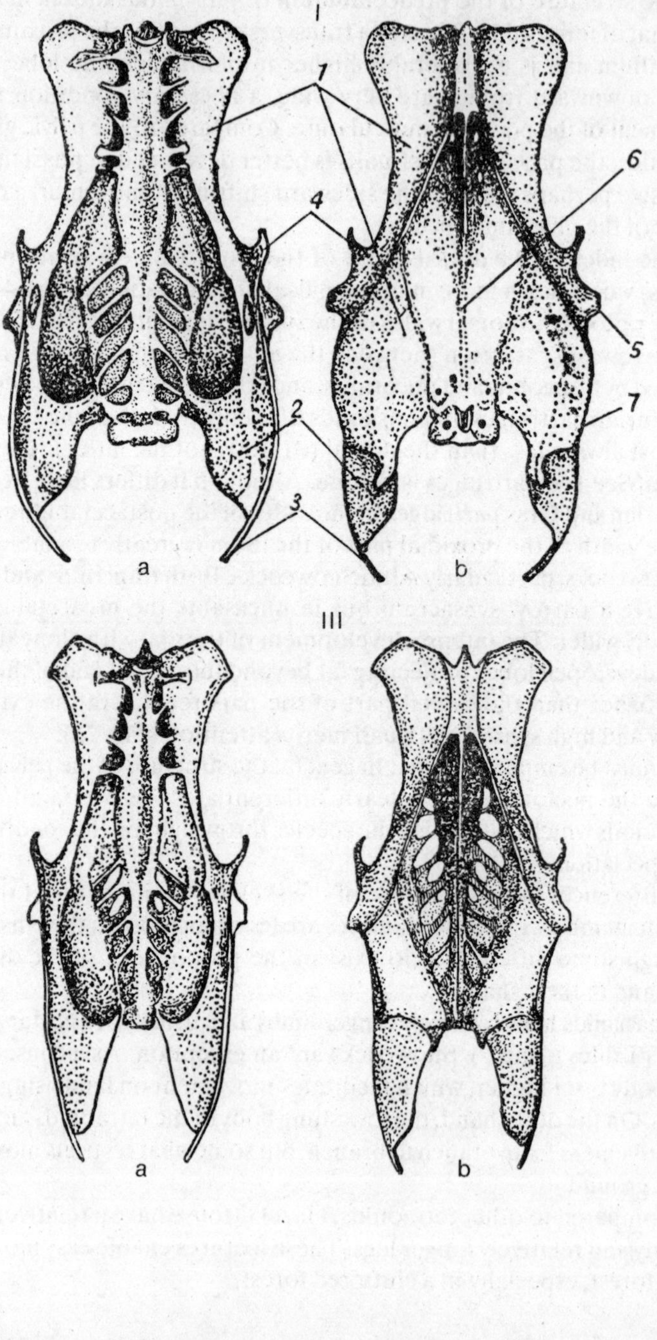

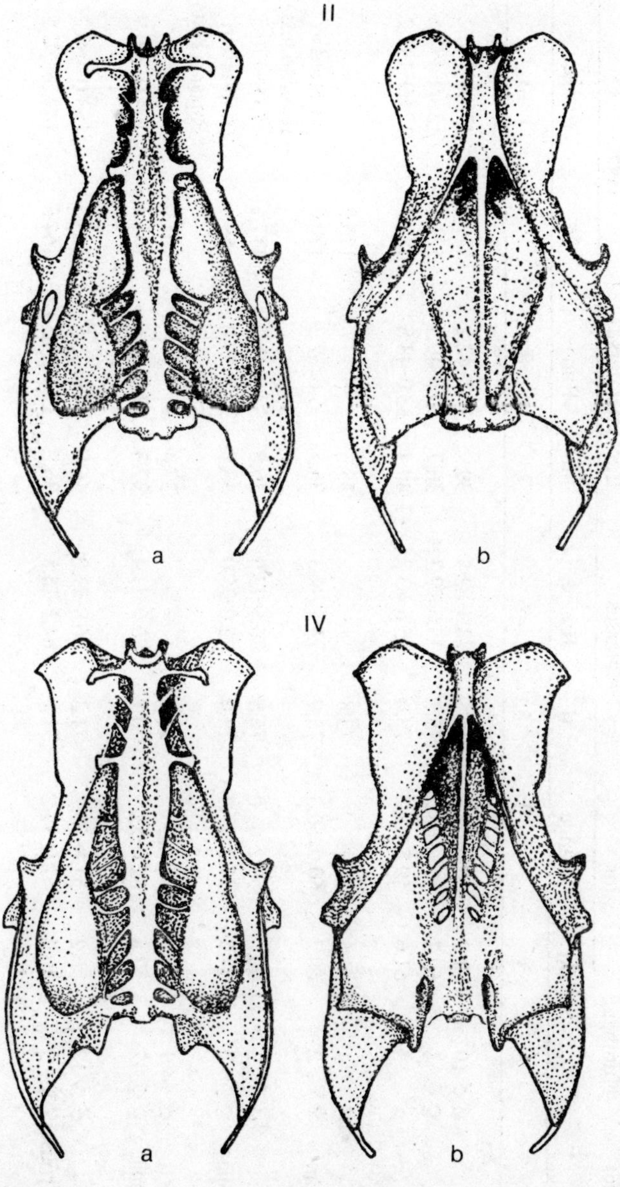

188–189 Fig. 22. Synsacra of phasianids: I—Himalayan Snowcock; II—Gray Partridge; III—Common Quail; IV—See-see Partridge: a—ventral view; b—dorsal view. 1—os ilium; 2—os ischium; 3—os pubis; 4—processus pectinalis; 5—antitrochanter; 6—crista transversa; 7—linea ilio lateralis.

190-191

Table 6. Ratio of skeletal components of the hind limbs to truncus, %

Bird species	Number	Femur		Tibiotarsus		Tarsometatarsus		Total	
		M	Range	M	Range	M	Range	M	Range
1	2	3	4	5	6	7	8	9	10
Willow	♂♂ 10	61.7	60.0–63.5	77.5	73.4–82.5	39.9	37.7–42.2	179.3	173.0–188.2
Grouse	♀♀ 4	60.3	57.6–63.2	75.7	75.0–79.3	36.7	30.8–40.5	172.6	158.9–181.5
	o 3	61.6	59.8–65.2	78.6	75.6–83.2	40.4	38.0–43.8	180.7	174.9–192.2
Rock	♂ 1	61.5	–	73.8	–	35.3	–	170.0	–
Ptarmigan	♀♀ 2	60.2	60.0–60.3	73.9	73.5–74.3	36.5	35.6–37.3	170.6	(omitted in Russian text)
Western	♂♂ 3	60.3	59.2–61.0	76.8	76.5–77.0	39.8	38.8–40.7	177.2	175.9–178.5
Capercaillie	♀♀ 3	57.5	57.2–57.7	73.0	72.0–75.0	38.9	38.8–39.0	169.4	(omitted in Russian text)
Black-billed	♂ 1	58.3	–	73.1	–	37.4	–	167.0	–
Capercaillie	♀♀ 2	57.5	57.0–57.9	70.7	70.0–71.4	37.3	37.0–37.6	165.4	164.0–166.9
Northern	♂♂ 12	56.2	51.8–58.2	71.3	66.8–77.0	36.8	34.0–38.5	163.4	153.0–169.2
Black Grouse	♀♀ 9	56.4	52.0–59.2	71.7	64.3–75.5	36.8	33.0–39.5	164.8	149.3–172.5
Siberian	♂ 1	59.5	–	74.0	–	38.4	–	171.9	–
Spruce	♀♀ 2	59.4	58.3–60.5	72.8	72.6–73.0	37.1	36.9–37.3	169.3	168.6–170.0
Grouse	o 3	60.9	57.3–62.6	76.3	71.0–81.5	36.8	36.4–37.2	174.5	165.0–181.3

1		2	3	4	5	6	7	8	9	10
Hazel	♂♂	10	62.3	59.4–66.2	81.3	78.2–84.0	41.7	36.4–45.1	185.3	177.7–190.0
Grouse	♀♀	8	62.8	61.2–65.5	82.2	80.0–84.0	42.7	41.2–43.8	187.7	183.4–193.3
Common	♂♂	17	62.3	59.3–69.2	76.3	70.2–83.0	45.9	41.8–48.4	184.2	176.7–192.4
Quail	♀	1	59.2	–	73.8	–	45.6	–	178.6	–
Gray	♂	1	68.5	–	85.0	–	51.8	–	205.3	–
Partridge	♀♀	5	66.1	64.2–68.4	85.1	81.5–89.0	50.0	48.7–51.3	201.0	196.7–203.0
	o	8	67.5	62.2–73.0	89.2	78.0–91.3	50.2	47.5–53.8	202.0	187.7–218.1
Daurian	♂♂	2	66.1	66.0–66.2	81.3	80.2–82.3	48.8	48.3–49.3	196.2	194.7–197.6
Partridge	♀♀	5	69.6	67.2–73.3	87.1	83.0–93.2	50.9	48.1–57.3	207.6	198.7–222.9
	o	2	67.1	66.8–67.3	85.8	85.2–88.0	51.6	50.8–52.5	204.6	202.8–204.6
Chukar	♂♂	16	66.1	63.3–69.6	89.1	84.5–95.5	51.1	47.0–55.4	206.7	198.5–220.5
	♀♀	9	65.3	64.5–68.0	87.9	79.0–90.9	49.9	48.3–51.2	202.0	188.5–209.1
See-see	♂♂	2	61.6	60.2–63.0	88.2	87.5–88.8	48.9	47.5–50.3	198.6	195.2–202.1
Partridge	o	2	60.3	–	85.0	82.5–87.5	47.8	47.5–48.2	193.2	191.0–195.3
Black	♂♂	4	66.6	64.4–68.5	88.3	86.8–90.8	57.4	56.2–58.8	211.1	208.4–218.1
Francolin	♀♀	3	66.7	65.7–67.1	90.4	89.8–91.0	59.7	59.3–60.3	216.8	215.8–217.4

(continued)

1		2	3	4	5	6	7	8	9	10
Himalayan Snowcock	♂♂	9	66.5	63.2–71.0	88.6	84.5–93.8	45.8	43.7–51.0	200.9	191.1–214.0
	♀♀	7	64.8	62.2–67.5	87.2	84.6–90.0	44.6	43.9–45.5	196.6	191.8–203.0
Altai Snow-cock	♂♂	2	70.8	69.9–71.1	90.1	89.0–91.2	44.9	42.8–47.0	205.8	201.7–209.9
	♀	1	61.3	–	79.5	–	41.0	–	181.8	–
Caucasian Snowcock	♀	1	62.4	–	82.0	–	42.9	–	187.0	–
Common Pheasant	♂♂	8	66.3	64.1–72.7	90.1	85.1–96.7	59.9	56.6–64.7	216.8	206.8–234.1
	♀♀	12	66.8	62.0–69.3	87.9	83.0–92.1	59.2	55.2–62.0	214.0	202.1–222.4

Differences in the structure of the limb bones are reflected in the
192 degree of heaviness of the bones, their relative sizes and structural details. In larger species (capercaillies and snowcocks) the skeletal bones are more massive, they are more slender in smaller species.

Small differences are observed in the structure of the femur but they are quite characteristic of each family; they arise from differences in the pelvic structure. In Tetraonidae, the femur is straighter, while in Phasianidae it is slightly bent and even somewhat twisted. This is greatly reflected in the disposition of the entire limb. In tetraonids the femur is set off from the body axis at a somewhat larger angle than in phasianids.

In such a position, the knee joint in tetraonids is shifted laterally farther from the body, and the tibiotarsus forms an acute angle with the trunk and is tucked under it. Only because of this position of the appendicular skeleton can birds with a broad synsacrum move on branches, stepping in tandem. In tetraonids (especially Northern Black Grouse) the significant sideward displacement of the femur has made it more mobile sideways but its movements in the body plane are restricted. Hence tetraonids (possibly not all) can move their knee joints rather freely sideways.

In phasianids movements of the femur are limited to a single plane, but such movements toward and away from the body are very free. The femur of ground birds is much more mobile. The maximum angle of movement is observed in quail (60°), and Chukar and pheasants (45°), and the maximum angles of movement with respect to the body axis are 110° and 105°, respectively. The highly mobile femur in quail and its maximum movement backward widen the pace of this bird which has a low-slung body. The degree of freedom of femur movement in Altai Snowcocks is next to quail, with a maximum angle of movement of 40°, while in Himalayan Snowcocks and Willow Grouse it is 30°. The maximum angle of movement of the femur from the body in Willow Grouse is the same as in quail, i.e., 110°, that in Altai Snowcocks is 100°, and in Himalayan Snowcocks, 90°. The femur movements in forest tetraonids are more restricted; the maximum degree of movement is 20° in Hazel Grouse and 10° in Northern Black Grouse, but the angles of withdrawal from the body are 90° and 75° respectively. The restricted femur movements in snowcocks apparently provide more stability in these large birds. The greater femoral mobility of the Willow Grouse compared to forest tetraonids (and in particular the wider angle of withdrawal from the body) and the preservationing of the common characteristics of synsacral structure found in other
194 forest tetraonids, should be considered consequences of its change to a terrestrial life style.

Table 7. Ratio of length of skeletal components of hind limbs, %

Bird species	Number	Ratio to total length of limbs					
		Femur		Tibiotarsus		Tarsometatarsus	
		M	Range	M	Range	M	Range
1	2	3	4	5	6	7	8
Willow Grouse	♂♂ 10	34.4	33.7–35.3	43.3	42.4–44.0	22.3	21.7–22.7
	♀♀ 4	34.9	34.3–35.7	43.6	42.8–44.2	21.5	20.6–22.3
	∘ 3	34.2	34.2–34.4	43.4	43.2–43.8	22.4	21.8–22.7
Rock Ptarmigan	♂ 1	36.2	—	43.1	—	20.7	—
	♀♀ 2	35.2	34.8–35.6	43.5	43.4–43.5	21.3	21.6–21.0 [sic]
Western Capercaillie	♂♂ 4	33.9	33.4–34.2	43.3	42.7–43.7	22.8	22.4–22.9
	♀♀ 6	33.5	31.9–34.1	43.5	42.7–44.7	23.0	22.7–23.4
Black-billed Capercaillie	♂ 1	34.8	—	42.8	—	22.4	—
	♀♀ 2	34.4	34.3–34.5	43.0	—	22.6	22.5–22.7
Northern Black Grouse	♂♂ 11	34.5	33.7–35.4	43.1	42.2–43.6	22.4	21.3–22.6
	♀♀ 11	34.5	34.1–35.0	43.2	42.9–43.6	22.3	22.0–22.5
Siberian Spruce Grouse	♂ 1	34.0	—	43.4	—	22.6	—
	♀♀ 2	35.0	34.3–35.7	43.0	42.6–43.3	22.0	21.7–22.3

1		2	3	4	5	6	7	8
Hazel Grouse	♂♂	10	33.4	32.5–33.8	44.0	43.4–45.6	22.6	20.5–23.2
	♀♀	8	33.3	32.5–33.8	43.7	43.3–44.1	23.0	22.5–23.6
Common Quail	♂♂	15	33.6	33.0–34.6	42.0	40.2–42.3	24.9	24.3–25.7
	♀	1	33.2	—	41.3	—	25.5	—
Japanese Quail	♂	1	34.5	—	41.6	—	23.9	—
Gray Partridge	♂	1	33.2	—	41.7	—	25.1	—
	♀♀	5	33.1	32.6–34.0	41.7	41.5–42.2	25.0	24.5–25.5
	o	7	33.5	33.2–33.9	41.6	41.4–42.0	24.9	24.7–25.4
Daurian Partridge	♂♂	2	33.7	33.5–33.9	41.3	41.2–41.5	25.0	24.9–25.0
	♀♀	4	33.3	32.6–33.8	41.7	41.2–42.3	25.0	24.2–25.5
	o	2	32.8	32.6–33.2	41.9	41.9–42.0	25.3	25.0–25.5
Chukar	♂♂	17	32.2	31.4–33.0	43.5	42.2–43.8	24.3	23.8–25.3
	♀♀	7	32.2	31.8–32.7	43.1	42.0–43.8	24.7	24.3–26.2
See-see Partridge	♂♂	2	31.1	31.0–31.2	44.3	43.8–44.7	24.6	24.3–25.0
	o	2	31.6	31.6–31.6	43.2	43.1–43.2	25.2	25.2–25.3
Black Francolin	♂♂	5	31.3	31.0–31.6	41.6	41.5–41.8	27.1	26.6–27.4
	♀♀	3	30.7	30.5–31.0	41.8	41.6–42.0	27.5	27.3–27.8

(continued)

1	2	3	4	5	6	7	8
Himalayan ♂♂	9	33.1	32.6–33.7	44.2	43.8–44.5	22.7	22.2–23.2
Snowcock ♀♀	7	33.2	32.6–33.9	44.3	43.9–44.5	22.5	21.9–23.0
Altai Snowcock ♂♂	2	34.4	34.1–34.6	43.9	43.6–44.1	22.8	21.3–22.3
♀	1	33.7	—	43.8	—	22.5	—
Caucasian Snowcock ♂	1	33.6	—	44.1	—	22.3	—
Common ♂♂	10	30.9	30.4–31.4	41.4	41.1–41.9	27.7	27.0–28.1
Pheasant ♀♀	10	31.1	30.7–31.5	41.3	40.7–41.8	27.6	27.2–27.9

Differences in the structure of the tibia have been observed in relative measurements (length and thickness) and in the structural details of its proximal end. It is relatively long in tetraonids and mountain phasianids, the thickness depending on the size of the bird. Differences in the structure of the proximal end of the bone depend on the degree of development of the ridges: crista tibialis anterior, crista tibialis posterior, and crista patellaris. The sizes of these ridges depend on the size of the muscles (M. tibialis anticus and M. flexores perforati) attached to them. These ridges are especially large in francolins, pheasants, and Gray and Daurian partridges; in Siberian Spruce Grouse, Willow Grouse, Rock Ptarmigans, and Chukar, the proximal end of the tibia is very smooth (Fig. 23).

The structure of the tarsometatarsus also varies in the species I studied. In all these galliform birds the calcaneal tubercle is well developed in the proximal part on the posterior side of the hypotarsus. According to Shtegman (1958), the hypotarsus is better developed in arboreal forms but is very smooth in ground birds. Its shape and size vary greatly in the species I studied; no direct relationship could be established between its size and its ground or arboreal-ground life style.

The hypotarsus in capercaillies is especially large. Its high protuberance is divided into outer and inner ridges by a deep furrow. A high crest extends from the inner ridge along the posterior side. The same type of
195 structure is also found in other tetraonids, but the crest in these cases is like a thin, narrow plate bordering a large opening (Fig. 24).

Among phasianids the bony crest of the calcaneal tubercle is absent in snowcocks, See-see Partridges, quails, and francolins. In Himalayan Snowcocks this tubercle is high and broad and divided by two deep longitudinal furrows into three, not two, ridges. There is a small longitudinal ridge on the posterior side of the tarsometatarsus. In other species two longitudinal ridges are distinctly demarcated. The calcaneal tubercle is particularly prominent in francolins; its inner ridge is longer than the outer and in structure the tubercle is intermediate between species with

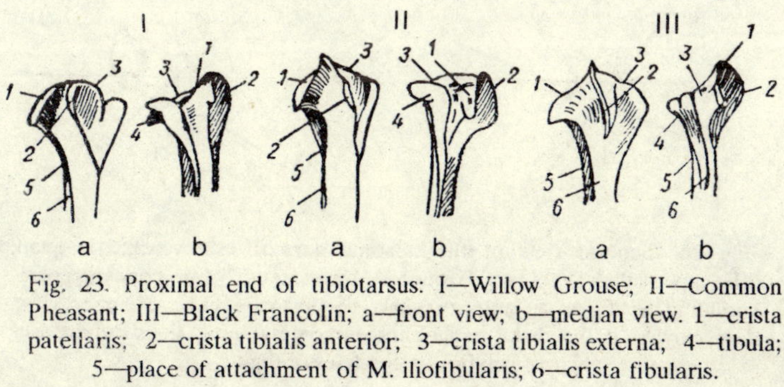

194 Fig. 23. Proximal end of tibiotarsus: I—Willow Grouse; II—Common Pheasant; III—Black Francolin; a—front view; b—median view. 1—crista patellaris; 2—crista tibialis anterior; 3—crista tibialis externa; 4—tibula; 5—place of attachment of M. iliofibularis; 6—crista fibularis.

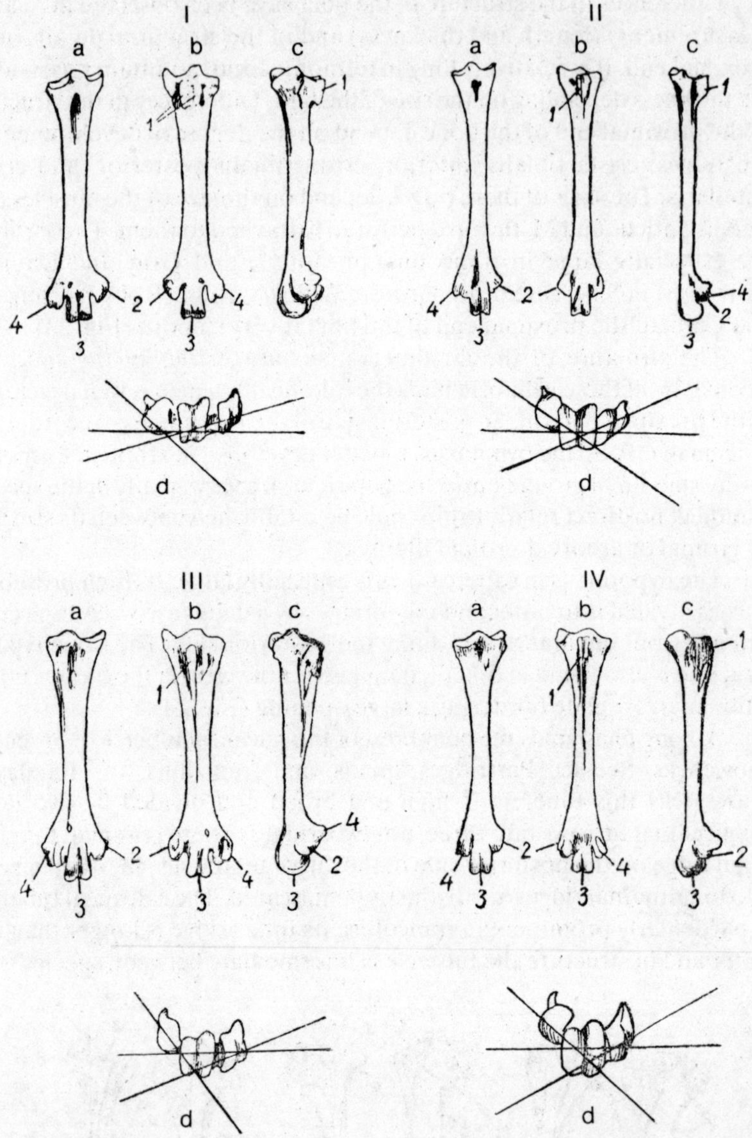

196 Fig. 24. General view of the tarsometatarsus: I—snowcock; II—quail; III—capercaillie; IV—Hazel Grouse. 1—hypotarsus; 2, 3, 4—condyles corresponding to digits; a—anterior view; b—posterior view; c—inner view; d—condyles at the distal end of the tarsometatarsus; lines depict axes of rotation of condyles of digits.

and without a ridge. The bony crest on the calcaneal tubercle is highest in pheasants, and only slightly smaller than in capercaillies. The posterior part of the tarsometatarsus in all species has longitudinal furrows for the attachment of the tendons of M. flexores digitorum. The notable variability in the structure of the hypotarsus and tarsometatarsus indicates the diverse dispositions of the muscle tendons which move the digits. Their function will be understood only after more detailed study.

The distal part of the tarsometatarsus, in particular the arrangement and shape of the articulate processes of the phalanges determining their position, is also highly variable. It has been observed that in birds leading an arboreal life the articulate process of the third digit is just as long as the second and fourth, while in ground birds it is much longer (Shtegman, 1958). In all the galliform birds I examined, both ground and ground-arboreal, the calcaneal tubercle of the third digit is much longer than that of the second and fourth (which lie at the same level) or the calcaneal tubercle of the fourth digit is smaller. The disposition and shape of the articulate processes greatly influence the separation of the digits. In some species I examined, these processes are wide set, but their axes of rotation diverge less than in species with close-set processes. In francolins and snowcocks the digits are comparatively close set and, despite the articulate processes being wide set, their axes of rotation diverge less.

The digits in forest-dwelling Galliformes such as capercaillies, Hazel, Northern Black, and Spruce grouse diverge greatly, and their second and fourth digits are wide set. The wide divergence of the digits in forest tetraonids is readily explained by the variety of functions they perform, and will be discussed below.

Weight indices have been used to characterize the musculature of the
197 hind limbs, namely the total weight of the musculature and the weight of individual muscles (Table 8). These data help characterize the degree of development of the groups of muscles performing similar functions and depict graphically the ratio of the weight of individual muscles in different species. It should be mentioned that in ornithological literature, weight indices of the muscles of locomotor organs have been widely used to compare their development in various species (Gladkov, 1949; Shtegman, 1950, 1952, 1953, 1954; Yakobi, 1960; Hartman, 1961; Greenewaldt, 1962; Kuz'mina, 1962; Shvarts, Smirnov, and Dobrinskii, 1968).

The muscle weight index is quite adequate to evaluate muscle power as a first approximation, more so since Kurochkin (1968) experimentally demonstrated the actual correlation between the relative weight and
198 relative volume of muscles: this correlation could be used to evaluate their relative power.

Table 8. Indices of hind limb musculature

Bird species	Number	Ratio of total weight of leg muscles to body weight, %	Number	Ratio of weight of individual groups of muscles to weight of leg musculature, %								
				Femur		Tibiotarsus		Tarsometatarsus			Phalanges	
				F	E	F	E	F	E	R	F	E
1	2	3	4	5	6	7	8	9	10	11	12	13
Willow Grouse	14	10.5	4	15.5	28.6	39.1	30.7	6.6	20.0	4.1	7.5	2.0
Rock·Ptarmigan	5	8.3	2	16.7	29.4	38.4	27.3	7.7	20.3	4.4	8.2	1.8
Capercaillie	8	9.8	2	15.6	27.8	38.6	28.5	7.2	21.5	6.0	10.2	3.0
Black Grouse	22	8.9	4	15.0	26.1	40.1	29.4	6.5	23.4	5.0	12.9	1.5
Siberian Spruce Grouse	4	5.5	2	16.7	25.5	39.3	28.6	5.3	25.0	5.2	12.3	2.2
Hazel Grouse	19	7.7	2	12.2	24.2	44.4	27.2	7.0	29.0	5.0	15.3	1.7
Common Quail	20	11.9	4	9.8	22.3	37.8	36.9	5.2	24.7	5.5	10.5	1.3
Common Partridge	11	12.7	2	13.2	25.1	40.0	32.6	7.5	23.2	4.0	10.3	2.2

1	2	3	4	5	6	7	8	9	10	11	12	13
Daurian Partridge	8	12.2	3	13.9	24.3	38.3	32.6	6.8	23.2	4.8	10.3	1.7
Chukar	31	14.7	4	13.3	24.5	36.1	35.9	6.8	21.3	5.6	8.7	1.7
See-see Partridge	7	12.7	1	14.3	27.2	39.1	33.0	6.6	22.4	3.7	6.4	1.4
Black Francolin	8	17.5	3	11.4	20.1	39.1	40.4	7.8	26.2	4.1	10.5	1.6
Himalayan Snowcock	20	18.6	3	15.2	23.5	38.3	33.7	5.4	23.3	5.8	10.5	1.0
Altai Snowcock	1	19.0	2	14.3	25.4	37.0	35.5	5.5	20.7	4.3	10.2	1.1
Common Pheasant	18	15.7	2	11.1	23.2	45.8	26.1	6.4	31.2	4.9	11.0	1.5

Note: F = flexor; E = extensor; R = rotator—General Editor.

Compared to tetraonids, all phasianids have rather highly developed leg musculature, particularly the snowcocks, francolins, and pheasants. Of all the birds studied, the best developed muscles were those moving the tibiotarsus, namely the M. ischioflexorius and M. caudatus ilioflexorius. It must be conceded that many flexor muscles of the tibiotarsus serve multiple functions, acting simultaneously either as extensors of the femur (M. iliofibularis, M. caudatus ilioflexorius, and M. ischioflexorius) or as extensors of the tarsometatarsus (M. gastrocnemius and M. peroneus sublimus). In other words, these muscles specifically contribute to the movement of the limb. The development of the musculature of the digits varies significantly, especially of the flexor muscles (the flexores perforati group). Compared to other limb muscles, the M. flexores perforati group are very highly developed in Hazel and Spruce grouse. However, the overall functional characteristics of the musculature presented in Table 8 do not provide a comprehensive picture of the degree of development of individual limb muscles. In fact the picture derived from the ratio of measurements of muscles performing similar functions is rather disparate.

For a better, more cohesive, picture of the relative development of the musculature of the hind limbs, the bar diagrams in Figs. 25 and 26 depict the ratio of individual muscle groups in percentages.

The almost identical structure of the leg musculature in taxonomically close species is at once striking—in Gray and Daurian partridges among

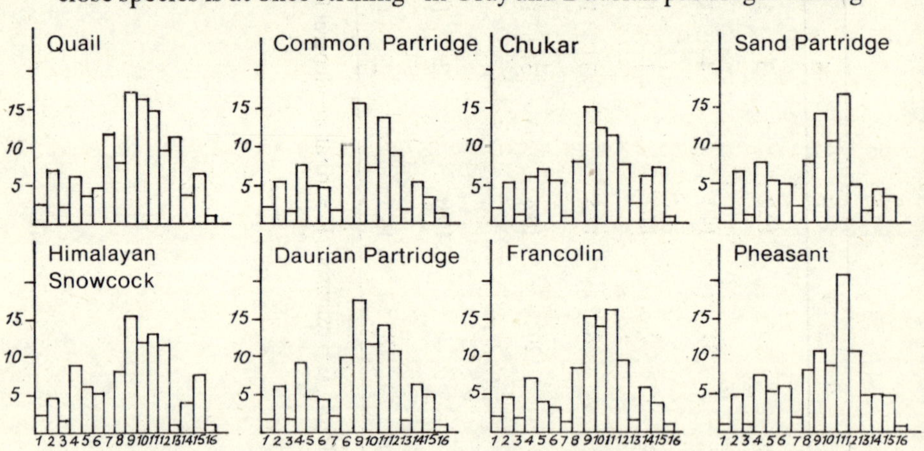

Fig. 25. Bar diagrams showing weight ratios of muscles of hind limbs in phasianids: 1—M. caudatus iliofemoralis; 2—M. pubi ischiofemoralis; 3—M. ischiofemoralis; 4—M. trochanterius; 5—M. sartorius; 6—M. iliofibularis; 7—M. ischioflesorius; 8—M. caudatus ilioflexorius; 9—M. iliotbialis; 10—M. femorotibialis; 11—M. gastrocnemius and peroneus sublimus; 12—M. flexores perforati; 13—M. extensor digitorum; 14—M. tibialis; 15—M. peroneus profundus; 16—M. ambiens.

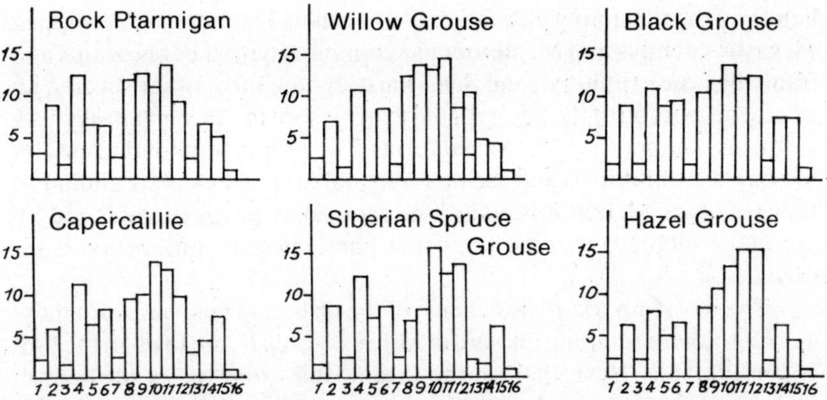

Fig. 26. Diagrams showing weight ratios of the hind limb muscles in tetraonids. Legend as in Fig. 25.

phasianids, and in Willow Grouse and Rock Ptarmigans among tetraonids. The greatest differences are seen in the members of the families which differ markedly in their life styles. Compared to phasianids, the more intensely developed M. trochanterius is typical of all tetraonids (except Hazel Grouses). Differences between the two families with respect to the size ratios of M. tibiotarsus extensor are rather marked. In tetraonids, M. femorotibialis is the largest muscle of the hind limbs while in phasianids, M. iliotibialis is always larger than M. femorotibialis. The ptarmigans in which these two muscles are almost equal occupy an intermediate position between forest tetraonids and phasianids, which might be the result of their terrestrial life style. The musculature of the hind limbs in the Hazel Grouse and Siberian Spruce Grouse is very poorly developed even though the individual muscles are differently developed in the two species. In Hazel Grouse M. gastrocnemius, M. peroneus sublimus, and M. flexores perforati, are best developed, while M. femorotibialis is the largest in spruce grouses, as in other tetraonids. In the development of individual groups of hind limb muscles, the spruce grouse are closer to Northern Black Grouse and even capercaillies than they are to Hazel Grouse.

200 In the various species of phasianids I examined, the development of the hind limb musculature varied more than in tetraonids. This could be due to the diversity in ecological conditions and to the gaits of these birds. In most birds M. iliotibialis is the largest muscle of the hind limbs. Common Pheasants and See-see Partridges are exceptions in which M. gastrocnemius and M. peroneus sublimus (the flexor of the tibiotarsus and the extensor of the tarsometatarsus) are the largest. These muscles are almost equally developed in the Black Francolin, i.e., it occupies an intermediate position between pheasants and Sand Partridges on the one

hand and the remaining phasianids on the other. The intensely developed M. gastrocnemius and M. pleroneus sublimus is typical of pheasants and francolins (fast runners), and Sand Partridges. Unlike other species, an inverse proportion of the flexors of the femur is seen in Chukar Partridge—M. sartorius is larger than M. trochanterius. The proportions of the muscles moving the tibiotarsus and the tarsometatarsus in Chukar are similar to those of quail: M. femorotibialis is larger than M. gastrocnemius and M. peroneus sublimus together; in other phasianids the proportion is reversed.

The most important movements of the phasianid hind limbs are those associated with scraping the soil, in which, obviously, the load on the toes is considerable. Nevertheless, the flexors and extensors of the toes in Hazel, Spruce, and Northern Black Grouse are better developed than in phasianids. But, on the whole, the leg muscles are relatively better developed in phasianids.

The relative development of individual muscles compared to the musculature of the whole limb explains quite well the gait of the birds; for example, M. gastrocnemius and M. peroneus sublimus are better developed in good runners such as pheasants and francolins, while in tetraonids and snowcocks, which have a less mobile femur, M. iliotrochanterius is better developed.

In describing the appendicular skeleton, it was mentioned that the development of crests on the proximal end of the tibiotarsus depends on the size of the muscles attached to it. The crests—crista patellaris and crista tibialis anterior—are especially well developed in francolins, pheasants, snowcocks, Gray and Daurian partridges—birds with strong muscles (M. flexores perforati and M. tibialis anticus). The very smooth head of the tibitarsus in Spruce and Willow grouse, and Rock Ptarmigans can be explained by the relatively poor development of the corresponding muscles.

The greatest differences in the hind limbs of these birds are seen in the structure of their feet (length and integument of the tarsometatarsus,
201 length and position of toes, their integument, and structure of the claws). This is quite understandable since the functions of the feet, the most mobile part of the limbs, are particularly diverse in birds living under diverse conditions.

The structure of the tarsometatarsus is a good taxonomic feature for separating these two families. In tetraonids, the tarsometatarsus is feathered to the base of the toes or to 2/3 its length, while in phasianids it is completely covered with scutes. In all the phasianids I examined the anterior region of the tarsometatarsus was covered with two rows of scutes. The covering on the posterior region differs widely in different species of phasianids. The feathered tarsometatarsus in tetraonids should be considered an adaptive feature which arose in response to severe climatic conditions. The tarsometatarsus and even the toes are especially well-feathered in Willow Grouse and Rock Ptarmigans which have penetrated more northward than other galliform birds. That the feathering of the tarsometatarsus is an adaptation to prolonged severe winters is demonstrated by the following observations. Among Hazel Grouses which I examined from the collection of the Zoological Institute, Academy of Sciences of the USSR, it is better feathered in birds from eastern Siberia. The long feathers in the specimens collected in winter from this region cover not only the bare surface of the tarsometatarsus but the toes to some extent. In the European forms the tips of these feathers reach only the base of the toes. The feathers of the tarsometatarsus are very long and dense in Black-billed Capercaillies in winter and cover the toes in such a way that only the claws are visible externally, while in Western Capercaillies the feathers cover no more than half the toes.

Seasonal changes in the integument of the feet and the shedding of claws are typical of tetraonids. In forest species the toes are covered with transverse scutes from above with a row of more or less rounded scales on the sides. In winter, a second row of horny scutes is seen under the lateral scutes in the form of narrow plates forming a fringe that is shed in spring; this fringe grows again in the second half of summer.

In winter, the underside of the toes is covered with a thick horny layer divided into an inner homogeneous and an outer one consisting of pointed spines or more rounded tubercles. The structure of the toes of all forest tetraonids is similar, but the shape and relative size of the horny structures reflect distinct species differences.

The toes of capercaillies are broad and flat, their dorsal surface is covered with fairly narrow scutes, and the rounded lateral scales diverge almost horizontally from their tips. The horny fringe under the lateral
202 scutes consists of two rows of narrow plates set sideways and somewhat downward; at the same time, the longer scutes of the outer row are somewhat incurved. The underside of the toes bears soft pads of varying thickness covered by a horny tubercular layer to form a rough surface. The thickest pads are at the base of the toes, the distal phalange of each digit, at the base of the claw, and at the joints of the phalanges. In places where the pads are thick, the horny fringes do not reach the bottom layer; conversely, in between the pads they extend beyond the bottom layer and, when walking, their tips touch the ground and prevent slipping.

The feet of Hazel Grouse are identical in structure, but the lateral scutes are extremely small and hence the toes are relatively narrow. The horny fringe consists of very thin, short plates set somewhat downward in such a way that they are almost invisible from above. The arrangement of the pads is the same as in capercaillies. The structure of the toes in Northern Black Grouse is also similar to that of capercaillies, but the horny fringe is slightly shorter and set more downward and the pads on the underside relatively thick. Spruce Grouse exhibit the same characteristics as Hazel Grouse, but with longer fringes set strictly downward.

In winter the tarsometatarsus of forest tetraonids is feathered more intensely than during the warmer season. This protects the feet from frostbite and, where the feathers fully cover the toes, as in Black-billed Capercaillies, the size of the supporting area is increased, facilitating movement on snow. The spines and tubercles on the underside of the feet reduce the chance of slipping and facilitate the bird's movement on frozen branches and probably on the snow crust. The increased horny layer (a poor heat conductor) also protects the toes from the harmful consequences of low temperatures. It has been observed that the horny layer on the underside of the toes is more intensely developed in arctic birds than in birds living in more southern regions (Madsen and Wingstrand, 1958). Opinions differ on the functional significance of the horny fringe in forest tetraonids. Most authors (Formozov, 1934; Teplov, 1947; Volkov, 1968) favor the view that the chief function of the horny fringes is to facilitate movement on frozen branches and they have hardly any role in movement on snow. On the other hand, Potapov (1969), who studied the structure of the podotheca in capercaillies and Black Grouse, considered that the horny fringes increase the supporting surface of the feet in these tetraonids and facilitate movement on snow and, possibly, help in digging into snow. Only the horny tubercles prevent slipping on frozen branches. According
203 to Potapov, the fringe bordering the toes at the point where the pads are very thick may not reach the surface of the branches, which precludes the possibility of the horny fringes playing a role in the movement of birds on branches. A cross section of the toe taken from such a place seemingly fully supports this view, but it does not reflect the complete complex structure of the pads, which changes when the toes bend and stretch. Furthermore, the surface of a branch is not strictly flat but rounded, and the thinner the branch the more rounded it is. Perched on a branch the bird clasps it tightly, resting on the spines on the pads, while the horny fringes in between help its hold on a thin branch. This acquires special significance when the bird perches on very thin branches.

In moving on a thin branch the middle toe is held on top of the curved surface of the branch, while the second and fourth fall slightly below and support the foot from the sides, as do the horny fringes.

The horny fringes increase the supporting area of the feet to some extent (Table 3) and thus facilitate movements on snow. However, this function would seem to be only secondary and in species with a short horny fringe set downward (Hazel and Spruce grouse), even this is doubtful. That the movement of forest tetraonids on loose snow is difficult has been pointed out in a number of works (Teplov, 1947; Donaurov, 1947; Dul'keit, 1960); they avoid movement on snow. They forage for a short period on trees and if there is no snow crust, spend most of the time in snow holes. Thus it must be considered that the horny fringes, together with the pads, fulfil a number of functions that assist tetraonids in life under the extreme conditions of prolonged severe winters. Depending on the structure of these horny formations, their functional importance may vary from species to species.

The foot structure differs notably in Willow Grouse and Rock Ptarmigans. In winter, the toes are feathered in addition to the tarsometatarsus; dense feathering covers the entire foot and gives it the appearance of a hare's foot. On the upper side of the toes the horny sheaths are retained only on the distal phalanges of the second, third, and fourth digits; the remaining surface is covered with keratinized feather sheathes with deep openings. In summer the feathery cover of the digits wears off and piliform
204 feathers and elastic shafts of broken feathers are retained at places. By the end of summer stumps begin to appear. The underside of the digits of Willow Grouse also reflects seasonal changes. In winter it is covered with a thick horny layer with slightly convex tubercles. In summer, distinct tubercles are retained only on the distal phalanges. The characteristic structure of the feet of Willow Grouse that distinguish it from the other tetraonids can be explained by the specialized adaptations of these birds to the extremely severe climatic conditions and their terrestrial life style. Feathers on the entire foot protect it against frostbite and simultaneously enlarge the supporting area substantially, thereby making movement on snow much easier. A summer molt also has a favorable effect on birds frequenting marshes.

However, the foot structure of Willow Grouse reveals certain features common to all tetraonids. As in other tetraonids, the shedding of claws and horny sheaths on the distal phalanges of the feet indicates their kinship. That the ancestors of the Willow Grouse, like forest tetraonids, had horny fringes on their toes is supported by the discovery of birds with small fringes on the margins of the toes, which can be considered an indication of atavism. The feet of these birds, without doubt, are more highly specialized, which assist their movement on snow in winter and, after molt, in wet areas in summer. Willow Grouse mainly use their broad, spadelike claws to burrow in snow.

Toes in phasianids are devoid of horny fringes and the lateral scutes are either totally absent or very poorly differentiated from the horny layer of the underside of the toes. Only some Gray Partridges have a much larger row of almost horizontal, lateral scutes which form a small serrate border on the margins of the toes. All phasianids have thick pads on the underside of the feet, which are especially large in snowcocks and Chukar Partridges apparently facilitate their movement on rocks. As in tetraonids, these pads are covered with a horny layer that thickens in winter. Internally it is homogeneous and externally differentiated into tiny prisms that, however, are never so hard and pointed as the spines of tetraonids.

The structure of the claws also has great functional significance. In phasianids, claws grow continuously, while in tetraonids they are shed in summer along with the podotheca. The shape of the claws in different
205 species of galliformes depends much on a number of factors such as mode of foraging, conditions of migrations, and so on.

In ptarmigans the claws are long, broad, flat, and highly curved at the tip in winter. Feet equipped with such claws enlarge the supporting area, which is important when moving on snow, while spadelike claws help the bird dig into the snow. Before the claw sheath is sloughed the rudiment of a new one has developed under it and attains normal size the following winter. In other tetraonids, even though the claws are shed periodically, their shape varies little from season to season. Flat and relatively small claws are typical of capercaillies; their claw plate is fairly broad and projects over the horny layer of the sole. In Hazel Grouse, the claws are narrow but fairly short, while in Spruce Grouse they are long and thin and undoubtedly provide a firm grip for the feet. The fairly long, sharp claws of Black Grouse are intermediate to those of capercaillies and of Spruce and Hazel Grouses. Apparently, the shape of the claws of forest tetraonids, like the structure of the toes, reflects the nature of the birds' movements, which are different on trees and on snow.

In phasianids the claws grow continuously as they wear from constantly raking the soil. The length of the claws may vary significantly in the same species depending on the hardness of the ground the birds have to rake. Figure 27 shows the longitudinal and transverse sections of the claws.
206 It is seen that the horny plate of the sole is always well developed in tetraonids while the sheath of the claw projects over its tip and sides. In pheasants and partridges the sole plate is less distinctly differentiated from the sheath of the claw, but is, nonetheless, distinctly visible. The structure of the claws in snowcocks and Chukar is somewhat typical. The horny sheath of their claws closely envelops the bone stipe. Even under high magnification it is difficult to distinguish the boundary between the claw and the sole plate. In snowcocks the horny cover of the claw is thick, especially at the tip, and hence the claws are fairly blunt. In Chukar the claws are thinner, and pointed in the sand partridges (Fig. 28).

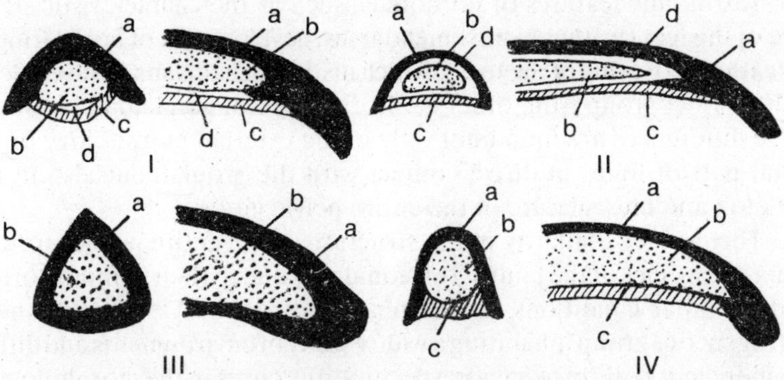

205 Fig. 27. Longitudinal and transverse sections of claws: I—capercaillies; II—ptarmigans; III—snowcocks; IV—pheasants; a—claw; b—bone stipe; c—pad; d—formation of a new layer of claw.

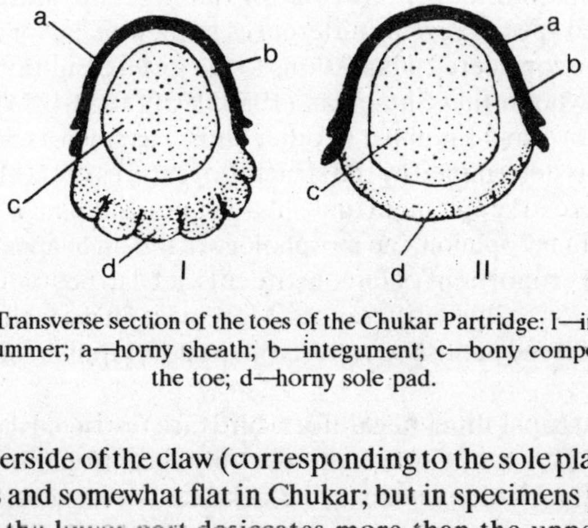

206 Fig. 28. Transverse section of the toes of the Chukar Partridge: I—in winter; II—in summer; a—horny sheath; b—integument; c—bony component of the toe; d—horny sole pad.

The underside of the claw (corresponding to the sole plate) is convex in snowcocks and somewhat flat in Chukar; but in specimens preserved for a long time, the lower part desiccates more than the upper, forming a furrow. This would seem to indicate that the underside of the claw differs from the upper sheath; quite likely, it is looser and softer. The characteristic structure of the claws (with a particularly thick and blunt tip and thicker sole) of birds living among rocks and stones is the result of intense wear on the horny sheath. Here similarity in the structure of the claws in species of diverse origin (snowcocks and Chukar) should be regarded as convergence, brought about by similar living conditions.

The structure of the hind limbs of galliform birds, carrying out most locomotor functions, in fact reflects all aspects of the life style of this group of birds (food procurement, courtship displays, and escape from enemies).

Taxonomic features of tetraonids, such as the characteristic structure of the feet (feathered tarsometatarsus, development of horny fringes,
207 or feathery covering of the toes), which distinguish this small but ecologically distinct group from other galliforms are not accidental. However, these differences are found not only in the external morphology of the distal part of limbs in direct contact with the ground, but also in the skeleton and musculature of the entire pelvic girdle.

The greater similarity in the structure of the entire pelvic girdle in different species of the family Tetraonidae is due to their common origin under similar conditions. In the phasianids of the USSR avifauna, a hetrogeneous group inhabiting a wide variety of environments and differing more markedly in behavior, specific differences in the morphology of the pelvic girdle are quite distinct.

FLIGHT AND STRUCTURE OF THE FLIGHT APPARATUS

An examination of the structure of the flight apparatus of various closely related species reveals differences in their behavior and life style that should be considered adaptations to the given conditions of life. It is from this viewpoint that Shtegman (1950, 1952, 1953,1954) studied the flight of galliform and a number of other birds. His studies reveal that each type of flight is determined by the structural features of the flight apparatus. I hope here to supplement and enlarge upon Shtegman's findings on Galliformes. In my opinion, the morphology of the flight apparatus and the structure and proportions of its constituents need to be studied in greater detail in relation to flight differences. Moreover, I have included here my recent material and published reports on the development of the pectoral muscles.

The flight capabilities of galliform birds are restricted in view of their sedentary life; migratory quails are the only exception. For these birds, flight across the Mediterranean and Black Seas is a very difficult proposition.

The galliform birds of the Soviet Union spend their entire lives either on the ground or on the ground and in trees. In both cases the hind limbs are the main locomotor organs; flight serves as a means of quick transport, usually for short distances. The chief purpose of flight in galliform birds is quick escape from enemies or hurrying to reach the nearest shelter, remaining airborne for the shortest time possible. According to observations made in England (Blank and Ash, 1956), Gray Partridges, if undisturbed, stay airborne for less than a minute per day. In their nesting sites quail rerely fly and remain hidden in grass. Yet it would be incorrect to assume that galliform birds fly only to escape enemies. The flight of Willow

Grouse from the Kolguev Island to the mainland or that of Rock Ptarmigans across the Brooks Range in Alaska, as well as their courtship displays, call for much greater energy than short rapid flights to escape an enemy. Gray Partridges, which rarely take to its wings during breeding season have been known to fly considerable distances in some localities and, at times, even perform migratory flights. Northern Black Grouse in northern Kazakhstan often fly to feeding grounds in the steppes several kilometers away.

The structural features of the wing obviously determine and limit the flight possibilities of birds. It should be remembered that they hardly utilize their full flight capabilities in day-to-day life except for sudden take-offs, prolonged migratory flights, or during some crisis.

It has been long known that Galliform are typical examples of poor fliers. However, Böker (1927) and later Shtegman (1950) considered it necessary to emphasize that the flight of Galliformes, or for that matter any other group of birds, should not be termed "poor" or "good" since each type of flight reflects specialization and definite adaptation to their living conditions.

A flapping flight with a large amplitude of wing strokes and an abrupt vertical take-off are characteristic of most Galliformes. The bird gains necessary speed during the take-off process.

These birds do not take to wing willingly, but prefer to crouch in case of danger, and suddenly take-off only when an enemy is close by. Even though sudden take-off accompained by raucous sounds is typical of all birds of this order, it is particularly characteristic of birds inhabiting thick shrubbery, tugai, or forest thickets (pheasants, Hazel Grouse, and francolins). Rapid take-off requires much energy and is undertaken when absolutely necessary.

Frequent and strong wing strokes alternated with gliding are typical of the flight of galliform birds. The duration of the active (frequent wing
209 strokes) and passive (gliding on steady wings) flights varies notably from species to species. It may also depend on the conditions of flight of each individual. In addition to the duration of active and passive flights, the maximum height of flight (capercaillies), maneuverability (pheasants and Hazel Grouse), and duration (quail and francolins) may also vary sharply.

Short flights up to 100–200 m are typical of most species, but this cannot be construed as a rule. Quails, Gray Partridges, Rock Ptarmigans, and Willow Grouse perform the most distant flights. Capercaillies and Northern Black Grouse are also capable of flying several tens of kilometers. In general, the flapping flight typical of Galliformes is uniquely revealed by each species and is the result of adaptation to the particular conditions under which they evolved.

Since the hind limbs are the chief organs of locomotion and flight plays only a secondary role, adaptive characteristics in the structure of the flight apparatus reflect the impact of habitat conditions (landscapes) rather than the phylogenetic relationship of the species.

Most species of Galliformes I studied differ little from each other in their flight characteristics while preserving individual features. As Shtegman (1950) pointed out, two highly aberrant types of flight can be distinguished among these birds with regard to flight pattern and morphological features. The flight of snowcocks exemplifies the first type—fast gliding with a rapid loss of height and total absence of projectile take-off. The other extreme is exemplified by quail, which fly fast and continuously. Projectile take-off is not seen among quail as in snowcocks, but the former freely gain the necessary height, often by flapping their wings.

Böker (1927) labels the flapping flight of Galliformes, which requires considerable muscular energy, as the most primitive. Shestakova (1950) emphasizes that adaptation to a new environment during evolution often takes place through flight differentiation. In light of these views, the aberrant forms of the flight of galliform birds of the USSR avifauna should be considered the most specialized type of primitive flight, resulting from important changes in life style. Only a more active type of flight (compared
210 to other Galliformes) could sustain migratory quail through its long journey.

High-mountain topography makes great demands on the locomotory organs of snowcocks. The need to negotiate areas which vary greatly in height is met with in two ways: snowcocks perform rapid downward flights by gliding on outstretched, steady wings and journey upward exclusively on foot. Such a highly specialized, economical flight could have developed only in high-mountain conditions, where the typical flapping flight is not suitable for upslope flight. Thus, the flight of snowcocks represents the highly modified flapping flight of galliform birds; in such flights the periods of gliding have been stretched to the maximum and alternate with very short periods of active wing strokes.

In studying the structural features of the flight apparatus, I paid special attention to the structure of the pectoral girdle and forelimbs, measurements of pectoral muscles, general shape of the body (relative sizes of wings, tail, and wingspan) and the wing structure (shape and proportions of its parts).

A large sternum supports the entire pectoral girdle. Other parts of the pectoral girdle such as the coracoids and ribs are attached to the sternum. The broad bony plate of the sternum protects the internal organs ventrally and simultaneously serves as the site for attachment of large pectoral muscles. The oblique muscles of the abdomen, which aid in respiration,

are attached to the posterior part of the sternum. The structure of the sternum of larger taxonomic groups of birds (orders) differs so markedly that it can even be used to study their phylogeny (Shtegman, 1958). Even at the level of families and still lower groups, however, adaptive characteristics reflecting the evolution of flight under certain conditions are seen in the morphology of the pectoral girdle. When the thoracic muscles are especially well-developed, the sternum is massive with a well-developed keel. Reduction in the musculature is usually accompanied by a decrease in the sternum size, and particularly a reduction in the height of the keel. In ground birds the sternum is usually very large and often quite broad. It has been noted that the broad sternum ensures dorsoventral attachment of the diagonal abdominal muscles, and such an arrangement (in ground
211 birds) functions more economically in the expiration process (Shtegman, 1958).

In the species I studied, the structure of the pectoral girdle exhibits certain typical characteristics. Very large pectoral muscles require a larger attachment area which, in turn, could well increase the weight of the sternum, and thus adversely affect flight. These contradictory requirements are met by the light weight of the sternum, in which much of the bony tissue is replaced by membranes. Paired notches in the caudal part of the sternum are so deep and broad that the body of the sternum eventually consists of a narrow bony plate with two pairs of thin lateral processes connected by fontanelles. The high keel—crista sterni—usually does not reach the hypocleidium of the clavicles. Between the sternum, coracoids, and furcula, there is a thick connective tissue membrane (membrana sternocoracoclavicularis) to which the proximal part of the thoracic muscles is attached. Such a structure of the pectoral girdle ensures a sufficiently large area for muscle attachment. Moreover, the large areas of the notch openings are apparently strong enough to withstand the strenuous, if intermittent, activation of the thoracic muscles.

In these species the length of the sternum relative to the truncus varies from 68.8 to 91.8%. The sternum is longest in Sand Partridges and shortest in Quail and Caucasian Snowcocks. I could establish no direct relationship between the weight of the pectoral muscles and the length of the sternum. At the same time, the general structure of the sternum, the height and length of the keel, and the length and position of the coracoids and clavicles vary in different proportions to ensure a larger area of attachment for larger muscles. Measurements of the pectoral girdle bones (Table 9) and diagrams (Figs. 29 and 30) drawn to the same scale reveal a general structural similarity in the skeletons of the pectoral girdle of tetraonids and phasianids as well as specific differences in species of the two groups. It may be noted that these differences are of no taxonomic importance, but

212-213

Table 9. Ratio of skeletal components of pectoral girdle, %

Bird species		Number	Ratio to truncus, %						Ratio to sternum %	
			Sternum		Coracoid + sternum		Keel		Keel	
			M	Range	M	Range	M	Range	M	Range
Willow	♂♂	9	79.6	76.0–81.7	114.9	111.5–120.1	72.9	69.5–74.5	89.4	81.5–93.2
Grouse	♀♀	4	78.6	77.0–81.1	112.2	108.5–120.0	70.5	65.5–74.4	89.4	86.0–91.5
	o	5	78.8	72.2–82.3	118.0	115.5–121.5	—	—	86.6	78.2–92.5
Rock	♂	1	78.5	—	115.6	—	67.8	—	86.0	—
Ptarmigan	♀♀	2	74.2	70.3–78.0	110.2	117.5–112.8 [sic]	67.5	66.8–68.3	91.4	85.8–97.0
Western	♂♂	6	84.1	81.3–87.6	124.8	121.0–129.0	82.1	79.0–87.2	98.8	97.3–100.3
Capercaillie	♀♀	3	81.8	78.9–85.0	119.5	—	78.1	77.2–78.8	95.8	93.0–98.0
Black-billed	♂	1	82.0	—	—	—	82.3	—	100.3	—
Capercaille	♀♀	2	82.8	81.1–84.5	119.6	117.8–121.5	79.2	77.9–80.5	95.7	95.6–95.8
Nortern	♂♂	11	79.0	75.3–82.8	116.2	110.3–120.5	76.6	72.4–80.3	96.6	92.0–100.0
Black Grouse	♀♀	9	81.3	76.5–84.5	116.3	108.5–123.0	74.6	73.9–80.5	95.8	91.5–101.1
Siberian	♂	1	85.1	—	122.0	—	76.0	—	89.1	—
Spruce	♀♀	2	88.7	88.6–88.8	125.0	—	81.1	81.0–81.2	91.7	95.1–91.8
Grouse	o	3	87.3	87.0–87.6	115.0	108.0–122.0	74.5	74.0–75.0	86.0	85.8–86.5
Hazel Grouse	♂♂	11	85.6	80.4–85.5	123.9	116.2–131.7	65.4	61.3–70.0	76.4	72.8–79.8
	♀♀	9	85.8	80.4–87.9	123.8	119.1–129.0	67.6	63.3–73.3	77.8	71.3–83.5
	o	3	87.5	84.9–88.3	119.5	—	60.9	60.4–61.3	76.8	71.3–80.3

Common Quail	♂♂	12	69.1	65.4–71.5	106.5	103.2-113.8	55.8	51.8-58.0	82.0	79.0–88.5
	♀♀	2	68.7	68.5-68.8	100.5	—	55.8	54.3-57.8	78.0	77.0–79.0
Japanese Quail	♂♂	4	74.5	72.1–83.6	113.0	107.2–120.0	—	—	—	—
Common	♀♀	7	84.3	82.0–86.8	124.9	118.0–126.5	68.9	66.7–71.0	81.3	79.5–83.5
Partridge	o	6	85.1	81.0–88.3	125.1	118.5–131.0	71.8	68.4–77.5	84.5	80.5–87.8
Daurian	♂♂	2	80.0	77.6–82.5	118.0	115.5–120.5	64.6	63.8–65.3	80.5	79.0–82.0
Partridge	♀♀	4	85.6	82.8–90.7	120.7	119.5–122.0	67.8	63.6–69.7	80.9	77.0–83.2
Chukar	♂♂	15	81.0	77.5–83.6	117.2	108.0–125.0	64.8	59.5–67.7	79.9	76.8–84.3
	♀♀	9	80.2	75.0–84.9	115.8	110.0–119.5	63.1	59.0–65.0	79.8	77.3–83.8
See-see Partridge	♂♂	2	91.8	90.1–93.5	122.5		—	—	80.5	80.1–81.1
Black Francolin	♂♂	4	83.5	80.1–88.5	121.4	116.5–131.0	68.8	67.5–70.8	81.0	77.0–85.1
	♀♀	4	82.8	82.0–83.5	118.2	115.5–121.0	67.4	65.6–69.4	79.5	77.5–83.3
Himalayan	♂♂	9	69.6	67.5–73.6	100.2	95.0–104.5	51.4	47.8–55.0	74.3	72.4–77.6
Snowcock	♀♀	6	69.4	67.8–70.3	99.9	95.5–105.0	51.6	48.5–55.3	73.3	70.4–76.8
Altai Snowcock	♂♂	2	75.7	72.7–78.5	107.3	104.2–110.5	56.5	59.3–58.7 [sic]	74.8	74.7–75.0
	♀	1	68.7	—	99.0	—	51.2	—	74.5	—
Caucasian Snowcock	o	1	64.8	—	—	—	49.3	—	73.0	—
Common	♂♂	9	84.6	78.7–92.2	119.7	113.4–128.0	75.2	70.3–81.2	87.4	84.0–91.0
Pheasant	♀♀	12	81.8	75.1–89.2	117.3	112.3–126.8	72.0	68.6–74.5	87.2	83.2–90.8

I

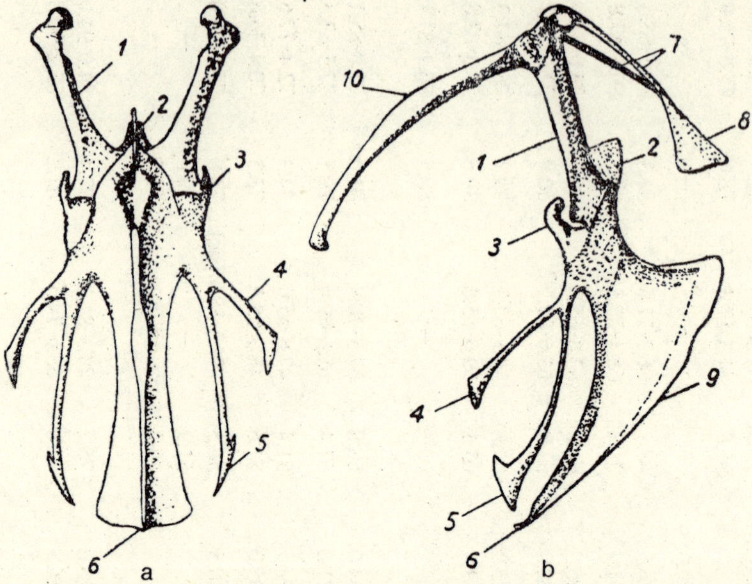

III

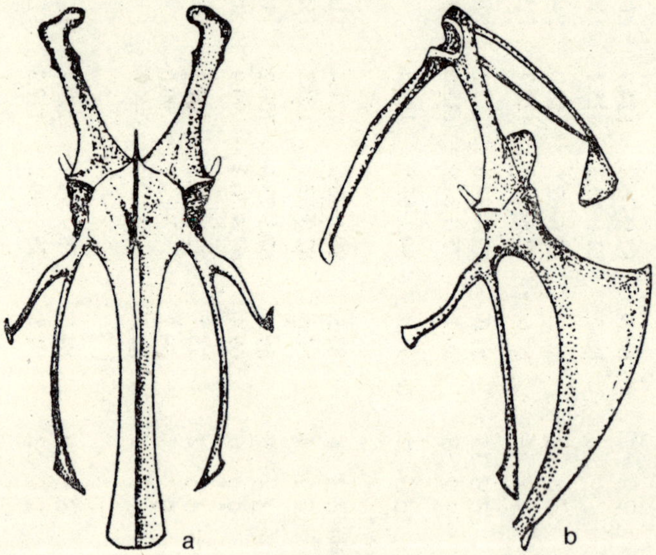

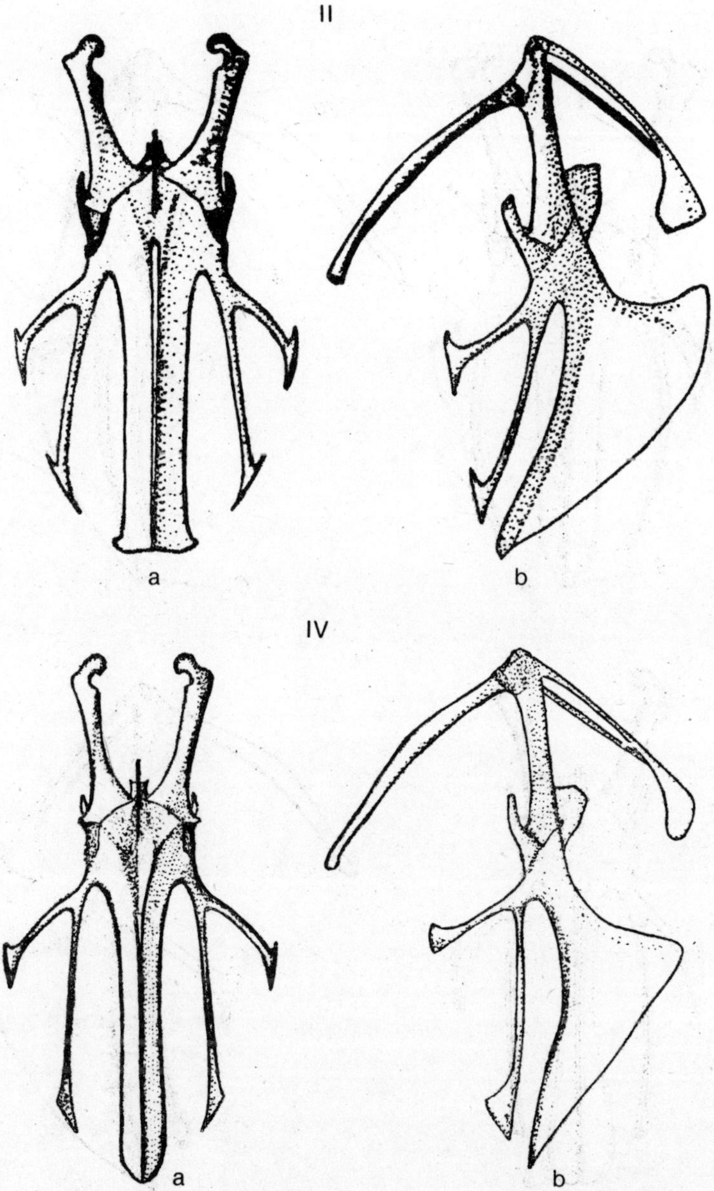

214-215 Fig. 29. Sternum of tetraonids: I—Western Capercaillie; II—Willow Grouse; III—Spruce Grouse; IV—Hazel Grouse; a—front view; b—lateral view. 1—coracoideum; 2—spina sterni; 3—processus lateralis anterior; 4—processus lateralis posterior; 5—processus intermedialis; 6—metasternum; 7—furcula; 8—hypocleidium; 9—crista sterni; 10—scapula.

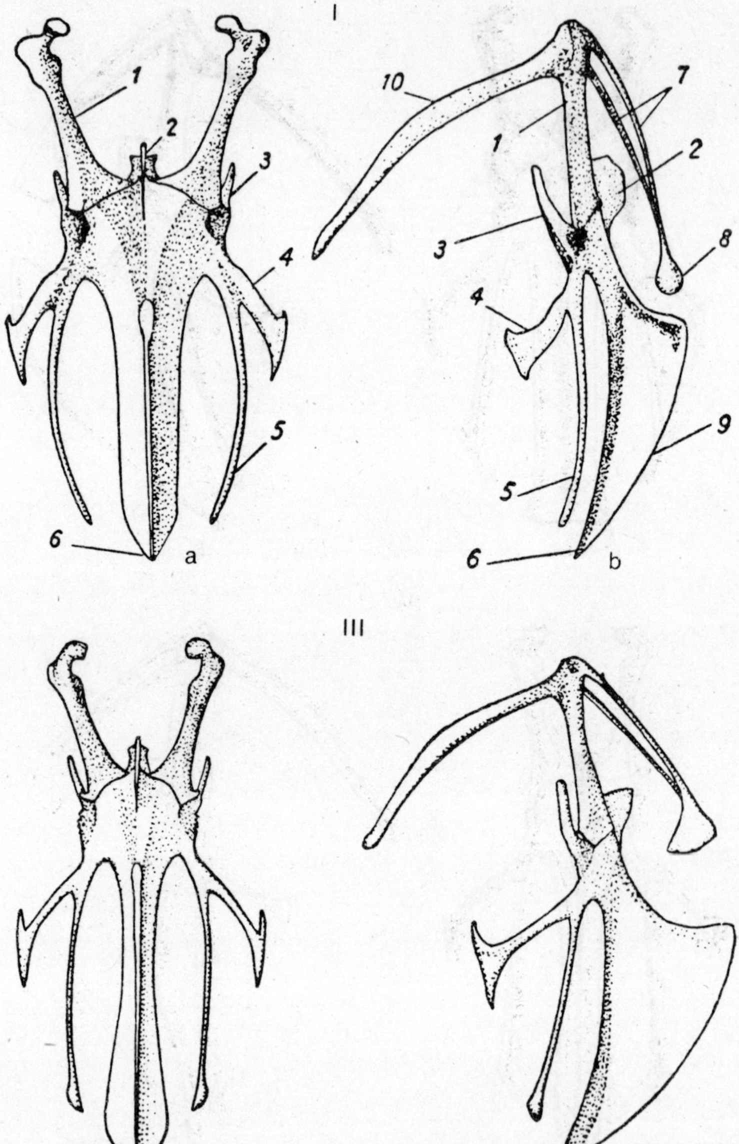
I
1
2
3
4
5
6
a
10
1
7
2
3
8
4
9
5
6
b
III
a
b

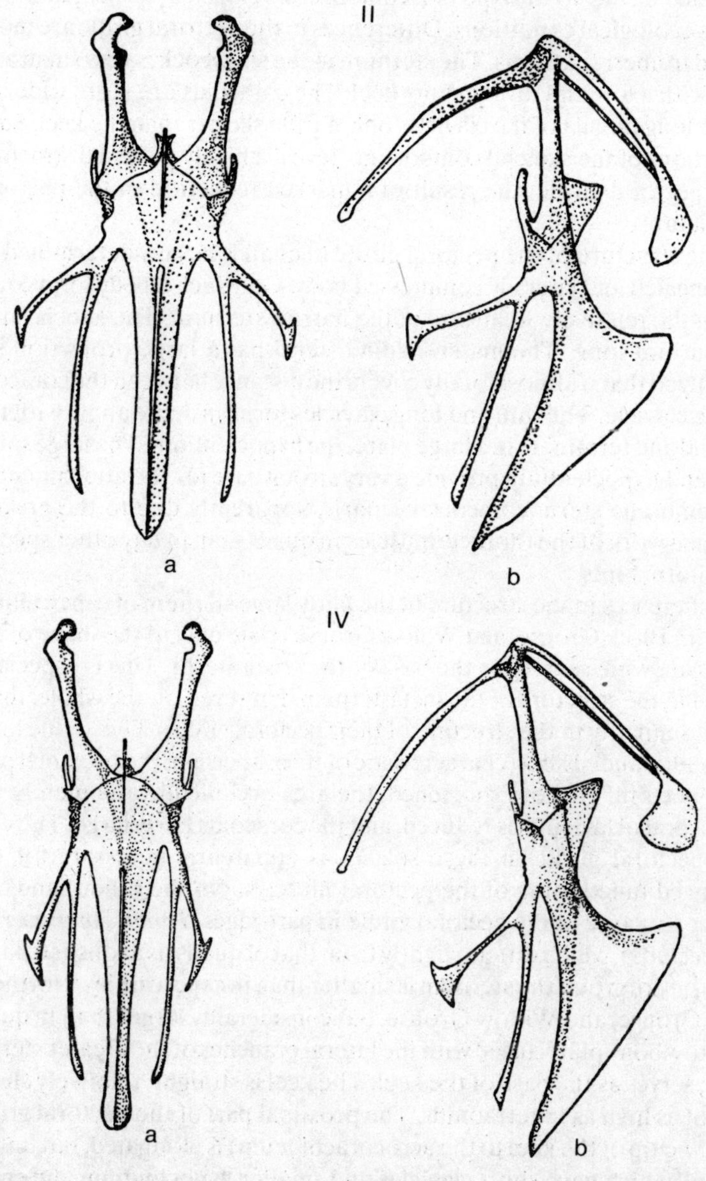

Fig. 30. Sternum of phasianids: I—Himalayan Snowcock; II—Gray Partridge; III—Common Pheasant; IV—Common Quail; a—front view; b—lateral view. Legend as in Fig. 29.

are related mainly to the type of locomotion developed as an adaptation to various ecological conditions. Differences in the pectoral girdle are more marked in aberrant forms. The sternum in the snowcock is very small and broad with a low and rather short keel. The coracoids are more wide-set and the long furcula of the clavicles only a trifle shorter than the keel. Such proportions of the skeletal components reveal a highly flattened structure of the pectoral girdle, the result of a marked reduction in the pectoral musculature.

218 The structure of the pectoral girdle of quail is largely determined by the elongated and laterally compressed body. Only such a body form could explain the relatively small size of the narrow sternum. The keel is fairly high but not long. The massive spina sterni has a large projection set ventrally so that it almost totally covers the distance between the coracoid and the clavicle. The thin and long clavicles form an acute angle with the coracoid and terminate in a large plate, the hypocleidium. The large spina sterni and hypocleidium provide a very strong base for the attachment of the membrane sternocoracoclavicularis, apparently due to the greater sustained work of the thoracic muscles in quails than in any other species of galliform birds.

Differences in the structure of the fairly large sternum of capercaillies, Northern Black Grouse, and Willow Grouse relate only to the shape of the bony plate which serves as the base of the crista sterni. This is especially marked in the structure of the metasternum. However, on the whole, there is great similarity in the structure of their pectoral girdles. Due to the long, high, and rounded keel (characteristic of these species), the proximal part of the pectoral girdle is shortened, the area occupied by the membrane sternocoracoclavicularis reduced, and the coracoids fairly large. The very large pectoral girdle in these species is apparently necessary for the prolonged functioning of the pectoral muscles. On the other hand, the simpler structure of the pectoral girdle in partridges (*Perdix*, *Ammoperdix*, and *Alectoris*), which differs slightly from that of quails, is at once striking. In this entire group the sternum is smaller than in capercaillies, Northern Black Grouse, and Willow Grouse, but considerably larger than in quail. A narrow bony plate, fused with the lateral branches of the greater sternal notch, serves as the base of the keel. The keel is straight, relatively short, and not as high as in tetraonids. The proximal part of the pectoral girdle (from the tip of the keel to the acrocoracoideum) is elongated, but unlike in quail, which have short clavicles and smaller hypocleidium and spina sterni.

Such a simplified skeletal structure allows partridges only relatively short flights. The low keel of Chukar and snowcocks is similar in shape. In Chukar, which have not lost the ability for rapid take-off and flight,

pectoral muscles are better developed and have a large area of attachment. The pectoral girdle of Sand Partridges, which fly often and far, is quite
219 different. The relative size of the sternum is larger than in any other species studied. Its keel is considerably higher than in Chukar and the hypocleidium is more ventral, which increases the area of attachment of the large thoracic muscles. In Gray Partridges, which fly even longer distances, the keel is slightly higher and longer than in the Sand Partridge. Thus the distance between the tip of the keel and the acrocoracoideum is more, but the area occupied by the membrane is less because of the large size of the furcula of the clavicles. In Gray Partridges the hypocleidium is considerably larger and, together with the large spina sterni, provides a stronger base for muscle attachment. This feature is more marked in quail.

The pectoral girdle of Hazel Grouse differs significantly. It resembles most the skeleton of Gray Partridges, but the unusual development of the pectoral musculature in this bird of limited flight has affected the structure. The ratio of the length of its sternum together with the coracoid to that of the trunk is higher than in all the species studied. Although the length of the sternum is slightly more in See-see Partridges, Hazel Grouse have longer coracoids which, on the whole, increase the length of muscle attachment.

The arrangement of the skeletal components of the pectoral girdle is unique. The short clavicles are greatly pushed ventrally and connected with the straight, though short, keel by a very long ligament. With such a position of the clavicles, the area of the membrane sternocoracoclavicularis is greatly increased. Such an extremely simple construction is undoubtedly less strong than in quail and Gray Partridges, in which this membrane occupies a more limited area in the proximal part of the pectoral girdle.

The skeleton of the pectoral girdle of Spruce Grouse falls in a somewhat intermediate position between that of Hazel Grouse and other tetraonids. The relative measurements of the pectoral girdle and the ratios of the length of the sternum and coracoids are the same as in Hazel Grouse, but the shape of the keel (its length and curvature) is as in other tetraonids.

The skeleton of the pectoral girdle of the pheasants and francolins, capable of the most spectacular projectile take-off, is somewhat typical. The high, curved, and long keel (especially in the Pheasant) and fairly short proximal part of the pectoral girdle (between the acrocoracoideum and the tip of the keel) are more similar to those of tetraonids than phasianids. Differences in the structure of the sternum and in the skeletal features of the entire pectoral girdle are largely caused by the sizes and working of the
220 pectoral muscles. It should be emphasized that details of the structure of

the pectoral girdle in the various species are undoubtedly adaptive in character and may not be useful to assess their phylogenetic relationships. A good example is the resemblance of the pectoral girdle of Hazel Grouse and Gray Partridges to that of other partridges of the family Phasianidae. Moreover, the sternum of pheasants shares some common features with that of tetraonids.

Of the large number of muscles of the pectoral girdle and wing which perform the entire complicated flight process, let us examine the common features of the pectoral muscles—M. pectoralis and M. supracoracoideus. The pectoral muscles in galliform birds are highly developed despite their extremely limited flight. The working of the pectoral muscles (M. pectoralis major—wing depressor, and M. supracoracoideus—wing elevator) largely determines the nature of flight, i.e., frequency of wing stroke, speed, and probably maneuverability also. Certain patterns have been observed in the development of the musculature of the fore- and hind limbs. In studying the Anseridae, Shtegman (1950) came to the conclusion that the development of musculatrues of the pectoral and pelvic girdles is strictly inversely proportional. True, such a pattern was later observed (Shvarts et al., 1968) only in some families. In the galliform birds I studied, the inverse ratios of the sizes of musculature of the pectoral and pelvic girdles was quite evident (Table 10). The pectoral muscles (M. pectoralis + M. supracoracoideus) together with the leg musculature constitute 30.8–43.2% of the body weight. The better the development of the pectoral muscles, the weaker the leg muscles. The weight of the pectoral muscles is usually 2 to 4 times greater than that of the leg muscles. Snowcocks are an exception: the proportion of the leg muscles in them is just the opposite. It should be noted that the index of the musculature of the locomotor organs underwent considerable individual variability, which is fully explained by assuming that the development of muscles depends greatly on the intensity of their use.

I have observed differences in the proportion of the musculature of the fore- and hind limbs of Chukar inhabiting low hills of varying topography. Those inhabiting ravines with more gentle but high gradients fly far less than those which live in intensely rugged terrain with narrow ravines and relatively low gradients. In the former population the index of the pectoral muscles is lower and that of the leg muscles higher than in the latter group.

222 At first glance the intense development of the pectoral muscles in birds with limited flight capability seems contradictory. This could be explained by the flight characteristics and structural features of their flight apparatus. A rapid flapping flight is less economical and the projectile take-off, typical of most species, requires considerable energy. Hence,

even the short-distance flights of galliform birds call for very strenuous, albeit brief, use of the pectoral muscles. It is possible that the strong pectoral muscles compensate to some extent for the disadvantages of their flight apparatus, namely, short wings (Harting, 1869; Magnan, 1921). Since the wings are small, a very high frequency of wing flapping is required to keep the birds aloft and, consequently, energy expenditure increases.

The relationship between the length of the wings and the development of the pectoral muscles and flight characteristics was studied by Gladkov (1949). He distinguished several groups in which the relationship between the pectoral muscles, the flight, and wing shape varied. He assigned galliform birds to the group in which birds with strong pectoral musculatures fly more and those with weaker muscles tend to adapt to movement on water or land.

My results reveal that a direct correlation does not always exist between the size of the pectoral musculature and flight characteristics in Galliformes. The pectoral muscles (M. pectoralis major and M. supracoracoideus) constitute 14.4–35.3% of the body weight of these birds. The lowest index was found in Altai Snowcocks, which have passive flight, and the highest in Hazel Grouse, whose flight capabilities are inferior to those of many other species of Galliformes. Quail, Willow Grouse, Rock Ptarmigans, Northern Black Grouse, and Western Capercaillies fly longer at a stretch. Pheasants and francolins adopt a more intense, projectile take-off. I can only conclude that the pectoralis major muscles probably provide Hazel Grouse with unique flight maneuverability in the forest, a feature not typical of other species.

Ornithological publications report the lowest ratio of pectoral muscle weight to body weight for *Lateralus albigularis* (7.8%) of the order Rallidae (Greenewaldt, 1960), and the highest (36.7%) for *Leototila rufinata* of the order Columbiformes (Hartman, 1966). A comparison of these data with mine makes possible the conclusion that the size of the pectoral muscles in the very small group of galliform birds I studied, is close to the limits known for birds in general. This can be explained by the
223 great reduction in musculature in Snowcocks due to its passive flight, and, on the other hand, the intense muscular development in other species (Table 11).

According to published reports, the weight index of the pectoralis major is far less variable than the overall weight of pectoral muscles. Gladkov (1949) pointed out that the average weight of this muscle is 1/6 the body weight. Greenewaldt (1960) even suggested that this muscle constitutes 15.5% of the body weight in all birds—from the smallest to the swan—with negligible variations from the mean. Hartman (1961) estab-

Table 10. Indices of pectoral and leg muscles, %

Bird species		Number	Ratio to body weight, %					
			Leg muscles		Pectoralis major and M. supracoracideus		Weight of major muscles of locomotor organs	
			M	Range	M	Range	M	Range
1		2	3	4	5	6	7	8
Willow Grouse	♂♂	11	10.2	7.5–11.2	26.1	22.9–30.4	36.3	31.2–41.6
	♀♀	6	10.9	8.4–12.5	24.8	21.4–28.5	34.9	29.8–40.2
Rock Ptarmigan	♂♂	2	8.2	7.7–8.6	25.8	25.5–26.1	34.0	33.2–34.7
	♀♀	3	8.4	8.1–8.8	26.2	25.3–26.9	34.6	33.4–35.4
Western	♂♂	4	11.2	10.4–12.2	23.7	22.8–24.5	35.0	33.2–35.5
Capercaillie	♀♀	4	8.4	8.0–9.0	26.6	24.5–28.6	35.0	32.5–36.0
Black-billed	♂	1	7.9	–	25.2	–	33.1	–
Capercaillie	♀	1	6.7	–	26.9	–	33.6	–
Northern	♂♂	16	9.2	7.0–10.5	25.5	23.5–26.9	34.7	30.5–37.0
Black Grouse	♀♀	6	8.1	7.0–9.4	27.3	25.4–28.4	35.4	32.4–37.3
Hazel Grouse	♂♂	13	7.9	6.5–9.3	35.3	32.9–39.1	43.2	40.0–46.1
	♀♀	6	7.0	7.0 [sic!]	35.1	33.0–37.0	42.1	39.1–43.6

1		2	3	4	5	6	7	8
Common Quail	♂♂	17	12.3	9.1–15.0	21.6	19.4–26.0	33.9	30.8–39.0
	♀♀	3	10.7	9.0–13.5	21.7	19.2–24.0	32.4	29.0–35.3
Gray Partridge	♂	1	12.0	—	26.7	—	38.7	—
	♀♀	4	12.5	11.0–13.8	28.8	27.4–29.9	41.3	39.0–43.7
Daurian Partridge	♂♂	4	11.5	9.6–13.2	27.9	25.1–31.7	39.4	34.7–44.9
	♀♀	6	12.9	12.4–13.6	29.2	26.7–30.5	42.1	39.9–42.9
Chukar	♂♂	15	15.4	10.8–20.0	23.3	20.3–28.2	38.7	31.1–43.8
	♀♀	16	14.1	11.3–18.6	23.1	20.3–25.1	37.2	33.1–43.8
See-see Partridge	♂♂	3	12.6	12.3–12.9	27.8	26.8–29.7	40.4	39.1–42.4
	♀	1	11.9	—	29.8	—	41.7	—
Black Francolin	♂♂	4	17.4	16.5–18.2	23.7	22.6–25.2	41.2	40.1–42.6
	♀♀	4	17.5	16.9–19.5	23.9	22.9–25.2	41.7	39.8–42.7
Himalayan	♂♂	11	18.9	16.6–23.0	14.8	13.3–16.5	33.7	31.4–37.9
Snowcock	♀♀	9	18.3	16.6–20.4	16.0	14.2–17.6	34.3	32.5–37.5
Altai Snowcock	♀♀	2	19.0	19.0–19.1	14.4	13.9–14.8	33.4	32.9–33.9
Common Pheasant	♂♂	4	16.2	15.2–17.9	23.5	21.3–26.7	39.7	36.5–44.6
	♀♀	14	15.7	12.1–19.9	22.8	19.8–26.1	38.5	34.5–46.0

Table 11. Indices of pectoral muscles (%) and heart (%)

Bird species		Number	Ratio to body weight, %				Ratio of supracoracoideus to pectoralis major, %		Ratio of weight of heart to body weight, %	
			pectoralis major		supracoracoideus					
			M	Range	M	Range	M	Range	M	Range
1		2	3	4	5	6	7	8	9	10
Willow Grouse	♂♂	13	20.4	17.8–24.3	5.7	5.1–7.0	27.6	25.1–30.3	12.7	9.7–15.2
	♀♀	7	19.1	16.2–22.5	5.7	5.2–6.5	29.6	26.7–33.2	12.1	10.0–15.5
Rock Ptarmigan	♂♂	2	20.2	–	5.5	5.2–5.8	27.4	25.8–29.0	23.5	23.0–24.0
	♀♀	3	20.6	20.0–20.9	5.6	5.3–6.0	27.3	26.6–28.7	19.0	17.0–21.0
Western Capercaillie	♂♂	4	19.3	18.6–19.9	4.4	4.2–4.6	23.1	22.2–24.1	9.7	7.7–10.9
	♀♀	4	21.1	19.2–22.2	5.5	5.3–6.4	25.3	23.1–24.4	10.4	9.0–11.7
Black-billed Capercaillie	♂	1	20.3	–	4.9	–	27.4	–	9.5	–
	♀	1	22.2	–	4.7	–	21.3	–	9.3	–
Northern Black Grouse	♂♂	19	20.7	18.9–21.9	4.9	4.1–5.6	23.8	21.7–29.8	10.7	8.3–14.4
	♀♀	7	21.5	19.9–22.9	5.6	5.0–6.0	26.9	25.1–30.5	9.7	9.0–11.6
Hazel Grouse	♂♂	13	26.9	24.9–30.0	8.3	7.9–9.1	30.5	27.0–33.0	3.9	3.0–5.5
	♀♀	8	26.0	24.2–28.0	8.1	7.2–9.0	31.0	28.9–33.0	3.5	2.6–4.0

1		2	3	4	5	6	7	8	9	10
Common Quail	♂♂	17	17.2	15.4–20.0	4.6	3.7–5.3	24.6	19.6–30.0	10.4	6.7–14.0
	♀♀	3	16.4	15.2–18.1	3.9	3.7–4.1	24.9	20.6–27.7	10.4	8.7–11.0
Gray Partridge	♂♂	1	20.1	—	6.6	—	32.7	—	8.5	—
	♀♀	4	22.0	21.4–23.0	6.9	6.6–7.6	30.9	28.8–33.9	8.8	7.0–10.9
Daurian Partridge	♂♂	4	21.3	19.3–24.1	6.7	5.8–7.6	31.9	31.2–33.6	6.7	6.1–7.4
	♀♀	5	22.0	20.7–23.2	7.2	6.0–7.8	32.2	28.9–34.6	6.9	6.4–7.2
Chukar	♂♂	15	17.9	15.6–20.7	5.2	4.6–6.2	30.0	24.1–36.8	4.0	2.8–4.8
	♀♀	16	17.8	16.0–19.4	5.3	4.1–6.3	29.9	24.4–37.4	4.0	3.6–5.2
See-see Partridge	♂♂	3	20.9	20.1–22.5	6.9	6.7–7.2	32.8	31.9–33.4	3.6	3.4–4.0
	♀	1	22.0	—	7.8	—	35.4	—	4.0	—
Black Francolin	♂♂	4	18.5	17.4–19.6	5.4	5.2–5.6	28.8	27.6–30.3	2.9	2.0–3.5
	♀♀	4	20.1	18.4–22.9	5.4	5.2–5.6	27.7	24.3–29.3	2.6	2.1–3.0
Himalayan Snowcock	♂♂	11	11.7	10.5–13.4	3.1	2.8–3.4	26.8	25.0–29.7	5.5	4.4–7.0
	♀♀	9	12.5	11.1–13.9	3.4	3.0–3.8	26.7	23.5–29.7	4.6	4.0–5.2
Altai Snowcock	♀♀	2	11.2	11.0–11.4	3.2	2.9–3.4	26.4	26.3–26.4	3.6	—
Common Pheasant	♂♂	4	17.6	16.0–19.9	5.8	5.3–6.8	32.6	31.5–34.0	4.4	3.0–4.0
	♀♀	14	17.4	15.2–20.3	5.4	4.6–6.2	30.7	27.1–33.6	3.2	2.2–3.7

lished that the range of this index is very wide: in herons—10 to 17%, terns—12 to 22%, and swallows—14 to 21%. In my data the variability of this muscle within the family, as well as in the group as a whole, differed little from the total weight of the pectoral muscles. In phasianids the index of the pectoralis major varies from 9 to 22.9% and in tetraonids from 19.3 to 26.9% (Table 11). According to published reports, the relative size of the supracoracoideus muscle varies greatly. Hartman showed that in the birds he studied this muscle constituted 1.8–30% of the entire flight muscles. According to my data, this muscle is always well developed in all galliform birds. To determine its value I used two indices: ratio to body weight and ratio to the weight of the pectoralis major. The supracoracoideus constitutes 3.1–8.3% of the body weight and 21.3–35.4% of the weight of the pectoralis major.

It is generally believed that the intense development of the supracoracoideus muscle is characteristic of birds with rapid take-offs. According to my data, however, this is not always so. The maximum development of this muscle (in relation to body weight) is seen in Hazel Grouse, followed by See-see, Gray, and Daurian partridges, and Siberian Spruce Grouse. The sharp projectile take-off is characteristic of pheasants, whose supracoracoideus muscle is of the same size as in francolins and Chukar (the latter resorts to such a take-off only in exceptional cases). If we examine the value of the supracoracoideus muscle in relation to the pectoralis major, it would appear that its ratio to the pectoral musculature, on the whole, is maximum in See-see Partridges, followed by Gray and Daurian partridges and Hazel Grouse; pheasants occupy the last place. The strong development of the supracoracoideus muscle in See-see partridges cannot be explained since information on its flight characteristics is very scanty.

The inconsistency between the development of the pectoral muscula-
226 ture and flight greatly depends on the histological structure of the muscles. Marked differences can be judged by their color which has long been the basis for distinguishing muscles as "white" and "red".

A complete description of the histological structure of the "white" and "red" muscles based on electron microscope studies has been given in *Avian Myology* (George and Berger, 1966). It was established that the color differences are due to the combination of two basic types of muscle tissues (fibers) with a third intermediate one. The working, as well as the color, of the muscles depends on the predominance of one or the other type. Red fibers (R) are slender and contain a larger amount of coloring matter (myoglobin) ensuring intense oxygen metabolism. These tissues contain large quantities of fat, the chief source of energy during contraction. The functional peculiarity of these tissues is their capacity for continuous work.

The thick white fibers (W) are almost devoid of myoglobin and fat, but are rich in glycogen, responsible for metabolism in these tissues. White fibers not only contract rapidly, but tire quickly.

The intermediate type of fiber (I) is midway between "R" and "W" in the histological structure as well as the physiological state. The combination of these three tissues in different proportions determines the physiological properties of the muscles and is one of the factors which determines the nature of the flight. A predominance of "W" fibers ensures rapid take-off but short flight. The "I" fibers constitute the major part of the muscles in birds with a hovering flight, especially in predators. A predominance of "R" fibers ensures active flight over considerable distances, as well as the tremolo flight of the humming-bird.

Histological studies on the pectoral muscles of galliform birds are almost nonexistent. In *Avian Myology* the authors only mention that M. pectoralis major of quail has the same histological structure as that of pigeons, i.e., it consists mostly of red tissues, a small percentage of the intermediate type, and only traces of white tissues.

According to Verbolovich (1961), who studied the myoglobin in animal tissues, "red muscles" with their high activity and high level of energy processes are one animal adaptation for oxygen utilization.

Based on color, the pectoral muscles of Galliformes can be divided into "red"—found in Western Capercaillies, Northern Black Grouse,
227 Willow Grouse, Rock Ptarmigan, and Common Quail, and "white"—in the other species studied. It is quite apparent that a rough division of the musculature into two types based on color alone does not suffice to explain the complex histological structure and functional features of the musculature of each species. At the same time, even the rough classification distinctly reveals its association with the capacity of birds for continuous flight, projectile take-offs, or relatively short flights.

It has been reported that the pectoral muscles of flying animals are far better developed than in terrestrial and aquatic animals (Oehme, 1968). The relative weight of the heart could serve as a main indicator of the agility of an animal. A number of experiments have demonstrated that the physical stress on the heart increases with increasing muscle contractures; the more the work done by the heart, the greater its hypertrophy. The relatively high weight of the heart in birds compared to other vertebrates can be explained by the greater energy stress during flight. Hence the heart index could be considered an indicator of the flight characteristics—duration, speed, and degree of economy of energy expenditure (Shtegman, 1950). A comparison of the heart index of birds of similar sizes (according to Shtegman)—Pheasants and Northern Black Grouse, and Gray Partridges and Hazel Grouse—supports this assumption. In the male Pheasant this index is 5.5‰ while in male Northern Black Grouse, it is 12‰. The great difference in the size of the heart is explained by their varying flight characteristics. In the short, low-economy take-off of Pheasants, although it requires considerable energy, it does not demand a large heart since the flights are short. In Northern Black Grouse, which have rapid and continuous flight, the stress on the heart is considerably greater and thus the heart muscles are better developed. The same explanation has been offered by Shtegman for differences in the heart index of Gray Partridges (10‰) and Hazel Grouse (4‰).

It must be emphasized that the heart index in every species is subject to variation due to a number of factors, primarily changes in body weight (caused by nourishment and the increased weight of female gonads during breeding). Furthermore, the heart index varies in birds in different geographic areas, in different populations, and even in individuals of different sexes and ages shot in the same area. These differences may also depend on their type of locomotion.

228 According to Kirpichev (1960), the heart index in Western Capercaillies from the eastern part of their range is 7.8–9.6‰, but it is 10.4–11.9‰ in Black-billed Capercaillies. The author explains the larger heart size of the latter as due to its more active flight and its ability to fly at considerable heights of several hundred meters, a feat not seen in Western Capercaillie.

According to Semenov-Tyan-Shanskii (1959), this value is much higher in Western Capercaillie of the Kola Peninsula—10.8‰ in the male and 12.4‰ in the female. He emphasizes that the long flight of capercaillies in autumn proceeds at great height. Despite the fairly large variation in the heart index, a wide but definite range of it is characteristic of this species.

Among the galliform birds I studied (Table 11), the heart index varied from 2.6 (Black Francolin) to 23.5‰ (Rock Ptarmigan). According to published reports, the minimum value is found among Rooks—3.0‰ (Shvarts, et al., 1968) and the maximum among Common Sandpipers—24.4‰ (George and Berger, 1966). Thus the heart index in the Galliformes of the USSR avifauna falls within the limits known for birds in general.

Considering that differences in heart size are associated with oxygen metabolism and flight durations, it must be conceded that the oxygen requirement in the Galliformes birds varies greatly. My data do not, however, support the assumption that the relative weight of the heart of birds is proportional to the size of the pectoralis major muscle (Shvarts, et al., 1968). This becomes clear if one recalls that in Galliformes there is no direct relationship between the weight of the pectoral muscles and the duration of flight. The relationship between the heart index and the physiological properties of the muscles is far more evident. In birds with

brightly colored muscles the heart index is higher. According to my data, this index varied from 9.3 to 23.5‰ in birds with "red" muscles and from 2.6 to 8.8‰ in species with "white" muscles. The particularly large heart of Rock Ptarmigans markedly distinguishes it from its kin, Willow Grouse. According to the material collected in Lapland Preserve, the index was 20.5‰ for Rock Ptarmigans and only 14.4‰ for Willow Grouse (Semenov-Tyan-Shanskii, 1959). This difference may be due to the fact that Rock Ptarmigans, which are mountain birds, fly more often and perform more work in the process (great ascents and struggle against strong winds). According to Dobrinskii and Nikul'tsev (1969), the higher heart
229 index of Rock Ptarmigans (18.2‰) may be explained by their life style, such as active locomotion and specific conditions of foraging in mountains.

In the birds caught in Taimyr, this index is slightly lower: March–April, 15.1‰ and July–August, 18.7‰ (Pavlov, 1973).

The Rock Ptarmigans I caught in the high mountains of Altai had an even greater heart index (17–24‰). Here it must be remembered that birds in high mountains often have a higher heart index than those inhabiting low hills or living near plains, a fact often recorded. But it is interesting to note that Small-billed Willow Grouse (*L. lagopus brevirostris*) from Altai also have a higher heart index (11.1 to 15.2‰; three specimens) than the larger Willow Grouse from northern Kazakhstan (8.2 to 14.6‰; ten specimens).

The very low heart indices of Black Francolins, Common Pheasants, Hazel Grouse, See-see Partridges, and Chukar may be due to their short flights. A slightly higher heart index has been observed in snowcocks, a passive flier in which the chief mode of locomotion is long, uphill walks along steep slopes of high mountains.

As mentioned above, body size influences the flight of a bird but the structure of the wing (size, width, curvature and notches of the primaries, size of the alula, plumage, etc.) is even more significant. I have described the general structure of the body in terms of the relative measurements of the wing and tail in relation to those of the trunk (Table 12). Most of the birds I studied have rather small wingspans and lengths of the wrist region of the wing. These values vary markedly in quail which has a highly elongated truncus. The length of the tail is even more variable (in quail and pheasants).

In studying the structure of the locomotor organs, special attention has been paid to the ratio of the limb bones. When correlating the wing skeleton of birds of various taxonomic groups with different types of flight, Böker (1927) distinguished a number of "characteristic morphological structures" and related them to definite types of flight. Thus he attempted to classify flight based only on the ratios of skeletal components of the

Table 12. Indices of some wing and tail measurements

Bird species		Number	Index (% of truncus)							
			Wingspan		Length of wing from ulna flexion		Length of alula		Length of tail	
			M	Range	M	Range	M	Range	M	Range
Willow Grouse	♂♂	10	643	540–695	191.4	164.0–207.8	85.8	73.8–93.0	141.3	131.8–150.50
	♀♀	4	620	608–628	195.5	184.0–206.9	83.7	78.0–89.8	113.5	106.8–116.3
	o	2	542	530–554	186.0	180.0–192.0	81.6	79.8–83.4	120.2	108.2–132.3
Rock Ptarmigan	♂	1	700	—	208.5	—	97.3	—	127.5	—
	♀♀	2	676	650–702	198.2	194.5–201.0	84.3	81.2–87.4	107.1	103.2–111.0
Western Capercaillie	♂♂	4	653	610–678	196.2	186.0–201.0	92.9	81.6–95.3	165.0	145.0–177.0
	♀♀	2	607	588–626	194.0	193–195.0	92.1	91.0–93.3	142.5	142.0–143.0
Black-billed Capercaillie	♂	1	699	—	200.0	—	92.3	—	207.5	—
	♀	1	668	—	193.7	—	96.2	—	166.0	—
Northern Black Grouse	♂♂	6	611	573–660	185.1	177.0–196.0	80.2	75.5–83.7	127.2	91.0–152.0
	♀♀	4	585	545–613	177.1	167.0–183.0	93.5	—	103.1	99.0–105.0
Siberian Spruce Grouse	♂	1	570	—	168.0	—	86.6	—	117.0	—
	♀♀	2	563	530–595	176.0	175.0–177.0	85.6	85.3–87.8	102.0	96.0–108.0
	o	1	622	—	178.5	—	87.0	—	110.5	—

Hazel Grouse	♂♂	7	596	575–635	187.4	181.3–192.0	84.2	79.4–91.0	143.4	137.0–153.0
	♀♀	6	588	553–610	182.4	178.0–196.0	84.4	80.6–87.5	137.0	129.5–147.5
Common Quail	♂♂	8	561	515–600	169.8	167.0–173.0	75.9	70.0–84.0	63.0	50.3–84.0
	♀	1	570	—	168.0	—	—	—	72.4	—
Common Partridge	♂	1	648	—	200.5	—	89.0	—	111.0	—
	♀♀	3	616	594–650	188.2	175.0–200.0	81.4	76.0–84.2	109.7	105.5–116.5
	o	3	628	565–670	192.7	189.0–195.0	85.6	82.6–87.8	108.8	107.4–110.2
Daurian Partridge	♀♀	3	634.3	593–665	201.3	188.0–208.0	88.3	85.0–91.5	121.5	116.0—127.0
Chukar	♂♂	8	573	511–602	174.5	169.0–180.0	88.6	84.0–91.5	103.3	96.0–113.0
	♀♀	5	573	540–591	171.3	164.5–179.0	85.5	85.0–87.2	106.0	92.0–113.0
See-see Partridge	♂♂	2	601	590–613	217.5	210.0–225.0	91.8	89.6–94.0	115.5	109.0–114.0
Black Francolin	♂♂	4	594	570–610	180.6	173.0–184.5	77.3	73.0–80.8	112.7	105.0–118.2
	♀♀	4	600	585–624	178.6	172.7–184.5	80.8	77.5–85.3	121.7	112.5–129.0
Himalayan	♂♂	8	587	533–643	183.5	175.0–197.0	100.6	92.2–115.5	122.6	111.5–135.0
Snowcock	♀♀	6	576	550–558	184.1	179.0–187.5	94.5	85.5–104.0	116.9	104.5–127.0
Altai Snowcock	♂♂	2	622.5	615–630	175.1	161.3–189.0	98.1	96.6–99.5	135.8	133.5–138.0
	♀	1	634	—	183.5	—	100.0	—	133.0	—
Common Pheasant	♂♂	3	646	640–650	188.0	186.0–194.0	88.5	86.5–90.5	407.0	338.0–425.0
	♀♀	10	575	533–670	186.5	173.0–204.0	84.7	76.5–95.5	308.0	273.0–375.0

232-233

Table 13. Ratio of skeletal components of wing and longest primary to truncus, %

Bird species		Number	Humerus		Ulna		Manus		Total length of skeletal components of wing		Number	Longest primary wing feather	
			M	Range	M	Range	M	Range	M	Range		M	Range
Willow	♂♂	10	61.0	58.3–62.0	58.6	57.0–60.1	56.8	55.0–58.7	176.5	170.3–180.7	3	167.3	172.0–162.4
Grouse	♀♀	5	61.9	59.0–63.3	59.4	57.1–61.7	58.4	55.6–61.4	179.7	171.7–186.3	3	159.0	149.4–161.4
	o	4	60.6	58.5–61.1	59.3	58.5–60.8	56.5	55.5–58.0	176.4	174.2–179.4	2	149.6	148.1–151.1 [sic]
Rock	♂	1	64.5	—	63.0	—	61.8	—	189.3	—	—	—	—
Ptarmigan	♀	1	61.2	—	57.5	—	60.0	—	178.7	—	—	—	—
Western	♂♂	5	67.3	66.0–69.5	68.2	66.8–71.4	61.0	59.3–63.0	195.5	193.1–197.7	1	162.9	—
Capercaillie	♀♀	3	64.8	64.0–66.1	64.1	64.7–65.1	59.5	57.5–62.4	188.9	185.8–193.6	3	162.6	158.2—163.3
Black-billed	♂	1	67.5	—	67.8	—	58.8	—	194.1	—	—	—	—
Capercaillie	♀♀	2	66.0	64.5–67.6	66.3	65.0–67.7	59.2	58.8–59.6	191.6	189.1–194.1	—	—	—
Northern	♂♂	12	59.3	56.0–62.2	56.7	52.8–59.2	54.6	51.5–57.5	170.5	163.8–178.0	2	148.0	142.0–154.0
Black Grouse	♀♀	9	59.6	55.0–62.0	56.9	52.0–59.8	54.0	49.7–56.7	169.9	157.3–178.7	1	161.0	—
Siberian	♂	1	58.8	—	55.7	—	54.0	—	165.5	—	1	139.4	—
Spruce	♀♀	2	59.6	59.5–59.8	56.4	55.8–57.0	56.0	55.0–56.0	172.0	171.8–172.3	2	134.6	132.6–136.5
Grouse	o	2	56.0	53.7–58.4	53.3	51.1–55.6	53.6	52.0–55.3	163.0	156.8–169.0	1	145.7	—
Hazel Grouse	♂♂	10	56.4	53.7–56.5	52.6	49.9–55.8	51.8	48.5–55.8	160.2	156.8–166.7	1	158.1	—
	♀♀	10	56.9	54.7–59.8	53.5	51.8–55.0	52.5	50.0–55.0	162.4	158.1–169.8	1	154.3	—
Common	♂♂	17	58.7	56.0–64.0	51.4	49.5–55.2	58.0	54.7–62.8	167.6	161.3–175.6	1	174.0	—
Quail	♀♀	2	59.8	58.1–61.6	51.3	49.0–53.6	57.6	54.5–60.9	168.8	161.6–176.1	—	—	—

Common Partridge	♂♂	4	60.2	59.3–60.7	54.6	53.9–55.6	55.9	53.5–57.2	170.7	168.0–172.3	2	141.4	132.5–150.6
	♀♀	4	56.9	55.0–58.5	52.9	50.4–55.3	53.9	52.2–55.0	163.7	157.6–166.3	4	161.4	158.1–165.0
Daurian Partridge	♂♂	2	57.7	57.5–58.0	52.1	52.0–52.3	54.2	53.2–55.2	163.9	162.7–165.0	—	—	—
	♀♀	4	61.2	59.0–66.5	56.3	54.3–60.1	56.3	54.0–58.5	173.8	169.8–175.1	2	161.2	148.6–173.0
Chukar	♂♂	19	56.4	54.6–60.5	52.6	50.3–55.9	55.5	53.7–61.2	164.6	159.0–168.3	3	133.8	133.6–134.0
	♀♀	10	56.4	54.6–58.0	52.9	50.8–54.2	54.9	53.8–57.3	164.2	161.7–169.9	3	133.4	128.1–142.9
See-see Partridge	♂♂	2	56.4	55.2–57.5	55.2	54.0–56.3	56.6	55.2–58.0	168.1	166.4–171.8	—	—	—
	♀♀	2	55.6	55.2–56.0	54.3	54.2–54.5	56.1	55.2–57.0	166.0	164.6–167.5	—	—	—
Black Francolin	♂♂	4	61.4	59.6–62.0	54.3	53.6–54.0	55.0	51.5–56.5	170.9	168.1–173.0	3	145.3	141.9–147.5
	♀♀	3	62.1	62.0–62.2	54.7	54.0–55.2	55.9	55.1–56.8	173.0	171.9–174.2	3	146.9	144.9–150.5
Himalayan Snowcock	♂♂	8	59.0	56.3–63.2	57.7	54.9–62.3	58.3	55.0–61.6	174.8	166.8–186.2	6	148.9	136.9–159.6
	♀♀	7	59.7	58.0–61.0	58.3	56.5–59.8	59.5	58.5–60.5	178.5	173.9–179.8	4	147.5	142.3–155.2
Altai Snowcock	♂♂	2	60.7	60.4–61.0	59.4	58.0–60.8	59.8	59.5–60.2	179.9	178.6–181.3	2	145.3	144.4–146.2
	♀	1	54.8	—	53.7	—	54.5	—	163.0	—	—	—	—
Caucasian Snowcock	♂	1	51.5	—	50.2	—	50.1	—	151.8	—	—	—	—
	♀	1	55.7	—	54.6	—	54.4	—	164.7	—	—	—	—
Common Pheasant	♂♂	8	60.6	57.6–65.3	53.6	50.9–58.2	53.7	50.6–57.5	167.2	161.2–181.0	2	153.2	149.3–157.2
	♀♀	12	60.7	58.0–63.6	54.2	50.4–56.5	54.0	51.3–56.7	169.0	159.9–176.4	3	153.9	149.6–161.3

wing. In spite of the fact that such a one-sided morphological approach to the classification of flight, based only on the structure of the skeleton,
234 cannot be adopted, this method of studying the ratios has proved useful, chiefly when studying the species within a given taxonomic group (Yudin, 1950; Yakobi, 1960).

In studying the wing proportions of galliform birds, I have given great importance to the ratios of various skeletal components of the wing and the longest primaries to the truncus length (Table 13). These indices were used by Yudiñ (1950) to describe the wing proportions in the family Falconidae. Indices calculated in relation to the truncus give the most accurate idea of the structural qualities of the wing, since along with the ratios of the different parts of the wing, these indices reflect the measurements of the whole wing in relation to the truncus. However, it should be noted that the sum of the skeletal parts given in the tables provides a somewhat exaggerated values of wing length, since skeletal components in a fully stretched wing fall not in a straight-line but at varying angles. In my data, the length of the wing skeleton varies from 163% (Hazel Grouse) to 195.5% (Western Capercaillie) of the length of the trunk skeleton. In all species the maximum length of the primary wing feathers is longer than the wing skeleton and constitutes 133.4% (Chukar) to 174.0% (Common Quail) of its length. The wing skeleton in birds of the family Falconidae, whose wings are not very long, varies from 248.5 to 309.9%, and the longest primary feather from 200 to 321% (Yudin, 1950).

As pointed out by Böker, the humerus is the longest part of the wing skeleton in all galliform birds. Capercaillies are an exception: the humerus is either equal to or slightly shorter than the ulna. In my material, See-see Partridges are another exception in which the manus is the longest. Differences have been observed in the proportions of the wing skeleton: in tetraonids the manus is the shortest bone, while in phasianids it is the ulna. A detailed discussion of the proportions of the skeletal components follows.

Proportions of the wing skeleton permit a comparison of its two functionally different parts: the proximal upthrust part of the wing consisting of the humerus and ulna and the distal extended or propulsion part consisting of the manus and the extremely long primary wing feathers (Table 14).

In the galliform birds studied, the proximal part of the wing was 107.4–134.5% of the truncus; the minimum upthrust surface was in Quail,
235 Chukar Partridge, and Hazel Grouse, and was maximum in Western Capercaillies. The distal part of the wing varied from 188.1–223.8%. The propulsion part of the wing was the shortest in Chukar Partridge, Siberian Spruce Grouse, and Black Francolins, and longest in Willow Grouse. On

Table 14. Relative measurements of upthrust and propulsion parts of wings

Bird species		Number	% of truncus		
			Proximal part of wing	Distal part of wing	Total wing
Willow Grouse	♂♂	3	120.8	223.8	344.6
	♀♀	4	118.8	211.0	330.7
Western Capercaillie	♂	1	134.5	223.2	357.7
	♀♀	3	130.1	220.9	351.0
Northern Black Grouse	♂♂	2	116.8	202.6	319.4
	♀	1	118.0	216.9	334.9
Siberian Spruce Grouse	♂	1	114.4	193.2	307.6
	♀♀	2	115.9	190.6	306.5
	○	1	114.1	192.0	306.1
Hazel Grouse	♂	1	109.0	211.0	320.0
	♀	1	108.6	207.7	316.3
Common Quail	♂	1	107.4	213.7	321.1
Common Partridge	♂♂	2	115.8	210.4	326.2
	○	4	112.8	216.4	329.2
Daurian Partridge	♀♀	2	119.0	217.4	337.2
Chukar	♂♂	2	107.5	188.4	295.9
	♀♀	3	109.0	188.1	297.1
Black Francolin	♂♂	3	115.5	202.4	317.9
	♀♀	3	116.9	203.1	320.0
Himalayan Snowcock	♂♂	6	117.8	206.6	324.4
	♀♀	4	117.7	206.8	324.5
Altai Snowcock	♂♂	2	119.6	205.2	324.8
Common Pheasant	♂♂	2	111.6	207.1	318.7
	♀♀	3	113.7	206.3	320.0

the whole, among the USSR avifauna Chukar Partridge and Siberian Spruce Grouse are particularly short-winged (295.9–306.1% of the truncus) while Western Capercaillies and Willow Grouse are long-winged (357.7 and 344.6%, respectively).

My data contradict those of Airumyan (1967), which were obtained by the same method for Chukar and Common Pheasant. If the total wing
236 length and the ratios of the distal and proximal parts of Chukar Partridge are close to my values, then the statement that the wing of pheasants, especially the distal part, is much shorter than that of Chukar Partridge is not correct. The length of the longest primary cannot be 77% of the length of the trunk skeleton. Airumyan's entire conclusion, therefore, that the Pheasant is exceptionally shortwinged is erroneous.

A comparison of the wing proportions of galliform birds with those of the family Falconidae given by Yudin (1950) revealed that the relative measurements of whole wings of galliform birds and the ratios of the proximal and distal parts are fairly constant. The overall wing length of the Falconidae is 425–620% of the truncus, while in Galliformes of the USSR avifauna it varies only from 295–362%.

The relative size of the skeletal components of the wing, calculated with reference to the truncus in birds of the same species and sex, generally varies within 2–3%; however, these differences are much higher in a number of instances. In female Willow Grouse, for example, the relative length of the humerus varies up to 4.3%, of the manus in female Western Capercaillies up to 4.9%, and of the humerus in female Daurian Partridges as much as 7.5%. Far greater variations have been observed in the relative sizes of the primaries, which can be explained by the state of the plumage (not fully grown or worn out feathers) and by individual differences in the length of the primaries. In certain specimens in my small collection, the size of the longest primary with reference to the truncus varied from 20 to 23 (Himalayan Snowcock) and even 30% (female Gray Partridges). Thus, when comparing the ratio of the skeletal components of the wings to the length of the truncus, I found "short-winged" and "long-winged" individuals among birds of the same species and sex. The relative length of the wing depends mainly on the size of the primary wing feathers; the ratio of the skeletal components with reference to the truncus is more constant, in spite of its rather wide range at times.

Engels (1938, 1942), when studying individual differences in the body ratios of birds, came to the conclusion that differences in the ratios of various skeletal components of different organs (femur—humerus) are more marked than between different parts of the same organ. To support
237 this view a number of authors have cited the very small individual

differences in the ratios of the skeletal components of the wing. Kattinger (1929) has shown that the degree of variability in the ratio of wing components of predatory birds does not exceed 1%. Yudin (1950) has recorded the range of variability in the family Falconidae as 0.3-1.2%; while for the Accipitridae, Yakobi (1960) cites a variability of 0.7%.

Individual differences in the ratios of the skeletal components of the wings of galliform birds, although trifling, are slightly higher than in predatory birds (Table 15). In most cases the range does not exceed 1-1.5%, but as an exception individual differences in birds of the same species and sex could be higher (2.4%—manus of male Quail and 2.7%—manus of male Willow Grouse). It should be noted that sizes of the humerus, ulna, and manus of several species of galliform birds studied are so close that individual differences of 1.5%, more so, up to 2.6% in some specimens, disturbs the proportions of different bones.

In all tetraonids except Western Capercaillies, the humerus is longer than the ulna and the ulna is longer than the manus. In my collection (5 specimens) the humerus was longer than the ulna in three capercaillies (1 male and 2 females), and the ulna was slightly longer than the humerus in the other two (1 male and 1 female), while the humerus and ulna of one male were equal.

Unlike tetraonids, the shortest bone in the wings of phasianids is not the manus but ulna; the humerus is longer than the ulna, and the ulna shorter than the manus. Sand Partridges are an exception whose manus is longer than the humerus and ulna, but the differences are only a few tenths of one per cent. It should be mentioned that the ratio of the various skeletal components of the wing varies less in tetraonids than in phasianids. For example, although the measurements of the humerus, ulna, and manus in snowcocks fall in a range typical of phasianids, they are so close that they differ by only a few tenths of one per cent. The longest humerus is found in the Pheasant, but the proportion of the manus and ulna in individual birds coincides. In some of the birds studied the ulna was shorter than the manus (1 male and 2 females); in most (1 male and 6 females) the proportion was reversed; and in one female the manus and ulna were of equal length.

The proportion of the skeletal components of the wings of quails
239 differs markedly. These birds have a very short ulna and a very long manus. Though the humerus is generally slightly longer than the manus, a reverse ratio was found in some birds. Thus the proportions of the skeletal components of the wing and the relative size of the functionally different proximal and distal parts hardly differ in the various species of galliform birds I studied. Nevertheless, minor differences in the relative lengths of the manus in the two families are quite distinct. Marked differences in the

Table 15. Ratio of length of skeletal components of wing to overall length of wing skeleton

Bird species	Number		Humerus		Ulna		Manus	
			M	Range	M	Range	M	Range
Willow Grouse	♂♂	10	34.6	34.2–35.2	33.3	32.4–34.7	32.1	30.1–32.8
	♀♀	4	34.5	34.3–35.2	33.1	32.9–33.3	32.4	32.2–32.9
	o	4	34.7	34.2–35.2	33.4	33.0–33.8	31.9	31.0–32.4
Western	♂♂	3	34.3	33.7–34.6	34.5	—	31.2	30.9–31.8
Capercaillie	♀♀	4	34.5	34.3–34.6	34.5	34.4–34.6	31.0	30.9-31.1
Black-billed	♂	1	34.7	—	34.9	—	30.4	—
Capercaillie	♀♀	2	34.5	—	34.6	—	30.9	—
Northern	♂♂	9	34.9	34.3–35.4	33.2	33.0–33.6	31.9	31.2–32.4
Black Grouse	♀♀	9	34.8	34.3–35.4	33.4	33.1–33.9	31.8	31.1–32.4
Siberian	♂♂	2	34.6	34.3–34.8	33.0	32.9–33.2	32.4	32.0–32.8
Spruce Grouse	♀♀	3	34.6	34.2–35.0	32.8	32.3–33.0	32.6	32.0–33.1
	o	2	34.6	33.7–35.0	32.6	32.5–32.7	32.8	32.3–33.2
Hazel Grouse	♂♂	9	34.7	33.8–35.5	32.7	32.0–33.3	32.6	31.3–33.2
	♀♀	8	34.9	34.5–35.6	32.8	32.4–33.1	32.3	31.3–32.8
Common Quail	♂♂	12	35.0	34.1–35.8	30.7	29.2–31.5	34.3	33.4–35.8
	♀♀	2	34.8	34.7–34.9	30.6	30.4–30.9	34.6	34.4–35.7
Gray Partridge	♂♂	4	35.2	35.1–35.8	32.1	31.8–32.4	32.6	31.8–33.4
	o	4	34.6	33.6–35.2	32.3	31.7–33.3	33.1	33.0–33.1

Daurian Partridge	♂♂	2	35.2	35.0–35.4	31.8	31.6–32.0	33.0	32.6–33.4
	♀♀	4	35.3	34.7–35.7	32.4	31.8–33.0	32.3	31.6–33.5
	o	1	35.6	—	32.3	—	32.1	—
See-see Partridge	♂♂	2	33.4	33.3–33.5	32.0	32.9–33.0	33.6	33.5–33.8
	♀♀	2	33.4	33.3–33.5	32.8	32.5–33.0	33.8	33.5–34.2
Chukar	♂♂	13	34.3	34.1–34.7	32.0	31.5–32.5	33.7	33.1–34.4
	♀♀	9	34.3	33.9–34.8	32.1	31.9–32.5	33.6	33.1–34.1
Black Francolin	♂♂	3	35.5	35.0–36.0	31.4	31.4–31.5	33.1	32.7–33.8
	♀♀	3	35.1	34.5–35.4	31.8	31.4–32.8	33.1	32.8–33.9
Himalayan Snowcock	♂♂	7	33.8	33.7–34.3	33.2	32.7–33.6	33.0	32.3–33.5
	♀♀	5	33.5	33.3–33.8	32.2	32.7–33.2	33.6	33.3–33.8
Altai Snowcock	♂♂	2	33.7	—	33.0	—	33.3	32.8–33.8
	♀	1	33.5	—	33.0	—	33.5	—
Caucasian Snowcock	♂	1	33.8	—	33.1	—	33.1	—
	o	1	33.9	—	33.1	—	33.0	—
Common Pheasant	♂♂	8	36.1	35.7–36.7	32.0	31.3–33.5	31.9	31.6–32.8
	♀♀	9	35.7	35.3–36.5	32.2	31.5–33.0	31.9	31.2–32.4

proportions of the skeletal components of the wing in Quail (elongated manus and shortened ulna) could explain its active flight.

In addition to the size of the wing and proportions of its different parts, its shape is no less important to flight (relative width, structure of tip, sharpness, trimness of primary wing feathers, relative size of alula, and compactness of primaries and their curvature). The wings of the galliform birds studied, although generally short, differ markedly in shape in different species. This applies particularly to the wing elongation, i.e., the ratio of the length of its average width. This index varies from 179 (Common Pheasant) to 270 (Common Quail). On the whole, birds of the family Tetraonidae have narrow wings. Variations in elongation are less (223—249) than in the Phasianidae, in which pheasants and quails represent the two extremes. The shape of the wing is greatly determined by wing projection, i.e., the ratio of the distance between the initial secondaries and the wing tip to its length expressed in percentage. Usually the narrower, pointed wing has a greater wing projection, but exceptions do exist.

The maximum wing projection is found in Common Quail (42%), Willow Grouse (41.1%), and Northern Black Grouse (40%); and the minimum in Black Francolins (18%), Spruce Grouse (21.4%), and Altai Snowcocks (25.9%).

The great similarity in shape, caused by nearly equal length (204 and 208) and wing projection (31.8 and 31.2%) in Himalayan Snowcocks and Chukar Partridges is at once striking. But Altai and Himalayan snowcocks differ in these very indices. Altai Snowcocks have a relatively narrow wing and smaller wing projection (25.9%) than the broader-winged Himalayan Snowcocks (31.8%), which hardly differ in this index from pheasant with the broadest-wings.

The distal part of the wing performs the main operation of flight, i.e., generating upthrust. That is why the shape of the primaries (their relative length and curvature, elasticity of the shaft, and vanes and notches in them) considerably influences flight characteristics. The wing formula in the
240 galliform species studied was fairly similar. The longest primary wing feather is the fourth, rarely the third or fifth, but the differences between them are so small that in most species the apex of the wing is actually formed by all three primaries. Far-migrating Quail, with a typically sharp wing (formula: 1 =2 > 3 > 4 > 5), are an exception.

The wing formula differs substantially in tetraonids and phasianids. The first primary in tetraonids is 70.3 (Northern Black Grouse) to 77.1% (Western Capercaillie) of the longest primary, while in phasianids this value ranges from 83.5 (Common Pheasant) to 95.4% (Himalayan Snowcock). The most truncated wings are found in mountain species—snowcocks,

Chukar, and See-see Partridges—in which the extreme first primary is only 4.6 to 7.0% shorter than the longest, i.e., the fourth.

The shape of the feather, its curvature, and the vane ruffling are no less important than the structural formula of the wing tip. Birds of different sizes, belonging to different taxonomic groups and with varying types of flight, often have outer and inner webs with slots of different shapes which look like digitate scalloped tips in an outspread wing. According to Shestakova (1953), birds use the ruffling in the wing tips to temporarily increase the upthrust. It has been recorded that the stretched wing of a galliform bird with its highly convex and hard primaries, looks like a whole row of very long wings, which develop independent aerodynamic forces (Storer, 1948).

In studies on the Accipitridae, Yakobi used the sum of the slots to describe the wing ruffling. Unfortunately, this method cannot be used for the Galliformes, since in most species the primary wing feathers are so highly falciform that linear measurements cannot be taken with accuracy. Hence I have given here only a description of the structure of the wing tip in various species.

The contour of the wing depends on the number of slots in the primaries, their depth, length, and the curvature of the feathers. In birds with relatively short, but deep slots on the outer and inner webs, the primaries are digitate only at the very tip of the wing.

The bird groups studied differ markedly in the structure of the wing tip. Quail, which perform far migrations, is the most outstanding in this
241 respect. Their wing is narrower than in other species and has a pointed, almost entirely unslotted apex which develops the speed necessary for distant migrations (Savile, 1957). Wings of the snowcocks and Chukar Partridges have relatively few slots, while those of Sand Partridges have more. The tetraonid wing is deeply incised and Common Pheasants, and Gray and Daurian partridges have wings with long but narrow slots; narrowed webs are especially marked if the feathers are curved. The wing contours of Black Francolins are somewhat typical. Broad slots reach almost to the middle of the outer webs of the second to sixth primaries, and the inner webs are not as narrow. Thus wing ruffling can be seen only in a fully outstretched position.

The wing structure in Spruce Grouse is quite unusual; it is close to that of Hazel Grouse in proportion, but shares nothing in common with the shape of the primary wing feathers. In both species the primaries are intensely arcuate. The curvature of the first primary feather of Hazel Grouse begins from the middle third of the feather; in the second and third, only the apical third is curved; and in the fourth to sixth, only the very

tip. The inner webs are broad and attenuate gradually and gently toward the middle. In a slightly folded wrist, the wing apex becomes almost entire, but in a fully stretched wing narrow clefts between the primaries separate the feathers rather sharply. The ability to freely change the wing ruffling
242 assists in flight maneuvers essential for life in dense vegetation. Spruce Grouse have a distinct curvature of the primary wing feathers (Fig. 31). The webs of the first and second primaries are curved throughout, starting from the tip of the wing. The curvature of the third and fourth is visible from the middle third of the feather; only in the fifth and sixth are just the tips curved. Thus the anterior margin of the wing is highly curved. But the most important difference in the wing of Spruce Grouse is the characteristic slots in the inner webs, due to which it acquires a totally different shape. Sharp and curved, the entire tips of the primaries somewhat resemble spines and are distinct even in the folded wing. The deeply-slotted wing of Spruce Grouse distinguishes them sharply from other galliform birds, especially from Hazel Grouse, which have broad, soft vanes. The functional significance of such sharply-ruffled wings is not clear. Moreover, data on the behavior of this interesting bird are almost nil. From the sharp differences in the wing structures of these two species inhabiting dense forests, it could be assumed that the flight of Spruce Grouse differs intensely from that of Hazel Grouse and probably is less maneuverable.

The curvature of the primary wing feathers is noticed in a large number of galliform species, but not uniformly. In cases where the primaries are curved to varying degrees, birds in flight are able to change the flat wing surface into a cleft surface more rapidly and with less strain.

1 2

Fig. 31. Primary wing feathers. 1—Spruce Grouse; 2—Hazel Grouse.

An outspread wing forms fairly wide gaps between the stretched digitate feathers; moreover, even a slight bending at the wrist gathers the feathers so close together that the wing tip becomes compact with very little change in the total area. Unrestrained changes in the structure of the wing tip in birds with curved primaries facilitate the change to gliding and allow greater flight maneuverability.

The primaries are less curved in snowcocks, Chukar Partridges, Western Capercaillies, Common Quail, Willow Grouse, and Rock Ptarmigan. The feathers of pheasants, Spruce Grouse, Hazel Grouse, Northern Black Grouse, and Sand Partridges are highly curved.

The structure of the wing of Galliformes with its intense development of the bastard wing (alula) is noteworthy. Various views have been expressed on its function. The alula of birds is comparable to the flaps of an airplane, whose role is to increase upthrust and decrease frontal resistance (Graham, 1932). It has been shown that the action of the alula increases the resistance necessary to brake during landing (Liliental',
243 1905). It has been observed that forest and bush birds which take-off and land constantly have a larger alula than birds of open expanses (Gladkov, 1949). Yakobi (1960) observed differences in the size of the alula relative to the flight characteristics of the Accipitridae. Increased foraging activity and a greater number of take-offs and descents at sharp angles increase the size of the alula. In changing from a flapping flight to soaring, the relative length of the alula decreases. According to Shestakova (1971), a large alula compensates to some extent for a shorter wing length by increasing the upthrust (in short-winged birds). This assumption is evidently confirmed by the especially large alulae of all short-winged galliform birds which are rather poor fliers. However, no definite correlation has been established between the wing length and size of alula among various species. In my data (Table 16), the average index of alulae* in galliform birds varied from 42.7 to 54.4 (range 39-57.5). The very low indices cited for Common Pheasants (29.3) and Chukar (41.2) by Airumyan (1967) are dubious. According to my data, the shortest alula is found in Northern Black Grouse, Willow Grouse, and Francolins—birds which vary widely in flight characteristics and wing shape. An especially large alula (51.1 to 54.4) is found in Chukar Partridges and snowcocks—birds which glide downward. Quite probably, the size of the alula, which performs several functions in flight, depends on a number of factors.

First of all, the size of the alula often depends on wing-loading and flight maneuverability, in addition to other characteristics such as wing

*Ratio of alular length measured from the wrist flexion to alar length—Author.

Table 16. Wing-load and size of alula

Bird species		Number	Body weight, g	Wing-loading, g/cm²		Number	Length of alula, % of wing length	
				M	Range		M	Range
Willow Grouse		9	709	1.15	0.88–1.43	15	43.8	41.2–45.5
Western Capercaillie	♂♂	7	4,044	1.73	1.49–1.86	8	47.2	43.7–50.7
	♀♀	5	2,030	1.52	1.45–1.61	5	46.8	44.0–48.3
Black-billed Capercaillie	♂	1	3,800	1.41	—	1	46.0	—
	♀	1	2,200	1.38	—	1	46.8	—
Northern Black Grouse	♂♂	8	1,322	1.22	1.00–1.66	7	42.9	40.0–45.8
	♀♀	3	966	1.22	1.19–1.24	2	42.7	—
Siberian Spruce Grouse		4	714	1.46	1.38–1.48	4	49.4	47.5–51.3
Hazel Grouse		15	375	0.90	0.68–1.09	23	45.2	39.0–48.5
Common Quail		8	91.2	0.59	0.52–0.67	18	44.4	39.4–48.5
Gray Partridge		7	363	0.89	0.83–0.95	12	44.6	41.8–47.4
Daurian Partridge		7	331	0.86	0.80–0.92	11	44.5	41.0–46.0
Chukar		9	506	1.34	1.17–1.47	19	51.1	47.8–55.8
See-see Partridge		7	178	0.75	0.67–0.82	6	44.1	42.2–45.8
Black Francolin		8	473	0.83	0.73–0.89	8	43.9	41.2–47.5
Himalayan Snowcock	♂♂	5	3,365	2.10	1.92–2.23	12	54.4	50.0–57.0
	♀♀	5	2,130	1.17	1.49–1.96	9	52.4	46.5–57.5
Altai Snowcock	♂♂	3	3,150	2.24	2.14–2.50	3	52.0	51.0–53.8
	♀	1	2,616	2.28	—	2	51.7	49.4–54.3
Common Pheasant	♂♂	3	1,425	1.38	1.27–1.54	3	47.8	46.4–49.2
	♀♀	11	942	1.11	0.82–1.38	13	45.7	41.4–50.0

structure and the nature of the flight. The maximum wing-load and, concomitantly, the largest alula, are seen in snowcocks. Western Capercaillies, although much larger in size with longer wings and less load, have smaller alulae. The next larger alulae are found in Chukar, which have a very high weight-load. In Common Pheasants and Black Francolins, the alula is much smaller, despite their shorter wings. Thus I could establish no direct relation between the size of the alula and wing elongation. Perhaps the very large alula in snowcocks compensates for its difficulty in take-offs from a flat terrain and acts as a brake when landing, by slightly
244 reducing the wing-load. The same explanation can be applied to Chukar, which live in similar localities and have a similar flight apparatus. It is generally assumed that, within related groups, the increase of weight-load
245 is proportional to an increase in bird size. At the same time, birds of the same size but from different taxonomic groups may differ markedly in this respect.

Within Galliformes the wing-load changes very sharply, but rarely conforms to the bird weight (Table 16). One must remember that wing-load is subject to considerable individual variability, depending on the plumpness of the bird and the condition of its wing feathers. From this view, one cannot help notice that increased wing-load corresponds to body size in tetraonids quite clearly; the Spruce Grouse is an exception. The high wing-load in this bird could well be due to the very intensely ruffled wing feathers.

Among phasianids, the high wing loading of mountain birds—snowcocks and Chukar Partridges—is noteworthy. In francolins the wing-loading is 0.84 g/cm^2 and is 1.34 g/cm^2 in Chukar Partridge, a bird of medium weight. This is more than that of Northern Black Grouse, that is, nearly 2.5 times lighter. The contrast in the wing-load of capercaillies and snowcocks is thus quite striking. Such an apparent disagreement between bird size and wing-load can be explained by the rapid flight of the mountain species. In rugged terrain flight could be in three directions: upward through a ravine, along a straight horizontal line, and downward through a ravine. Chukar resort to vertical flight very rarely and only in moments of crisis. Snowcocks have lost this ability entirely. Upward flight along slopes is very difficult for the birds and involves a great expenditure of energy. Hence Chukar use this mode of flight very rarely and only over short distances; snowcocks never resort to this type of flight. Flights from one slope to another along a horizontal line or downward through a ravine are completely passive. Rapid gliding down a ravine with quick loss of altitude is accomplished by taking advantage of the bird's weight and its high wing-load. This type of flight is possible only in mountain habitats with steep drops.

To conclude, the characteristic flapping, short flights of galliform birds exhibit marked variations among the species brought about by the structural features of their flight apparatus, which evolved under diverse natural conditions. Apart from this, no common morphological characteristics could be established in the pectoral girdles of the two families.

Ecological Groups of Galliform Birds

246 Based on the information relating to the ecology, ethology, and morphology of locomotor organs presented in the preceding chapters, five rather distinct ecological groups could be distinguished among the galliform birds of the USSR. These include species whose distribution is closely associated with definite biotopes, a fact manifested in the nature of their feeding and structure of their locomotor organs.

The family Tetraonidae can be divided into two distinct ecological groups: inhabitants of open biotopes (Willow Grouse) and forest tetraonids. Phasianids of the USSR can be divided into three groups: inhabitants of mountains, thick undergrowth, and open expanses. Obviously these ecological groups are far from identical. Adaptive features not only to particular biotopes, but also to specific plant groups have often been observed in the family Tetraonidae. The evolution of the group as a whole exhibits a basic trend—the utilization of woody-shrub species which remain above the snow in winter—which leads to a clear-cut differentiation of food species during this season. Diet changes are reflected in the structure of the digestive organs, beak, and pelvic girdle.

The distribution of phasianids (foraging on the ground) along the northern boundary of the range of the family is determined by the depth, duration, and distribution of the snow cover. For all species inhabiting various biotopes, the snow cover is an extremely important factor in their distribution and winter life. Birds of varying origin inhabit similar biotopes or live together and exhibit common adaptive characteristics brought about not by phylogenetic relationships, but by convergence.

247 *Forest tetraonids:* Distribution of this group is associated with boreal forests of Eurasia and North America. The habitat of each species depends primarily on the availability of tree species, the chief source of food in winter.

Western Capercaillies inhabit pine or cedar forests, while Black-billed Capercaillies occupy larch or Erman's birch forests. Hazel Grouse live in mixed spruce and deciduous stands where they find food and shelter. The range of Spruce Grouse is limited to areas of boreal taiga; their chief winter food is needles of Yeddo spruce and Manchurian fir. Northern

Black Grouse prefer sparse birch groves, but penetrate even coniferous forests, especially felled and burned-out areas, and steppe shrubbery.

The ground-arboreal mode of life of this group of birds developed due to seasonal change of food. For seven to eight months Western Capercaillies, Hazel Grouse, and Spruce Grouse feed predominantly in trees, consuming the shoots, needles, catkins, and berries of both trees and shrubs. Western Capercaillies, Northern Black Grouse, and Hazel Grouse inhabiting the same forests often live together in summer, but their winter biotopes are usually different. Winter habitats reveal the food association of each species with definite tree species; this, in my opinion, is the proof of their origin in these forests. This view is supported by Kozlova (1960): "... when the winter and summer habitats of a species are not the same, the winter habitat may characterize its more antiquated biotopic associations." The locomotion of the hind limbs of this group of birds is markedly complex since the surfaces on which they move in the different seasons vary notably (soil, snow, and tree branches).

In winter these birds live on trees, wander on the snow crust (very rarely on loose snow), and make snow holes. Most of the species partly breed in trees in spring. During the snowfree period they feed on the ground, but excavating soil is limited to construction of their nest hole and raking. The morphology of the pelvic girdle of forest tetraonids reveals common characteristics associated with their arboreal mode of life and also specieswise differences, some of which are the result of living in the crowns of various trees.

All forest tetraonids have a flat, dorsoventrally somewhat compressed body, broad and flat synsacrum, and relatively short limbs (the tarsometatarsus is particularly short). Such a proportion of the skeletal components makes for a low-slung body which ensures greater equilibrium while
248 moving along branches. Compared to phasianids their digits are wider apart, a fact explained by the complex movement associated with an arboreal mode of life. The leg muscles weigh relatively less; larger birds such as Western Capercaillies and Northern Black Grouse have stronger leg muscles. Relatively better development of M. trochanterius (anterior, posterior, and medius) and M. flexores perforati (flexors of the phalanges) is characteristic of all tetraonids. The latter muscle of Hazel Grouse is especially well developed and is nearly the largest muscle of the hind limbs. In terms of the development of leg muscles, Hazel Grouse stand out with well-developed M. gastrocnemius and M. peroneus sublimus in addition to the strong musculature of the digits.

Small size and high-set body enable Hazel Grouse to move freely among the dead wood, stumps, and felled tree trunks in a cluttered forest. In contrast, the low-slung body of large Western Capercaillies impede their movement and ability to negotiate even the smallest obstacles.

The characteristic structural features of the hind limbs of forest tetraonids are the horny formations on the toes, spines and tubercles on the underside of the soles, and fringes along their sides. The functional significance of these structures varies. Spines on the sole and horny fringes prevent slipping from tree branches and on the snow crust. The horny fringes enlarge the tread and also facilitate movement on snow. Furthermore, when brought close together, the toes form a scoop—a feature useful for making holes in snow.

Association with forest biotopes is distinctly seen in the type of flight and structure of the flight apparatus in Hazel Grouse and, apparently, also in Spruce Grouse. The projectile take-off of Hazel Grouse is short, but extreme maneuverability in the forest is possible due to its short, highly curved wings, with a small wing projection, arched primaries, and fairly long tail.

The highly convex, obtuse, short wing of Spruce Grouse resembles that of Hazel Grouse, but the exceptionally deep slots of the primaries (the functional significance of which is still not clearly understood) reveal an essential difference between the two species, arising probably from differences in flight characteristics.

In flight characteristics and wing shape, not only Northern Black Grouse but also Western Capercaillies may rightly be considered (strange as it may seem) birds of open biotopes. Their comparatively long, narrow
249 wings are hardly suitable for flight in the forest. It is known that Western Capercaillies and Black Grouse fly little in the forest and particularly avoid thick vegetation. At the same time, they can undertake distant migrations, rising high above the forest.

Marked differences in the flight characteristics of Hazel Grouse and Western Capercaillies, often living side by side, are explained by their present-day habitats and evolution in different types of forests.

Tetraonids of open biotopes: This group of birds includes ptarmigans of the genus *Lagopus* which inhabit vast open expanses in high latitudes as well as boreal forests. Many morphological features of this group of birds and forest tetraonids are so similar that they lend credence to the assumption that ptarmigans led a ground-arboreal mode of life in the past and became ground birds only after settling in forest-less areas.

Their main winter food is branches and buds of shrub willow, birch, and alder (Willow Grouse) and; herbaceous plants in less snowy regions and catkins and buds of shrub alder, birch, and willow (Rock Ptarmigan).

Such a diet calls for free movement on snow and a more specialized function of the hind limbs compared to forest tetraonids. The general structure of the skeleton of the hind limbs in ptarmigans differs little from that of forest tetraonids. A flat synsacrum, straight femur, and relatively short limbs are as typical of ptarmigans as of the ground-arboreal species.

The musculature of ptarmigans, highly varying in proportions, occupies an intermediate position between other tetraonids leading a ground-arboreal life and phasianids. The largest muscles of the hind limbs—M. caudatus ilioflexorius, M. iliotibialis, M. femorotibialis, and M. gastrocnemius—are almost equal in size. The M. trochanterius (flexors of the femur) are as well developed as in other tetraonids. This is probably responsible for the limited mobility of the femur which, in turn, shortens the stride somewhat. On the other hand, M. flexores perforati (flexors of digits) are more poorly developed in Willow Grouse than in forest tetraonids, perhaps due to their behavioral pattern.

The life of these birds on the ground is thus reflected in the musculature of their hind limbs, but has not markedly changed their skeleton. The maximum changes occurred in the distal parts of their feet. Constant
250 movement on snow led to the complete disappearance of the horny fringes on the toes, feathering of the toes, and the growth of claws. These changes notably increased the area of the tread, formed "snow-shoes," and facilitated scraping and movement on snow.

In the nature of flight—range and considerable height (seen during courtship displays and migrations)—ptarmigans are closer to capercaillies than any other forest tetraonid. The two groups also have much in common in the structure of their flight apparatus: relatively long, narrow, wings, with slightly curved primary wing feathers, and structurally similar sternums. This similarity between birds inhabiting different biotopes can only be explained by their close phylogenetic relationships.

The habitats of phasianids are more diverse. Unlike tetraonids, in which evolution mostly proceeded in the direction of food specialization, adaptations to the topographic features and vegetation density are better manifested in the heterogeneous ecological groups of phasianids. The similarity in their behavior and morphology is often the result of convergence caused by life under similar conditions and not due to phylogenetic relationship.

Phasianids of hilly terrains: This group includes species of diverse origin, clearly manifested in their vertical distribution. Snowcocks, isolated under the severe climatic conditions of Central Asia, inhabit high mountains while Sand Partridges, a group closely related to species inhabiting the arid lands of West Asia, inhabit low mountains. Xerophytic mountain ranges of the Mediterranean represent the homeland of Rock Partridges while Chukar, which are the most adaptable species of this genus, inhabits mountains from the foothills to alpine meadows. The mode of foraging is the same in all species of this group: they feed on various parts of herbaceous plants, including the roots, bulbs, and rhizomes. Seasonal changes in food composition are restricted to the availability of insects.

General adaptive features in the structure of the locomotor organs are found in all mountain galliform birds, but marked differences due to their different origin and different topographic conditions of habitat and to some extent the size of the birds have also been observed.

Compared to forms from open lands, mountain phasianids have a broader synsacrum; this is particularly evident in Sand Partridges, which have a broad, short synsacrum with highly developed lobes in the proximal part of the ilium. In snowcocks, the synsacrum is relatively larger and
251 heavier but is somewhat narrow in Chukar. The hind limb proportions of snowcocks are almost the same as in tetraonids—long femur and short tarsometatarsus—resulting in relatively short but high strides. Chukar and Sand partridges with their high-set bodies are more mobile. On the whole, the structure of the skeleton and hind limb musculature of snowcocks reveals certain features which ensure the equilibrium of this large but relatively slow-moving bird. Vast slopes and alpine plateaus provide a wide vista for snowcocks to detect danger from a distance and help it escape on foot leisurely or fly into a nearby ravine. Because narrow ravines are the typical habitat of Chukar, they have developed the ability to run fast and clamber with dexterity over rocks and stones. This kind of movement calls for greater mobility of the whole limb. The structure of the leg musculature of Sand Partridges reveals features common with Black Francolins, but differs from all other birds in its extremely weak toe muscles. The structure of the feet of the galliform birds I studied exhibited many common features; all the same there were notable differences. The digits of snowcocks are relatively thicker, but those of Chukar and Sand partridges are thinner and longer. The digits of Chukar are wider apart, which probably explains its astonishing dexterity in clambering over rocks. The sole of the toes of all these species consists of separate soft pads, most developed at the base of the toes. Their surface is covered with a horny layer, thicker in winter, with prominent tubercles. The soft sole of the toes enhances the bird's grip and reduces slipping, thus facilitating movement on stones and rocks.

Similarities in the pelvic girdles of mountain species are expressed in the low-slung body and structure of the distal part of the foot (soft sole of toes and structure of claws). Marked differences relate chiefly to the structure of the synsacrum.

The need for considerable vertical ascent exerts great influence on the flight and structure of the flight apparatus in mountain species. In snowcocks, vertical diurnal migrations are performed by rapid flights down the ravines and ascents on foot up steep slopes. With such alternating movements the main load is shared by the hind limbs; since the flight is passive, little energy is expended. Snowcocks have lost the ability to take-

off vertically from a flat surface as well as to perform active flight. Aberrant types of flight are reflected in the structure of the flight apparatus. The
252 small but thick sternum, with a very low keel and long furcula, is the site of attachment of weak muscles. Short, obtuse, and flat wings with very small slots in the primary feathers carry a heavy load facilitating rapid gliding, while limiting active flight and take-off from flat surfaces.

The descent of Chukar resembles that of snowcocks and is also effected by their high wing-load relative to the bird size. However, because of their more powerful pectoral muscles, Chukar continue to retain the ability for active flight. These birds can fly short distances as well as up the slopes and, in moments of crisis, are even capable of vertical take-off.

For Sand Partridges, gliding downward in low hills is less convenient. Typically the range of flight of this bird is considerably longer than that of Chukar. The structure of its flight apparatus is intermediate to Gray and Chukar. Their pectoral musculature is developed as in Gray Partridges and far better than in Chukar. The structure of the pectoral girdle also combines features of these two species. The curvature of the dorsal part of the sternum and its high keel bring the Sand Partridge closer to Gray Partridges, while the fairly short furcula and small hypocleidium more resemble the Chukar.

Adaptive features of locomotion and the structure of the locomotor organs are dictated by life in mountains. These are more pronounced in the highly specialized Snowcock inhabiting high mountains. On the other hand, the Sand Partridge, a resident of low hills, has much in common, both in behavior and structure of the locomotor organs, with birds of the plains—Black Francolins and Gray Partridges. Chukar with their much wider vertical distribution occupy an intermediate position in this respect.

Phasianids of dense undergrowth: The distribution of pheasants and francolins is associated with various types of dense vegetative cover: intrazonal regions in deserts (tugai, reeds, and oases), brush of woody and shrub species in hilly regions, and the dense vegetation of cultivated areas. Protective conditions in their habitats are of primary importance in the evolution of these species. Life in dense vegetation is reflected in the
253 behavior of these birds. Only life in dense brush could produce their exceptional agility on land and the very limited but typical flight pattern causing the structural similarity of locomotor organs in these species of totally different origin.

The great similarity in the diet of these birds (francolins and pheasants), characterized by the maximum content of animal food and more uniform plant components and differing markedly from that of other galliform birds, is very striking. Morphological features are similar in many respects in these birds which differ much outwardly. In both species the laterally compressed body facilitates movement in dense undergrowth. The body is highly compressed in francolins and hence the synsacrum is extremely narrow. The synsacrum of pheasants is slightly broader, especially its distal part, but still narrower than in the other phasianids studied (except for quail). The ratios of the skeletal components of the hind limbs in pheasants and the francolins are almost identical and differ from the other galliform birds in a short femur and very long tarsometatarsus. The limbs are relatively longer than those of other phasianids, giving them a high-set body; this becomes quite noticeable when they run fast as well as when moving among thick undergrowth. The musculature of the Francolin and Pheasant differs noticeably. Broadly speaking, the leg muscles in francolins are better developed than in pheasants and are second only to those of snowcocks. The relative development of individual muscles in pheasants differs markedly from other members of the family in the exceptionally powerful M. gastrocnemius and relatively weak M. iliotibialis. In francolins these muscles are equally developed. Thus, in the development of individual limb muscles, francolins appear to occupy an intermediate position between pheasants and Sand Partridges on one hand, and Gray Partridges on the other.

In the structure of the foot there is much in common between francolins and pheasants, but the digits of francolins are relatively small, not so thick, and armed with small claws. The toes of pheasants are much bigger, with fairly long and obtuse claws. There is no marked difference in the structure of the sole pads of the toes in the two species. The horny layer is better developed in pheasants (winter specimens) and has more convex regular tubercles than francolins. The much thicker horny layer on the pads of pheasants probably formed in response to the severe winters they have to endure in the northern parts of their range. Black Francolins, typical members of the subtropics, inhabit regions with less snow and mild winters.

The short flight of pheasants and francolins could well be described as an "escape flight." They fly out of the thicket only *in extremis* and after a
254 short distance settle down again in the thicket. The rapid vertical and raucous take-off characteristic of galliform birds is particularly distinct in these species. This type of flight is also reflected in the structure of their flight apparatus. The narrow sternum (due to the medially flattened body) with its long, high keel provides sufficient area for the attachment of the well-developed pectoral muscles to the flat trunk.

Broad, short, convex wings, with a small wing projection and deep slots in the primaries, and rather strong pectoral muscles facilitate the projectile take-off that calls for a considerable expenditure of energy and short flights.

The great similarities in the behavior of francolins and pheasants and in the structure of their locomotor organs may be regarded as a convergence of two phylogenetically distant forms living in similar biotopes.

Phasianids of open biotopes: Open grasslands in the plains and gentle mountain slopes--meadows, steppes, and fields—are the habitats of these widespread species: Common Quail, Japanese Quail, *C. coturnix,* Gray Partridges, and Daurian Partridges. If the examination of other ecological groups of tetraonids inhabiting any biotopes readily revealed common ecomorphological features as a consequence of common origins (Tetraonidae) or convergence then in species inhabiting open grasslands such similarities could not be observed.

The two species of quail and two species of pheasants (the species pair in each of these birds have such a close resemblance they were at times considered subspecies) differ markedly in ecology and morphological features.

The migratory habit has left its imprint on the entire behavior of quail, including significant changes in their flight apparatus caused by distant migrations. An elongated, pointed wing without slots in the primary feathers facilitates flight for considerable distances, but rules out sudden vertical take-offs. Notable structural features of the pectoral girdle are the large keel of the sternum and the position of the clavicle forming a wide angle with the coracoid, making for a large area of attachment for the pectoral muscles. The large spina sterni and hypocleidium ensure a very strong attachment of the membranous sternocoraco-clavicularis, the major support of the thoracic musculature in the proximal part of the pectoral girdle. However, in the life of quail, flight is relatively infrequent
255 and they seldom fly at their nesting and wintering sites. Thus there are sharp seasonal changes in the functioning of the locomotor organs of quail.

Movement in the dense grass particularly affected the body shape of these small birds. The trunk of quail is not only highly compressed medially, but highly elongated; consequently the synsacrum and thoracic cavity are very narrow.

In the larger partridges that live under similar conditions as quail the body shape is different. It is somewhat flattened dorsoventrally, the synsacrum is broader than in other phasianids, and its distal part is broader than the distance between the antitrochanters. The bones of the hind limbs are relatively longer in partridges than in quail, probably due to the elongated body of the latter and the fact that this index is calculated in reference to the trunk length. Very high mobility of the femur seems to largely compensate for the short limbs and somewhat increases the length of the stride.

The ratios of the skeletal components of the hind limbs (without digits) of quail and partridges are almost identical and these birds are intermediate to fast runners (Black Francolin and Common Pheasant), which have long tarsometatarsus and short femur, and mountain species (snowcocks and Chukar), which have a short tarsometatarsus and long femur.

The relative weight of the muscles of the hind limbs of quail and partridges is the same; however, the size ratios of individual muscles differs markedly. In Quail, M. tibialis, M. femorotibialis, and M. gastrocnemius are very large and nearly equal in size, while the flexors of the femur and M. sartorius are less developed than in most phasianids. In Gray Partridges the strongest muscle is the M. iliotibialis, while M. femorotibialis and M. gastrocnemius are considerably weaker; M. trochanterius and M. ischiofemoralis, like M. tibialis, are much stronger than in quail.

The structure of the feet of quail and partridges differs significantly. Quail have very long, thin digits and the third toe is nearly as long as the tarsometatarsus. In summer, the ventral surface of the digits is covered with large soft pads with indistinct furrows. The digits of partridges are relatively short and the middle toe much shorter than the tarsometatarsus. The structure of the sole exhibits sharp seasonal changes. In summer large but flat pads grow on the underside of the toes; these leathery pads, resembling those of quail, have indistinct furrows. In winter the horny
256 layer is much thicker and consists of horny, highly prolate tubercles. These tubercles are set higher in Gray Partridges than in Daurian Partridges and their tips project forward, somewhat like the horny spines on the soles of tetraonids. The significant differences in the general structure of the body and of the locomotor organs can be attributed to deep differences in the life styles of quails and partridges; furthermore, they point to the completely different evolution of these species, which at present live together only during the breeding season.

* * *

A comparison of the ecological groups of tetraonids and phasianids and their association with definite biotopes leads us to some definite conclusions. Food specialization was apparently the basic direction in the evolution of tetraonids. A ground-arboreal mode of life—the result of the need to subsist on coarse twigs in winter—led to adaptations of the beak, digestive organs, pelvic girdle, and the integument of the digits.

No common features could be seen in the structure of the flight apparatus that could be regarded as typical of species of a single ecological group. The flight characteristics as well as morphology of the pectoral girdle differ markedly in various species and are wholly determined by habitat conditions (flight of Western Capercaillies high above the forest and maneuvered flight of Hazel Grouse in dense forest).

The basic direction of evolution among phasianids is not food specialization, but adaptation to different biotopes. This is well reflected in differences in locomotion and locomotor organs in species living under different conditions. Characteristic locomotion on foot and by flight, as well as common morphological features of the locomotor organs, are typical of species of each ecological group. The most distinctive are species with an aberrant type of flight (Common Quail and snowcocks), which also differ markedly in the structure of the pectoral girdle. The characteristic gliding of snowcocks could only have developed in high-mountain conditions calling for great vertical movement. This ability is developed to a lesser extent in Chukar. The medially flattened and highly elongated trunk of quail resulted from life in a dense grass canopy. Additionally, distant migrations led to changes in the entire flight apparatus, manifested in morphological changes of the pectoral girdle, "red muscles" (not characteristic of other phasianids), and the unique shape of the narrow wings with a wholly acuminate apex.

Conclusion

257 In the avifauna of the USSR the order Galliformes is represented by two families—Tetraonidae and Phasianidae. The total number of their species is not large—21 (or 7.6% of the entire order). However, the extensive geographic ranges and fairly high population of most of these birds place them among the top positions of the feathered game.

The diverse group of phasianids in the avifauna of the USSR consists mainly of the northernmost widespread species, which have phylogenetic links with the Southeast, and West Asia, Africa, and southern Europe. Snowcocks are the only exception; they originated in the severe conditions of the high mountains of Central Asia.

Ground life year round is characteristic of all species of the family, a feature preserved even in the northernmost species inhabiting regions with prolonged winters, when ground food is mostly covered with snow.

The isolation of Tetraonidae from Phasianidae was brought about by the cooling of the climate, the formation of distinct seasons, the displacement of tropical flora by deciduous species and, as a consequence, the formation of the taiga. Tetraonids represent a well-differentiated group that stands out in the entire order Galliformes in its ecological and morphological adaptations to severe climatic conditions.

The taxonomic position of tetraonids to date has been considered differently: either as a subfamily of Phasianidae or as an independent family. In my opinion, and that of the majority of Soviet ornithologists, tetraonids represent an independent family. The evolution of tetraonids proceeded along the course of adaptation of thermophilic forms to more severe climates. Changes in diet occurred concomitantly under the new
258 conditions and led to distinct seasonal dietary adjustments, e.g., the consumption of a limited assortment of widely available coarse twigs and other twig food during the snowy period. This, in turn, influenced the behavior of the birds leading to morphological adaptations in the procurement and assimilation of various types of food.

The common conditions of formation of the homogeneous group of tetraonids are also reflected in the extremely similar food composition of different species. This is manifested in seasonal changes of diet, the use of twig food in winter, and especially in the great similarity of their species composition during the snowfree period.

The feeding of phasianids is characterized by indistinct seasonal change of food items. Unlike tetraonids their food is more varied in winter when birds that do not feed on twigs face great difficulty in food procurement and feed on whatever herbaceous vegetation is available. During the snowfree period the food composition of different species varies significantly, which is explained by their evolution under varying conditions and also by great differences in their present-day habitats. The most important difference in the diet of tetraonids and phasianids is that the former never feed on the underground plant parts while the latter have no desire for twigs.

To date, 1,282 plant species have been identified in the food of Galliformes of the Soviet Union. Higher plants are represented by 105 families. Tetraonids feed on 735 plant species belonging to 80 families and phasianids on 645 species belonging to 79 families.

The similarities and differences in the plant food components of tetraonids and phasianids are revealed by the fact that of the total plants identified, 51.9% of the families are common to both, but their species composition varies; only 7.6% of the plants are common in the diets of both groups.

For most galliform birds animal food is less significant than plant food. In the quantity of animal food consumed, tetraonids differ less than phasianids; among the latter, snowcocks feed almost exclusively on plants while animals are the chief food of francolins and pheasants for much of the year.

The biochemical composition of the winter food of tetraonids and phasianids differs markedly. The abundantly available twigs are less nutritive, while the limited food resources of phasianids have a higher calorie value. During the snowfree period, when notable differences in diet are seen, the nutritive value is compensated for, which is so very important for the various biological processes occurring in this period (reproduction, growth of young, and molting). The consumption of twigs by tetraonids led to changes in their beak structure, as reflected in the formation of
259 characteristic "scissors" on the sides, which help shear off twigs. Coarse food of low nutritive value but with high cellulose content led to sharp changes in the morphology of the blind gut and apparently also in the physiology of digestion; the latter aspect has hardly been studied so far. The consumption of bulk food cut short the time required for fattening and brought about sharp changes in the activity of the birds, which had a favorable effect on their energy balance. The availability of food, albeit of low calorie value, together with the protection offered by the snow cover

favored the colonization by tetraonids of vast territories with particularly severe winters and heavy snowfall.

If snow cover plays a favorable role in the life of tetraonids, for most phasianids, which are ground feeders, it has only an adverse effect. It deprives them of food and hinders their movement. Only a few northernmost species of this family inhabit regions with prolonged winters; unlike tetraonids, the thickness of the snow cover limits the northward extension of their range.

Tetraonids habitually forage on tree tops, a behavioral characteristic reflected in the structure of the entire pelvic girdle. Compared to phasianids, this group of birds has relatively shorter limbs, especially the tarsometatarsus, resulting in a low-slung body. The broad, low sacrum produces a spacious body cavity. Such a structure of the limbs and synsacrum ensures better equilibrium, a feature essential when hopping from branch to branch. At the same time, the spacious body cavity can accommodate a voluminous intestine.

Ground life is a secondary adaptation in the genus *Lagopus*. In the structure of the pelvic girdle, ptarmigans have to a great extent preserved certain features in common with forest tetraonids, but they occupy an intermediate position between them and phasianids in weight ratios of the muscles of the hind limbs. Feathered toes and broad flat claws indicate the specialization of the toes for movement on loose snow and for digging in it; such morphological features are, however, less suitable for movement among tree branches.

Differences in the structure of the hind limbs of various species of phasianids can be explained by their diverse habitats and modes of foraging (digging the soil for food, noticed in all species).

The main purpose of the limited flight of galliform birds is quick
260 movement over short distances within their habitats. Hence the structure of their flight apparatus, unlike the pelvic girdle, distinctly reveals an adaptation to specific biotope conditions rather than common features of phylogenetic affinity between individual species and groups of birds. The most primitive flapping flight of galliform birds requires large expenditure of energy, requiring powerful pectoral muscles and a high sternal keel for their attachment. Despite an overall similarity in the type of flight, species differences are well defined, particularly in the highly aberrant forms.

Unlike other galliform birds, migratory quail have a narrow, relatively long wing, a proximally stronger, light-weight pectoral girdle, and a smaller M. pectoralis minor. Such a structure of the flight apparatus facilitates distant migrations, but deprives them of the ability for projectile take-off.

An overall reduction of the pectoral muscles and small M. pectoralis minor are characteristic of the pectoral girdle in snowcocks. The main differences in the sternum are its small size and especially low keel. Short wings with rather faint slots experience a high wing-load. The flight of snowcocks is restricted to rapid downward gliding. They are totally incapable of a vertical take-off. The characteristic type of flight of snowcocks could only have arisen under alpine conditions with great height differences, in which downward movement is by flight and ascent on foot.

Among the galliform birds of the Soviet Union, five ecological groups have been distinguished which are adapted to definite biotopes and climatic conditions.

The adaptation of tetraonids to various biotopes is manifest primarily in food associations revealed in the utilization of the twigs of plants of various levels; this has led to the formation of two ecological groups—forest tetraonids and inhabitants of open landscapes.

Phasianids display lesser differences in the mode of foraging. The ecological groups of this family have a more pronounced adaptation to topographic features or sheltering conditions of environment, which is reflected in the adaptive features of the locomotor organs. The ecological groups of phasianids include species of diverse origin; hence similarities in the adaptive features of these birds should be regarded as due simply to convergence. Of the three ecological groups of this family, adaptations of the locomotor organs are better manifested in birds inhabiting mountains.
261 Limited flight leads to the reduction of the pectoral muscles and skeletal changes in the pectoral girdle (low keel). Greater work by the hind limbs leads to a hypertrophy of their musculature.

For the second group of phasianids inhabiting thick undergrowth, the habitat primarily serves as a shelter. Life in rather dense, contiguous vegetation is reflected in the general body structure (medially flattened) and locomotor organs. Mobility and dexterity so necessary in thick undergrowth are provided by the shape of the high-set body. The structure of the pectoral girdle, rather well-developed musculature, and broad wings with deep slots in the primary feathers facilitate vertical take-offs from thick vegetation.

Open grasslands such as meadows, steppes, and fields are inhabited by the third, less homogeneous group of phasianids (Quail and Gray Partridges). These are widespread species and the least selective in choice of habitat. The grass canopy provides them sufficient shelter and abundant food. Species of this group have penetrated farther north, than any other group of phasianids, into regions with a fairly deep snow cover. Apparently this has become possible only because of their ability for distant migrations.

The extensive range of the forest tetraonids in the Soviet Union and the circumpolar and circumboreal distribution of Willow Grouse should not necessarily be regarded as indications of the remarkable adaptability of this species. In the vast expanse of the boreal forests and high latitudes, both climate and flora are fairly uniform, and each species finds for itself suitable conditions in these vast territories. Similarity of habitats in various parts of the vast range indicates a strict preference for specific conditions. The narrow food specialization of each species during winter and a sedentary mode of life, manifest in the seasonal changes of biotopes within a very limited territory, underscore the low ecological adjustment of most of these species.

The severe climatic conditions under which the tetraonids evolved gave rise to highly specialized species, with a rather wide distribution and high population density only in suitable natural conditions, in which some definite species of food plants are available.

Willow and Northern Black Grouse are relatively more adaptable species of this family; occupying more diverse habitats in varied natural
262 zones. Moreover, Willow Grouse feed everywhere (except in regions along the southern boundary of its range) on the shoots of shrubs in winter, while the diet of Black Grouse in this season varies greatly in different parts of its range.

The distribution of phasianids as well as tetraonids depends on the wintering conditions: while the availability of twigs is the decisive factor for tetraonids, it is the climate (thickness of the snow cover) on which phasianids, as ground feeders, depend for food.

In the avifauna of the Soviet Union the diverse group of phasianids includes both highly specialized species (snowcocks) and highly adaptive ones (pheasants, Gray Partridges, and quail), a fact proved by the expanding range of the latter, closely following the changes of biotopes caused by clearing forests, the expanding agriculture, and also by their successful naturalization.

Variations in the population of species of these two families in countries where great changes in natural cenoses have taken place reveal that the galliform birds of the Palearctic react differently to the changeover of natural biotopes into "cultivated" lands. The more specialized tetraonids usually do not adjust to sharp changes in their main habitats, and hence their population drops sharply and, at times, their range shrinks.

Spruce Grouse are the most specialized species of this family judging from their particularly narrow food specialization. The biology of these interesting birds has not yet been thoroughly studied. Low populations and limited range call for protecting this bird and conserving its natural habitats. Special attention should also be paid to protecting the endemic

Caucasian Black Grouse, which have small numbers and a highly restricted range. The remaining species of this family could be widely used as game birds now and in the future. However, it should not be forgotten that each species is strictly confined to specific biotopes and that their fecundity is relatively low. Large-scale felling of forests and extension of agriculture could shrink the range and reduce the population of most tetraonids, as has happened in western Europe. Only creation of necessary conditions for the wildlife through rational management of forest husbandry can, to a great extent, help conserve the rich wildlife.

It is known that frequent felling of forests and making clearings have promoted the dispersal and population rise not only of Northern Black
263 Grouse, but Hazel Grouse as well. Western Capercaillies adapt rather easily to changes in conditions of the summer habitats when their major breeding sites are protected (Doppel'maier, 1950, 1951; Osmolovskaya, 1966). Attempts at domestication and captive breeding of this species have not been successful (E.A. and E.V. Krutovskii, 1953; Nemtsev, et al., 1973; Kutovaya, 1976). Creating favorable environmental conditions year round and providing natural foods, especially for young birds in winter, their high mortality, and the difficulty of brood care hardly make the commercial breeding of Western Capercaillies feasible. Furthermore, attempts at introducing tetraonids have not been successful and those reared under semicaptive conditions could not adjust to new habitats. In my opinion, therefore, the only way to raise a sufficiently high population for game purposes is to protect the original habitats of these birds, even in the forests.

The high fecundity, ease of domestication, and widely practiced naturalization of a number of species of phasianids indicate the ready adaptability of many of them. All these qualities can be exploited by man, now and in the future, by evolving new breeds, domesticating and breeding birds under a variety of conditions including on "cultivated" lands.

The Palearctic phasianids are particularly adaptable and very fecund. Among these are pheasants and Gray Partridges, whose geographic ranges have been advancing with the advancing cultivation (felling of forests, expanding agricultural lands, irrigation of arid lands, etc.). Judging from the successful naturalization of these species and their wide use as game in western Europe and North America, Gray Partridge and pheasants represent the main stock of game birds in the "cultivated" zones. In many south European countries two species of Rock Partridges—*Alectoris rufa* and *A. graeca*—are very important in the rearing of game birds. A third species of this genus—Chukar—is especially promising as a game bird in the southern hilly regions of the Soviet Union. However, the potential possibilities of Daurian Partridges and quail for this purpose should not be ruled out.

Further enrichment of the fauna, its rational exploitation and the conservation of rare species should be based on a sound knowledge of their biology. However, the study of Galliformes of the Soviet Union opens a
264 mixed picture. Among tetraonids, the biology of Northern Black Grouse, Hazel Grouse, Western Capercaillies, and Willow Grouse has been studied exhaustively, while information on the life styles of Spruce Grouse and Caucasian Black Grouse is nearly non-existent, and we know very little about the biology of Rock Ptarmigan. Phasianids of the Soviet Union in general, even valuable game species such as Gray Partridges and pheasants, have not been adequately studied. Not much is known about the biology of the less numerous species of Sand Partridges and snowcocks and much remains vague about the life styles of such common and promising species as Chukar, Daurian Partridges, and quail.

A study of the biology of the lesser known species of Galliformes of the Soviet Union should be pursued for two reasons. First, a better knowledge of the biology of rare species confined to a narrow geographic range ensures a properly organized program for their conservation. Second, a detailed study of the life styles of the more common species would no doubt be invaluable in their rational exploitation of this group of birds for the good of mankind.

Bibliography

265 Abdusalyamov, I.A. 1957. Materialy po biologii gimalaiskogo i tibetskogo ularov na Pamire (Biology of the Himalayan and Tibetan snowcocks of Pamir). *Trudy Tadzh. Un-ta Seriya Estestv.* Nauk, vol. 2.

Abdusalyamov, I.A. 1957. Novye dannye po ptitsam Vostochnogo Pamira (New data on birds of the eastern Pamir). *Dokl. AN TadzhSSR,* vol. 22.

Abdusalyamov, I.A. 1964. Ptitsy Gornogo Zeravshana (Birds of Zeravshan Mountains). Dushanbe.

Abdusalyamov, I.A. 1971. Fauna Tadzhikskoi SSR (Fauna of Tadzhik SSR). Dushanbe, vol. 19, pt. I, Birds.

Abramov, K.G. 1962. Nablyudeniya nad dikushei v Priamur'e (Observations on Spruce Grouse in the Amur region). *Ornitologiya,* No. 4.

Afanas'ev, A.V. 1934. Okhotnichii promysel v raione khrebta Dusse-Alin' (k severu ot Dul'nikanskogo perevala) [Game management in the Dusse-Alin' range (north of Dul'nikansk Pass)]. *Trudy SOPS AN SSSR. Seriya Dal'nevost.,* No. 2.

Airumyan, K.A. 1967. Morfologicheskie adaptatsii fazana i keklika k spetsificheskim usloviyam obitaniya (Morphological adaptations of pheasants and chukar to specific environmental conditions). *Aiastani Gensapanakan Antes., Biol. Zhurn. Armenii,* vol. 20, No. 2.

Aleksandrova, I.V. and L.I. Krasovskii. 1960. Ob osennikh statsiyakh serykh kuropatok v Podmoskov'e (Autumn habitats of Gray Partridges near Moscow). *Byul. Mosk. Ob-va Ispyt. Prirody,* vol. 65, No. 4.

Aleksandrova, V.D. 1940. Kormovaya kharakteristika rastenii Krainego Severa (Dietary values of plants in the Far North). *Trudy Nauchno-Issled. In-ta Polyarn. Zemledeliya, Zhivotnovodstva i Promysl. Khoz-va,* No. 2.

Alferaki, S. 1882. Zametki o serykh kuropatkakh (Notes on Gray Partridges). *Priroda i Okhota,* July.

Alai, Salim and D.S. Ripley. 1962. In: *Handbook of the Birds of India and Pakistan Together with Those of Nepal, Sikkim, Bhutan, and Ceylon,* Bombay-London-New York.

Andreev, A.V. 1971. Osobennosti ekologii tundryanoi kuropatki *Lagopus mutus* na Komandorskikh ostrovakh [Characteristics of ecology of Rock Ptarmigans (*Lagopus mutus*) on the Komandor Islands]. *Zool. Zhurn.*, vol. 50, No. 8.

Andreev, A.V. 1975. Zimnyaya zhizn' i pitanie tundryanoi kuropatki (*Lagopus mutus*) na krainem severo-vostoke SSSR [Winter life and feeding habits of Rock Ptarmigan (*Lagopus mutus*) in far northeastern Soviet Union]. *Zool. Zhurn.*, vol. 54, No. 5.

Andreev, B.A. 1953. Ptitsy srednego Vilyuya (Birds of Central Vilyuysk). Yakutia.

Archer, G. and E. Godman. 1937. *The Birds of British Somaliland and the Gulf of Aden.* London, vol. 11.

Arkharov, I. 1939. Kormovoe znachenie berezovoi cerezhki dlya ryabchika i tetereva i ee khimicheskii sostav (Nutritive values and chemical composition of birch catkins for Hazel Grouse and Black Grouse). *Boets-Okhotnik,* No. 9.

Aubekerova, P.A. 1965. Novye dannye ob iskopaemykh ptitsakh Kazakhstana (New data on fossil birds of Kazakhstan). In: *Novosti Ornitologii.* Alma-Ata.

Austin, O.L. 1972. Birds of Korea. *Bull. Museum Comparative Zoology, Harvard College.* Cambridge, vol. 101, No. 1.

Averin, Yu.V. 1938. Kavkazskii teterev (Caucasian Blackcock). *Trudy Kavkazskogo Gos. Zapovednika.* Moscow, No. 1.

Averin, Yu.V. 1948. Nazemnye pozvonochnye Vostochnoi Kamchatki (Land vertebrates of eastern Kamchatka). *Trudy Kronotskogo Zapovednika.* Moscow, No. 1.

Averin, Yu.V. 1948. Kamennyi glukhar' na Vostochnoi Kamchatke (Black-billed Capercaillie in eastern Kamchatka). *Okhrana Prirody,* No. 5.

Bakeev, N.N. and V.I. Osmolovskaya. 1964. Chislennost' teterevinykh ptits Urala i Zaural'ya (The Abundance of galliform birds in the Urals and Trans-Urals). *Byul. Mosk. Ob-va Ispyt. Prirody. Otd. Biol.*, vol. 69, No. 3.

Baker, E. and C. Stuart. 1930. *Game Birds of India, Burma, and Ceylon.* Bombay-London.

Bannerman, D.A. 1953. *The Birds of West and Equatorial Africa.* London, vol. 1.

Bannikov, A.G. 1955. Zametki o fazane v nizov'yakh r. Samura (On the pheasants in the lower reaches of the Samura River). *Uchen. Zap. Mosk. Gor. Ped. In-ta,* vol. 38, No. 3.

Bannikov, A.G. 1957. Zametki o zimnikh kormakh ryabchika i daurskoi kuropatki v Kentee (On the winter food of Hazel Grouse and Daurian Partridge in Kentei). *Uchen. Zap. Mosk. Gor. Ped. In-ta.*, vol. 65, No. 6.

Bannikov, A.G. 1958. Ekologicheskie osobennosti i ochagi formirovaniya vysokogornoi fauny mlekopitayushchikh Evrazii (Ecological characteristics and formation centers of alpine mammalian fauna in Eurasia). In: *Problemy Zoogeografii Sushi.* L'vov.

Barancheev, L.M. 1954. Okhotnich'e promyslovye ptitsy Amurskoi oblasti (Game Birds of the Amur District). Blagoveshchensk.

Barancheev, L.M. 1964. Izmenenie areala (*Phasianus colchicus* Pall.) v Amurskoi oblasti [Changes in the geographic range of Common Pheasants (*Phasianus colchicus* Pall.) in the Amur District]. In: *Okhrana Prirody na Dal'nem Vostoke.* Vladivostok, No. 2.

Barancheev, L.M. 1965. Chislennost' i elementy ekologii kamennogo glukharya v Amurskoi oblasti (Numbers and ecological factors of Black-billed Capercaillies in the Amur District). *Ornitologiya,* No. 7.

Barancheev, L.M. 1965. Migratsii man'chzhurskikh fazanov v Amurskoi oblasti (Migration of Manchurian pheasants into the Amur District). In: *Migratsii Ptits i Mlekopitayushchikh.* Moscow.

Bates, B.S.P. and L.H.N. Lower. 1932. *Breeding Birds of Kashmir.*

Baziev, Zh. Kh. 1962. O kavkazskoi gornoi indeike (The Caucasian Mountain Turkey). Nal'chik.

Baziev, Zh. Kh. 1965. Ulary Kavkaza kak primer prisposobleniya ptits k usloviyam zhizni v vysokogor'e (Caucasian Snowcock as an example of the adaptation of birds to life in high mountains). Avtoref. Kand. Diss., Moscow.

Baziev, Zh. Kh. 1966. Lin'ka Kavkazskogo ulara (Molt in Caucasian Snowcocks). *Zool. Zhurn.,* vol. 45, No. 1.

Baziev, Zh. Kh. 1968. Migratsii otdel'nykh populyatsii kavkazskogo ulara (*Tetraogallus caucasicus* Pall.) i interpretatsiya ikh proiskhozhdeniya [Migrations of some populations of Caucasian Snowcocks (*Tetraogallus caucasicus* Pall.) and an interpretation of their origins]. *Zool. Zhurn.,* vol. 47, No. 1.

Baziev, Zh. Kh. 1972. Evolution of the genus *Tetraogallus* Gray, 1834. *Acta Ornithologica, Warszawa,* vol. 13, No. 5.

Beebe, W. 1918–1922. *Monograph on the Pheasants.* Edinburgh-London, vols. 1-4.

Bëme, R.L. 1960. Vozniknovenie ornitofauny vysokogornykh landshaftov Kavkaza (Origin of the ornithofauna of the high mountains of the Caucasus). *Ornitologiya,* No. 3.

Bent, A.C. 1932. Life histories of North American gallinaceous birds, *United States National Museum Bulletin,* Washington, vol. 162.

Berezovskii, M. and V. Bianki. 1891. Ptitsy Gan'suiskogo puteshestviya G.N. Potanina (Birds of G.N. Potanin's Kansu expedition). In: *Materialy po Ornitologii Kitaya, Glavnym Obrazom Yuzhnoi Provintsii Gan'su.* St. Petersburg.

Bianki, V.L. 1898. Obzor vidov roda *Tetraogallus* Gray (Review of species of the genus *Tetraogallus* Gray). *Ezhegodn. Zool. Muzeya Akad. Nauk*, No. 3.

Bianki, V.L. 1907. Mongoliya i Kam (Mongolia and Kama). In: *Materialy dlya Avifauny Mongolii i Vostochnogo Tibeta.* St. Petersburg, vol. 5.

Bibikov, D.I. and M.N. Korelov. 1961. K ornitogeograficheskoi kharakteristike Tarbagataya (Ornithogeographic characteristics of Tarbagatai). *Trudy In-ta Zool. AN KazSSR*, vol. 15.

Birulya, A.A. 1907. Ocherki iz zhizni ptits polyarnogo poberezh'ya Sibirii (An outline of the life of birds of the Polar seacoast of Siberia). *Zap. Akad. Nauk po Fiz.-Mat. Otdeleniyu, Seriya* 8, vol. 18, No. 2.

Birulya, A.A. 1910. Zoologicheskie rezul'taty russkikh ekspeditsii na Shpitsbergen. Biologicheskie nablyudeniya nad ptitsami Shpitsbergena (Results of the Russian Zoological Expedition in Spitsbergen. Biological observations on birds of Spitsbergen). *Ezhegodn. Zool. Muzeya Akad. Nauk*, vol. 15, Nos. 1-2.

Blank, T.H. and I.S. Ash. 1956. The concept of territory in the partridge *Perdix perdix perdix*, *Ibis*, vol. 98, No. 3.

Blokhin, N. 1929. Iz Sarapul'skogo okruga (From the vicinity of Sarapulsk). *Ural'skii Okhotnik*, Nos. 11-12.

Bodenstein, G. and F. Neubauer. 1954. Steinhuhn oder Rothuhn, *Ornithol. Mitt.*, vol. 6, No. 4.

Boev, Nikolai. 1959. Pernatá zver Bulharska, jeji vyuziti a ochrana. *Sb. Ceskol Akad. Zeméd. Ved. Lesn.*, vol. 5, No. 5.

Boev, N. and G. Konstantinov. 1954. Yarebitsata, Sofiya.

Bogdanov, M.N. 1868. Biogeograficheskii ocherk tetereva polevogo (Biogeographic history of Black Grouse). *Trudy I S"ezda Russkikh Estestvoisp. Otd. Zool.*

Bogdanov, M.N. 1874. Etyudy russkoi okhoty (Studies of Russian game hunting). *Illyustrir. Zhurnal Okhoty i Konnozavodstva*, Nos. 3-4.

Bogdanov, M.N. 1879. Ptitsy Kavkaza (Birds of the Caucasus). *Trudy Ob-va Estestvoisp. pri Imp. Kazansk. Un-te*, vol. 8, No. 4.

Böker, H. 1927. Die biologische Anatomie der Flugarten der Vögel und ihre Phylogenie, *Journ. Ornithol.*, vol. 75, No. 2.

Borovitskaya, G.K. 1961. K voprosu ob izuchenii pitaniya kurinykh ptits Buryatii (Studies on the feeding of Galliformes of Buryat). In: *Kraevedchesk. Sbor. Buryatsk. Fil. Geogr. Ob-va SSSR*, No. 6.

Borovitskaya, G.K. 1965. K zimnei zhizni daurskoi kuropatki (Winter life of Daurian Partridge). In: *Novosti Ornitologii.* Alma-Ata.

Brodkorb, P. 1964. Catalogue of fossil birds. Part 2 (Anseriformes through Galliformes), *Bull. Florida State Museum. Biological Sciences*, vol. 8, No. 3.

Brown, R.H.J. 1960. Flight. Chapter XX in *Biology and Comparative Physiology* edited by A.J. Marshall. London, vol. 2.
Brüll, H. 1960. Der derzeitige Stand subtiler Äsungsstudien des Flugwildes, *Meded. Inst. Toegep. Biol. Onderz. Natuur,* No. 50.
Budnichenko, A.S. 1965. Ptitsy iskusstvennykh lesonasazhdenii stepnogo landshafta i ikh pitanie (Birds of the forest plantations of the steppes and their feeding). In: *Ptitsy Iskusstvennykh Lesonasazhdenii.* Voronezh.
Burchak-Abramovich, N.I. 1962. Novye dannye o Binagadinskikh ptitsakh (New data on birds of Binagadinsk). *Ornitologiya,* No. 4.
Burchak-Abramovich, N.I. and V.I. Tsalkin. 1971. K izucheniyu ptits nizov'ev r. Amur-Dar'i (iz materialov arkheologicheskikh raskopok drevnego Khorezma) [Study of birds of the lower reaches of the Amu-Darya River (from archeological collections of ancient Khorezm)]. *Uzbeksk. Biol. Zhurn.,* No. 1.
Burchak-Abramovich, N.I. and V.I. Tsalkin. 1971. K poznaniyu ornitofauny yuga Ukrainy, Kryma i Podon'ya (po arkheologicheskim materialam) [Information on the avifauna of the southern Ukraine, Crimea, and the Don regions (based on archeological data)]. *Byul. Mosk. Ob-va Ispyt. Prirody, Otd. Biol.,* vol. 76, No. 5.
Buturlin, S.A. 1905. O geograficheskom rasprostranenii nastoyashchikh fazanov (Geographic distribution of pheasants). *Polevaya i Ruzheinaya Okhota,* No. 1.
Buturlin, S.A. 1935. Polnyi opredelitel' ptits SSSR (A Complete Guide to Birds of the Soviet Union). Moscow-Leningrad, vol. 2.
Buturlin, S.A. 1935. Turach (Francolin). *Boets-Okhotnik,* No. 2.
Chandra-Bose, D.A. and J.C. George. 1965. Studies on the structure and physiology of the flight muscles of birds. 13: Characterization of the avian pectoralis, *Pavo,* vol. 3, No. 1.
Chandra-Bose, D.A. and J.C. George. 1965. Studies on the structure and physiology of the flight muscles of birds. 14: Characterization of the avian supracoracoideus, *Pavo,* vol. 3, No. 1.
Chunikhin, S.P. 1960. Kavkazskii ular v Kabardino-Balkarii (The Caucasian Snowcock in Kabardino-Balkaria). *Ornitologiya,* No. 3.
Chunikhin, S.P. 1964. Kavkazskii teterev (The Caucasian Blackcock). *Okhota i Okhotnich'e Khoz-vo,* No. 8.
Courtenay, L.M. and P.A. Clancey. 1960. The occurrence of migratory Palaearctic Quail in South Africa, *Ostrich,* vol. 31, No. 4.
Dal', S.K. 1944. Pozvonochnye zhivotnye Sarai Bulakskogo khrebta (Vertebrates of the Sarai Bulaksk Range). *Zool. Sbornik, AN ArmSSR,* No. 3.
Dal', S.K. 1949. Ocherk pozvonochnykh zhivotnykh Aiodzorskogo khrebta (Brief account of the vertebrates of the Aiodzorsk range). *Zool. Sbornik, AN ArmSSR,* No. 6.

Danilov, D.N. 1938. Kaloriinost' osnovnykh kormov belki (Caloric values of important protein foods). *Zool. Zhurn.*, vol. 17, No. 4.

Danilov, D.N. 1958. Osnovnye kormovye rasteniya promyslovykh zverei i ptits (Major food plants of game animals and birds). *Zool. Zhurn.*, vol. 37, No. 8.

Danilov, N.N. 1965. Teploizolyatsionnye svoistva perovogo pokrova i adaptatsiya ptits k temperaturnym usloviyam (Heat insulation properties of the feather coat and the adaptation of birds to temperature conditions). In: *Novosti Ornitologii*, Alma-Ata.

Danilov, N.N. 1966. Puti prisposobleniya nazemnykh pozvonochnykh zhivotnykh k usloviyam sushchestvovaniya v subarktike (Adaptations of land vertebrates to subarctic habitats). *Trudy In-ta Biologii Ural. Fil. AN SSSR*, vol. 2, *Ptitsy*, No. 56, Sverdlovsk.

Danilov, N.N. 1975. Ural. i Zaural'e (The Urals and Trans-Urals). In: *Teterevinye Ptitsy.* Moscow.

Delacour, J. 1951. *The Pheasants of the World.* New York.

Dement'ev, D.P. 1943. K ekologii fazana na Issyk-Kule (Ecology of pheasants in Issyk-Kul'). *Trudy Kirgizsk. Fil'. AN SSSR*, vol. 1, No. 1.

Dement'ev, G.P. 1940. Ptitsy (Birds). In: *Rukovodstvo po Zoologii.* Moscow, vol. 6.

Dement'ev, G.P. 1948. Redkie ptitsy—pamyantniki prirody SSSR. Turach (Rare birds—memorials in the Soviet Union. Francolins). *Okhrana Prirody*, No. 3.

Dement'ev, G.P. 1952. Kharakteristika otryada kurinykh Rasores ili Galliformes (Characteristics of the order Rasores or Galliformes). In: *Ptitsy Sovetskogo Soyuza.* Moscow, vol. 4.

Dement'ev, G.P. 1952. Ptitsy Turkmenistana (Birds of Turkmenistan). Ashkhabad.

Dement'ev, G.P. 1958. K voprosu ob istorii fauny ptits Sovetskogo Soyuza (On the history of the avifauna of the Soviet Union). *Uchen. Zap. MGU, Ornitologiya*, No. 197.

Dement'ev, G.P. 1964. Klass Aves-Ptitsy (Class Aves—Birds). In: *Osnovy Paleontologii. Zemnovodnye, Presmykayushchiesya, Ptitsy.* Moscow.

Dement'ev, G.P. and I.P. Shimbareva. 1941. Ptitsy i zima 1939/40 g (Birds and the winter of 1939/40). *Zool. Zhurn.*, vol. 20, No. 2.

Dement'ev, G.P. and A.N. Shokhin. 1939. K avifaune verkhov'ev r. Kolymy (Avifauna of the upper reaches of the Kolyma River). *Trudy Zool. Muzeya MGU*, No. 5.

Devyatkin, V.A. and D.A. Obodovskaya. 1955. Oblepikha (Sallow-thorn). *Priroda*, No. 9.

Dezhkin, V. 1964. Fazan v Severnoi Amerike (Pheasants of North America). *Okhota i Okhotnich'e Khoz-vo*, No. 6.

Dobrinskii, L.N. 1972. Nekotorye morfofiziologicheskie osobennosti beloi i tundryanoi kuropatok v svyazi s kharakterom ikh pitaniya. Optimal'naya plotnost' i optimal'naya struktura populyatsii zhivotnykh (Some Morphological and Physiological Features of Willow Grouse and Rock Ptarmigan Relative to Their Feeding Habits. Optimal Density and Optimal Structure of Animal Populations). Sverdlovsk.

Dobrinskii, L.N. and A.P. Nikul'tsev. 1969. Opyt ispol'zovaniya metoda morfofiziologicheskikh indikatorov pri izuchenii populyatsionnoi ekologii teterevinykh ptits (An experiment in the use of morphophysiological indicators in studying the population ecology of Tetraonidae). *Zool. Zhurn.*, vol. 48, No. 4.

Dolbik, M.S. 1967. Pitanie glukharya i ryabchika v Belorussii (Capercaillie and Hazel Grouse Feeding in Belorussia). In: *Itogi Ornit. Issledovanii v Pribaltike.* Tallin.

Dolgushin, I.A. 1947. Materialy po faune ptits Severnogo Pribalkhash'ya i Kazakhskogo nagor'ya (Data on the avifauna of the northern Balkhash region and the Kazakh uplands). *Izvestiya AN KazSSR. Seriya Zool.*, No. 6.

Donaurov, S.S. 1947. Ryabchik v Pechorsko-Ilychskom Zapovednike (Hazel Grouse in Pechoro-Ilych Preserve). *Trudy Pechorsko-Ilychskogo Gos. Zapovednika,* No. 4, pt. 1.

Doppel'maier, G.G. 1950. Vzaimootnosheniya okhotnich'ego i lesnogo khozyaistva RSFSR (Interrelations of game management and forestry in the RSFSR). *Okhrana Prirody,* No. 10.

Dorofeev, P.I. 1970. Tretichnye flory Urala (Tertiary Flora of the Urals). Leningrad.

Dorogostaiskii, V.G. 1913. O gnezdovanii nekotorykh ptits (On the nesting of some birds). *Ptitsevedenie i Ptitsevodstvo,* vol. 4, No. 2.

Dubinin, N.P. 1953. Ptitsy lesov nizhnei chasti doliny r. Ural. Ch. I (Birds of the forests in the lower valley of the Ural River. Part I). *Trudy In-ta Lesa AN SSSR,* vol. 18.

Dubinin, V.B. 1956. Per'evye kleshchi (Analgesoidea). Ch. III. Someistvo Pterolichidae (Feather mites of the order Analgesoidea. Part III. Family Pterolichidae). In: *Fauna SSSR. Paukoobraznye.* Leningrad, vol. 6, No. 7.

Dubinin, V.B. 1958. O parazitologicheskikh kriteriyakh v sistematike ptits (Parasitological criteria in bird taxonomy). *Uchen. Zap. MGU. Ornitologiya,* No. 197.

Dubrovskii, A.N. 1930. Materialy k izucheniyu pitaniya tetereva (*Lyrurus tetrix*) v Pol'skom raione Novgorodskogo okruga [Data on Black Grouse (*Lyrurus tetrix*) feeding in the Pol'sk region of Novgorod]. *Trudy po Lesnomu Opytnomu Delu,* No. 5, Moscow.

Dul'keit, G.D. 1960. Zimnyaya zhizn' ptits v taige Severo-Vostochnogo Altaya (Winter life of birds in the coniferous forests of northeastern Altai). *Trudy Problemnykh i Tematicheskikh Soveshchanii ZIN AN SSSR,* No. 9.

Dul'keit, G.D. 1975. Altai i Sayani (Altai and Sayans). In: *Teterevinye Ptitsy.* Moscow.

Dul'keit, G.D. and L.M. Shul'pin. 1937. Ptitsy Shantarskikh ostrovov (Birds of the Shantar Islands). *Trudy Biol. Nauch. In-ta-pri Tomskom Un-te.*

Dunaeva, T.N. and V.V. Kucheruk. 1941. Materialy po ekologii nazemnykh pozvonochnykh tundry yuzhnogo Yamala (Belaya kuropatka) [Data on the ecology of land vertebrates of the tundra in southern Yamal (Willow Grouse)]. In: *Materialy k Poznaniyu Fauny i Flory SSSR. Novaya Seriya, Otd. Zool.,* No. 4 (19).

Dzerzhinskii, F. Ya. 1974. K funktsional'noi morfologii chelyustnogo apparata glukharya (Functional morphology of the jaws of Capercaillie). *Ornitologiya,* No. 11.

Dzerzhinskii, F. Ya. 1974. O znachenii biomekhanicheskogo analiza v rekonstruktsii filogenii ptits (na primere kurinykh) [Importance of biomechanical analysis in the reconstruction of the phylogeny of birds (on the example of Galliformes)]. In: *Materialy 6-i Vsesoyuzn. Ornitol. Konfer.* Moscow, pt. 1.

Edminster, F.C. 1954. *American Game Birds.* New York.

Efremov, P.G. 1968. Adaptivnye osobennosti glukharya v svyazi s drevesnym obrazom zhizni i pitaniem (Adaptive features of Capercaillie in relation to an arboreal mode of life and feeding). *Uchen. Zap. Mariiskogo Ped. In-ta,* vol. 30.

Egorov, O.V., Yu.V. Labutin, and A.A. Mezhennyi. 1959. Materialy po biologii kamennogo glukharya v Yakutii (Biology of Black-billed Capercaillie in Yakutia). *Trudy In-ta Biol. yakutskogo Filiala Sibirskogo Otdeleniya AN SSSR,* No. 6.

Engels, W.L. 1938. Variability in bone length and limb proportions in the coot (*Fulica americana*), *Journ. Morphology,* vol. 62.

Engels, W.L. 1940. Structural adaptations in thrushes (gen. *Taxostoma*) with comments on interspecific relationships. *Univ. Calif. Publ. Zool.,* vol. 42.

Fedorako, B. 1957. Seraya kuropatka v Predural'e (Gray Partridge in the Cis-Urals). *Okhota i Okhotnich'e Khoz-vo,* Nos. 6–7.

Fedorov, A.A. 1949. K biologii gornoi kuropatki Kopet-Daga (Biology of Rock Ptarmigan of Kopet Dag). *Priroda,* No. 6.

Fedorova, O.A. 1958. Khimicheskii sostav kormovykh rastenii, sena i sochnykh kormov Krainego Severa (Chemical composition of food plants, hay, and succulents of the Far North). *Trudy Nauchno-Issled. In-ta Krainego Severa,* vol. 6.

Fedosenko, A.K. 1966. Materialy k zimnemu pitaniyu i aktivnosti semirechenskogo fazana (On the winter food and habits of Semirechensk Pheasant). In: *Okhrana i Ratsion. Ispol'zov. Resursov Dikoi Zhivoi Prirody.* Alma-Ata.

Fedyushin, A.V. 1958. Belaya kuropatka—potentsial'nyi vreditel' plodovykh sadov (Willow Grouse—a potential pest of orchards). *Priroda,* No. 3.

Fedyushin, A.V. and M.S. Dolbik. 1967. Ptitsy Belorussii (Birds of Byelorussia). Minsk.

Fetisov, A.S. 1934. Materialy po pitaniyu tetereva (*Lyrurus tetrix* L.) v yugo-vostochnoi chasti Pribaikal'ya (Data on the feeding of *Lyrurus tetrix* L. in the southeastern part of the Baikal region). *Zool. Zhurn.,* vol. 13, No. 2.

Filonov, K.P. 1961. Zima v zhizni ptits Barguzinskogo zapovednika (Winter in the bird life of Barguzinsk Preserve). *Trudy Barguzinskogo Zapovednika,* No. 3.

Flora SSSR (Flora of the Soviet Union). 1949. Botan. In-t. im. Komarova AN SSSR, Moscow-Leningrad, vol. 15.

Folitarek, S.S. 1939. Materialy po pitaniyu beloi i tundryanoi kuropatok (Data on the feeding of Willow Grouse and Rock Ptarmigan). In: *Nauchno-Metod. Zap. Gl. Upr. po Zapvedn.,* No. 5.

Folitarek, S.S. 1939. Nekotorye dannye po pitaniyu ryabchika (Some information on the feeding of the Hazel Grouse). *Voprosy Ekologii i Biotsenologii,* No. 4.

Folitarek, S.S. and G.P. Dement'ev. 1938. Ptitsy Altaiskogo gosudarstvennogo zapovednika (Birds of the Altai State Preserve). *Trudy Altaiskogo Zapovednika,* No. 1.

Formosov [Formozov], A.N. 1939. The snow cover as an environmental factor and its importance in the ecology of mammals and birds. *Byul. Mosk. Ob-va Ispytat. Prirody,* vol. 48. No. 4.

Formozov, A.N. 1934. Materialy k biologii ryabchika (*Tetrastes bonasia volgensis* But.) po nablyudeniyam na severe Gor'kovskogo kraya (Data on biology of Hazel Grouse (*Tetrastes bonasia volgensis* But.) based on observations in northern Gor'kii). *Byul. Mosk. Ob-va Ispytat. Prirody, Otd. Biol.,* vol. 43, No. 1.

Formozov, A.N. 1935. Kolebaniya chislennosti promyslovykh zhivotnykh (Population Dynamics of Game Animals). Moscow.

Formozov, A.N. 1940. Prakticheskoe znachenie ptits (Economic importance of birds). In: *Rukovodstvo po Zoologii* edited by G.P. Dement'ev. Moscow-Leningrad, vol. 6, *Ptitsy.*

Formozov, A.N. 19 6. Snezhnyi pokrov v zhizni mlekopitayushchikh i ptits SSSR (Snow cover in the Life of Mammals and Birds of the Soviet Union). Moscow.

Formozov. A.N. 1959. O dvizhenii i kolebanii granits rasprostraneniya mlekopitayushchikh i ptits (Shifting and changing distribution ranges of mammals and birds). In: *Geografiya Naseleniya Nazemnykh Zhivotnykh i Metody ego Izucheniya.* Moscow.

Formozov, A.N. 1961. O znachenii struktury snezhnogo pokrova v ekologii i geografii mlekopitayushchikh i ptits (Importance of the snow cover structure in the ecology and geography of mammals and birds). In: *Rol' Snezhnogo Pokrova v Prirodnykh Protsessakh.* Moscow.

Formozov, A.N. 1964. Urozhai yagod brusniki, cherniki i ikh znachenie v zhizni ryabchika (Harvest of cowberry and bilberry and their importance in the life of Hazel Grouse). In: *Geografiya Plodonosheniya Lesnykh Drevesnykh Porod, Kusturnikov i Yagodnikov, Znachenie ikh Urozhaev v Narodnom Khozyaistve i Zhizni Fauny.* Moscow.

Formozov, A.N. 1964. Yadovitye perepela (Poisoned quails). *Priroda,* No. 9.

Formozov, A.N. 1970. Snezhnyi pokrov i kurinye ptitsy (Snow cover and galliform birds). *Okhota i Okhotnich'e Khoz-vo,* No. 5.

Fürbringer, M. 1888. *Untersuchungen zur Morphologie und Systematik der Vögel.* Jena, Amsterdam.

Gabuzov, O. 1958. Borodataya kuropatka pod Moskvoi (Daurian Partridge in the environs of Moscow). *Okhota i Okhotnich'e Khoz-vo,* No. 3.

Gagina, T.N. 1958. K poznaniyu massovykh neperiodicheskikh migratsii dikikh kurinykh v Sibirskoi taige (Mass nonperiodic migrations of Galliformes in the Siberian taiga). *Izvestiya Irkutsk. S.-Kh. In-ta,* No. 8.

Gaidar, A.A. 1969. Nekotorye dannye po osennemu pitaniyu ryabchika v verkhov'yakh r. Leny (Some information on the autumn feeding of Hazel Grouse in the upper reaches of the Lena River). *Trudy VNIIZhP,* No. 22. Kirov.

Gaidar, A.A. 1974. Ekologo-morfologicheskaya kharakteristika ryabchika Vyatskoi taigi (Ecological and morphological features of Hazel Grouse in the Vyatsk taiga). Avtoref. Kand. Diss., Sverdlovsk.

Gajic, I. and V. Jovic. 1969. Aktivnost facana u toku 24 casa, *Jelen,* No. 8.

Gavrin, V.F. 1956. Ekologiya teterevinykh Belovezhskoi pushchi (Ecology of tetraonids in the Belovezhsk dense forests). Kand. Diss., Alma-Ata.

Gavrin, V.F. 1969. Plotnost' i dinamika populyatsii teterevinykh v podzone smeshannykh lesov (Density and population dynamics of tetraonids in the subzone of mixed forests). In: *Estestvennaya Proizvoditel'nost' i Produktivnost' Okhotnich'ikh Ugodii SSSR.* Kirov, pt. I.

Gazan'yan, M.K. 1954. K biologii kamennoi kuropatki v Azerbaidzhane (Biology of Rock Partridges in Azerbaidzhan). *Trudy Azarbaidzh. Un-ta. Seriya Biol.,* No. 6.

George, J.C. and A.J. Berger. 1966. *Avian Myology.* New York-London.

Gibet, L.A. and G.A. Voronov. 1965. Raspredelenie i chislennost' borovoi dichi v Kalininskoi oblasti (Distribution and population of wildlife in the pine forests of Kalinin District). *Ornitologiya,* No. 7.

Gizenko, A.I. 1955. Ptitsy Sakhalinskoi oblasti (Birds of Sakhalin District). Moscow.

Gladkov, N.A. 1947. O svyazi mezhdu anatomicheskoi konstruktsiei kryla i kharakterom poleta ptitsy (Interrelation between the anatomical structure of the wing and flight characteristics of birds). *Izvestiya AN SSSR,* No. 1.

Gladkov, N.A. 1947. Ob ustanovlenii ekologicheskikh tipov poleta ptits (Determining the ecological types of birds flight). *Uspekhi Sovrem. Biologii,* vol. 23, No. 3.

Gladkov, N.A. 1949. Biologicheskie osnovy poleta ptits (Biological Principles of Bird Flight). Moscow.

Gladkov, N.A. 1952. Rod fazany *Phasianus* (The genus *Phasianus*). In: *Ptitsy Sovetskogo Soyuza.* Moscow, vol. 4.

Gladkov, N.A. 1952. Zhivotnyi mir zony zashchitnogo razvedeniya (Animals in a protected zone). *Prioroda,* No. 8.

Gladkov, N.A. and V.B. Grinberg. 1932. Materialy k ornitofaune r. Chu (Bird fauna of the Chu River). *Byul. Mosk. Ob-va Ispyt. Priorody, Novaya Seriya,* vol. 41, Nos. 3–4.

Godfrey, W.E. 1966. The birds of Canada. *National Museum of Canada Bulletin* 203, *Biolog. Series,* p. 73.

Gofman, D.N. 1944. Ob antsestral'nykh priznakakh v operenii *Lyrurus* i *Tetrao* (K voprosu o filogeneze sem. Tetraonidae) [Ancestral features in the plumage of *Lyrurus* and *Tetrao* (phylogeny of the family Tetraonidae)]. *Zool. Zhurn.,* vol. 23, Nos. 2–3.

Gorodkov, B.N. 1938. Rastitel'nost' Arktiki i gornykh tundr SSSR (Plants of arctic and hilly tundras in the Soviet Union). In: *Rastitel'nost' SSSR,* pt. 1.

Gorodkov, B.N. and E.S. Korotkevich. 1957. Pitanie beloi kuropatki v zone arkticheskikh pustyn' (Feeding of Willow Grouse in the barren Arctic lands). *Zool. Zhurn.,* vol. 36, No. 9.

Grachev, N.I. and M.A. Kuz'mina. 1947. Materialy po pitaniyu syrdar'inskogo fazana (On the feeding of Syr-Darya pheasants). *Izvestiya AN KazSSR, Seriya Zool.,* No. 36.

Grachev, V.A. 1960. O lin'ke semirechenskogo fazana (*Phasianus colchicus mongolicus* Brandt.) [Molt of Mongolian pheasants (*Phasianus colchicus mongolicus* Brandt.)]. *Trudy In-ta. Zool. AN KazSSR,* vol. 13.

Graham, R.R. 1932. Safety devices in wings of birds. *Journ. Royal Aeronautik,* vol. 36, No. 253.

Gray, A.P. 1958. *Bird hybrids.* Farnham Royal, Bucks, England, Commonwealth Agricultural Bureaux.

Greenewaldt, C.H. 1960. *Dimensional Relationships for Flying Animals.* Greenville, Delaware (Printed privately).

Gregory, M.M. 1921. *A Manual of the Birds of Australia.* London, vol. 1.

Grekov, V.S. 1965. Perepel na yuge Odesskoi oblasti (Quail in the southern Odessa District). In: *Novosti Ornitologii,* Alma-Ata.

Groebbels, F. 1926. Die phisiologischen Grundlagen des Vogelfluges. Anlage 9, *J. fur Ornithol.,* vol. 74, No. 2.

Groebbels, F. 1932. *Der Vögel. Atmungswelt und Nahrungswelt.* Berlin, vol. 1.

Hall, B.P. 1963. The francolins, a study in speciation, *Bull. Brit. Museum (Natural History) Zool.,* vol. 10, No. 2.

Hartert, E. 1905. *Die Vögel der Paläarktischen Fauna, Galli,* Berlin, vol. 1, No. 3.

Harting, P. 1869. Observation sur letendue relative des ailes et le poids des muscles pectoraux chez les animaux vertebre volants, *Arch. Néerlandaises,* vol. 4.

Hartman, F.A. 1961. Locomotor mechanisms of birds. Smithsonian Inst., Misc. Collection.

Heinroth, O. 1927. Über weisse und rote Musculatur bei den Vögeln, *Ornith., Monatsber,* vol. 35.

Helminen, M. and J. Viramo. 1962. Animal food of Western Capercaillies (*Tetrao urogallus*) and Northern Black Grouse (*Lyrurus tetrix*) in autumn, *Ornis Fennica,* vol. 39, No. 1.

Howard, H. 1950. Fossil evidence of aviain evolution, *Ibis,* vol. 92, No. 1.

Hudson, G.E. and P.J. Lanzillotti. 1964. Muscles on the pectoral limb in Galliform birds. *The American Midland Naturalist,* vol. 71, No. 1.

Ioganzen, G.Kh. 1934. Ptitsy Komandorskikh ostrovov (Birds of the Komandor Islands). *Trudy Tomsk. Gos. Un-ta,* vol. 86.

Irisov, E.I. and V.M. Totunov. 1970. K ekologii kurinykh Yugo-Vostochnogo Altaya (Ecology of the galliform birds of southeastern Altai). *Izvestiya Altaisk. Otd. Geogr. Ob-va,* No. 11.

Ivankin, V.K. 1935. Sostav i pitatel'naya tsennost' trav i sena iz del'ty Severnoi Dviny (Composition and nutritive value of grass and hay from the delta of the North Dvina River). *Zap. Vsesoyuzn. Kommun. S-kh. Un-ta,* Leningrad.

Ivanov, A.I 1951. Ptitsy SSSR (Birds of the Soviet Union). Moscow-Leningrad, pt. 1, Galliformes.

Ivanov, A.I. 1969. Ptitsy Pamiro-Aleya (Birds of Pamir-Alai). Leningrad.

Ivanov, V.G. and Zh.Kh. Baziev. 1961. K biologii Kavkazskogo ulara (*Tetraogallus caucasicus* Pall. v usloviyakh Kabardino-Balkarii [Biology of Caucasian Snowcocks (*Tetraogallus caucasicus* Pall.) in Kabardino-Balkariya]. *Uchen. Zapad. Kabar., Balkarsk. Un-ta,* No. 10.

Ivanter, E.V. 1962. K biologii ryabchika v Karelii (Biology of Hazel Grouse of Karelia). *Ornitologiya,* No. 4.

Ivanter, E.V. 1963. Teterev v Karelii (Northern Black Grouse of Karelia). *Ornitologiya,* No. 6.

Ivanter, E.V. 1965. Ekologiya i chislennost' glukharya v Karelii (Ecology and population of capercaillies in Karelia). In: *Novosti Ornitologii,* Alma-Ata.

Ivanter, E.V. 1974. Glukhar' i belaya kuropatka v Karelii (Western Capercaillie and Willow Grouse in Karelia). *Ornitologiya,* No. 11.

Izmailov, I.V. 1967. Ptitsy Vitimskogo Ploskogor'ya (Birds of the Vitim Plateau). Ulan-Ude.

Janda, J. 1956. Potrava despelych koroptvi (*Perdix perdix* L.), v CSR, *Vest. Ceskosl. Spolec. Zool.,* vol. 20, No. 2.

Janda, J. 1959. Potrava mladych koroptvi (*Perdix perdix* L.), *Zool. Listy,* vol. 8, No. 4.

Jewett, S.G., W.P. Taylor, W.T. Shaw, and J.W. Aldrich. 1953. *The birds of Washington State.* Univ. Wash. Press.

Johansen, H. 1956. Revision und Entstehung der arktischen Vögelfauna. Teil 1, Einführung und Revision der Gaviae. Galli, *Acta Arctica,* No. 8.

Johansen, H. 1957. Rassen und Populationen des Auerhuhns (*Tetrao urogallus*), *Viltrevy,* No. 1.

Johnsen, S. 1929. Draktskiftet hos lirypen i Norge, *Bergen Mus. Aarbok,* No. 2.

Kaftanovskii, Yu.M. 1951. Chistikovye ptitsy Vostochnoi Atlantiki (Alciformes of the Eastern Atlantic). Moscow.

Kalabukhov, N.I. 1940. Adaptivnoe znachenie chuvstvitel'nosti organizma k vneshnim faktoram (Adaptive significance of the sensitivity of organisms to the environment). *Zhurn. Obshch. Biol.,* No. 2.

Kalabukhov, N.I. 1946. Sokhranenie energeticheskogo balansa organizma kak osnova protsessa adaptatsii (Maintenance of energy balance in organisms as the basis for adaptation). *Zhurn. Obshch. Biol.,* No. 6.

Kapitonov, V.I. 1962. Ornitologicheskie nablyudeniya v nizov'yakh Leny (Ornithological observations in the lower reaches of the Lena). *Ornitologiya,* No. 4.

Kaplanov, L.G. 1938. K biologii dikushi—chernogo ryabchika (*Falcipennis falcipennis*) [Biology of Siberian Spruce Grouse (*Falcipennis falcipennis*)]. *Vestnik Dal'nevost. Fil. AN SSSR,* No. 32.

Kartashev, N.N. 1952. Rod Perepely—*Coturnix*, Rod Pustynnaya kuropatka—*Ammoperdix*, Rod Kamennaya kuropatka—*Alectoris*, Rod Turach—*Francolinus*, Rod Ulary—*Tetraogallus*, Rod Kuropatki—*Perdix* (Genera *Coturnix*, *Ammoperdix*, *Alectoris*, *Francolinus*, *Tetraogallus*, and *Perdix*). In: *Ptitsy Sovetskogo Soyuza*. Moscow, vol. 4.

Kartashev, N.N. 1974. Sistematika ptits (Avian Systematics). Moscow.

Kattinger, E. 1929. Sexual- und Subspecies-Unterschiede im Skelettbau der Vögel, *Journ. Ornithol.*, vol. 77, No. 1.

Kelso, L. 1935. O pishche seroi kuropatki (Food of Gray Partridges). *Boets-Okhotnik*, No. 2.

Kendeigh, S.C. and G.C. West. 1965. Caloric values of plant seeds eaten by birds. *Ecology*, vol. 46, No. 4.

Kessler, K.F. 1840. O nogakh ptits v otnoshenii k sistematicheskomu deleniyu etogo klassa (Legs of Birds as a Criterion of Taxonomy). St. Petersburg.

Khankhasaev, V.K. 1969. K ekologii i chislennosti daurskoi kuropatki (Ecology and population of Daurian Partridge). In: *Ornitologiya v SSSR*, Ashkhabad, bk. 2.

Khanmamedov, A.I. 1955. K biologii keklika (kamennoi kuropatki) v Azerbaidzhane (Biology of Chukar Partridges in Azerbaidzhan). *Trudy In-ta Zool. Akad. Nauk AzSSR*, vol. 18.

Khanmamedov, A.I. 1956. Turach v Azerbaidzhane, ego rasprostranenie, ekologiya i khozyaistvennoe znachenie (Francolins in Azerbaidzhan, their distribution, ecology, and economic importance). *Trudy In-ta Zool. AN AzSSR*, vol. 19.

Khanmamedov, A.I. 1965. Materialy k biologii kurinykh v severo-vostochnoi chasti Azerbaidzhana (On the biology of Galliformes in the northeastern part of Azerbaidzhan). *Trudy In-ta Zool. AN AzSSR*, vol. 12.

Khanmamedov, A.I. 1969. K ekologii zakavkazskoi seroi kuropatki v Azerbaidzhane (Biology of Transcaucasian Common Partridge in Azerbaidzhan). In: *Ornitologiya v SSSR*, Ashkhabad, bk. 2.

Kirikov, S.V. 1937. Sravnitel'no-ekologicheskii ocherk yuzhno-ural'skogo dubravnogo i borovogo glukharya (Comparative ecology of capercaillies of oak and pine forests). *Izvestiya AN SSSR, Seriya Biol.*, No. 3.

Kirikov, S.V. 1952. Ptitsy i mlekopitayushchie v usloviyakh landshaftov yuzhnoi okonechnosti Urala (Birds and Mammals of the Southern Extremity of the Urals). Moscow.

Kirikov, S.V. 1952. Rod glukharie *Tetrao* (The Capercaillie genus *Tetrao*). In: *Ptitsy Sovetskogo Soyuza*. Moscow, vol. 4.

Kirikov, S.V. 1968. Ekologicheskie osobennosti i istoriya naseleniya glukharya i tetereva na yuge Urala i Priural'ya (Ecological characteristics and history of capercaillies and Northern Black Grouse populations in the southern Urals and the Ural region). In: *Resursy Teterevinykh Ptits v SSSR*. Moscow.

Kirikov, S.V. 1975. Yuzhnaya taiga (Southern taiga). In: *Teterevinye Ptitsy.* Moscow.

Kirpichev, S.P. 1960. Kamennyi glukhar' (The Black-billed Capercaillie). *Okhota i Okhotnich'e Khoz-vo,* No. 5.

Kirpichev, S.P. 1960. Ob izmenchivosti velichiny i slozheniya glukharei (Size variability and composition of the Capercaillie). *Ornitologiya,* No. 3.

Kirpichev, S.P. 1961. Temno-seryi glukhar' (The Dark Forest Capercaillie). *Okhota i Okhotnich'e Khoz-vo,* No. 7.

Kirpichev, S.P. 1965. Opyt razvedeniya glukharei (Experimental breeding of capercaillies). *Okhota i Okhotnich'e Khoz-vo,* No. 5.

Kishchinskii, A.A. 1962. K kharakteristike zimnei zhizni ptits Koryakskogo nagor'ya (Winter life characteristics of birds of the Koryaksk upland). In: *Materialy 3-i Vsesoyuzn. Ornitol. Konfer.* L'vov, bk. 2.

Kishchinskii, A.A. 1968. Ptitsy Kolymskogo nagor'ya (Birds of the Kolyma Uplands). Moscow.

Kishchinckii, A.A. 1974. Arktoal'piiskaya avifauna i ee proiskhozhdenie (North alpine bird fauna and their origins). *Zool. Zhurn.,* vol. 53, No. 7.

Kishchinskii, A.A. 1974. K ekologii beloi kuropatki na Novosibirskikh ostrovakh (Ecology of Willow Grouse on the Novosibirsk Islands). *Ornitologiya,* No. 11.

Kishchinskii, A.A. 1975. Novosibirskie ostrova (Novosibirsk Islands). In: *Teterevinye Ptitsy.* Moscow.

Kloiber, A. and G. Rokitansky. 1954. Ein Fasanbastard der freien Wildbahn aus Aigen in Mühlkreis, *Jahrb. Oberösterr. Musealvereines,* No. 99.

Knysh, N.P. 1974. O vozmozhnosti gruppovoi termoregulyatsii u serykh kuropatok (The possibility of group thermoregulation in Gray Partridges). In: *Materialy 6-i Vsesoyuzn. Ornitol. Konfer.* Moscow, pt. 2.

Kogan, Sh.A. 1950. Materialy po pitaniyu keklika v Kopet-Dage (Feeding of Chukar in Kopet-Dag). *Izvestiya Turkm. Fil. AN SSSR,* No. 4.

Kokshaiskii, N.V. 1964. Osobennosti lëtnykh kachestv tsapel' s tochki zreniya teorii podobiya (Flight characteristics of herons from the viewpoint of the theory of analogy). In: *Problemy Ornitologii,* L'vov.

Kokshaiskii, N.V. 1966. Nekotorye soobrazheniya o bionicheskikh issledovaniyakh v ornitologii (Some views on the bionics of birds). In: *Mekhanizmy Poleta i Orientatsii.* Moscow.

Kokshaiskii, N.V. 1971. Ob issledovaniyakh poleta ptits (Studies of bird flight). In: *Itogi Nauki, Seriya Biologiya, "Zoologiya Pozvonochnykh,"* 1969, *Voprosy Ornitologii,* Moscow.

Kolesnikov, I.I. 1956. Fauna nazemnykh pozvonochnykh Kyuren-Daga (Land vertebrate fauna of Kuren-Dag). *Trudy SAGU,* No. 86.

Konev, G.I. 1956. Kedrovye orekhi—korm ryabchikov (Cedar nuts as food for Hazel Grouse). *Priroda,* No. 9.

Korelov, M.N. 1956. Fauna pozvonochnykh Bostandykskogo raiona (Vertebrate fauna of the Bostandyk region). In: *Prioroda i Khozyaistvennye Usloviya Gornoi Chasti Bostandyka.* Alma-Ata.

Korelov, M.N. 1956. Materialy k avifaune khrebta Ketmen' (Tyan'-Shan') [Data on the avifauna of the Ketmen Range (Tien Shan)]. *Trudy In-ta Zool. AN KazSSr,* vol. 6.

Korma SSSR, sostav i pitatel'nost' (Fodder in the Soviet Union, Its Composition and Nutritive Value). Moscow, 1944.

Kost, Wolfgang. 1967. Wachtel (*Coturnix coturnix*) "wassert" in Roten Meer, *Vogelwarte,* vol. 24, No. 1.

Kosygin, G.M. 1962. Osobennosti zimnego pitaniya tetereva i vorona v Lenskom raione Yakutti (Peculiarities of winter feeding of Black Grouse and crows in the Lena region of Yakutia). *Ornitologiya,* No. 5.

Kotov, V.A. 1959. Gornaya indeika v Kavkazskom zapovednike (Mountain turkey in the Caucasian Preserve). *Trudy Kavkazskogo Gos. Zapovednika,* No. 5.

Kovshar', A.F. 1966. Ptitsy Talasskogo Alatau (Birds of Talassk Alatau). Alma-Ata.

Kozlova, E.V. 1930. Ptitsy Yugo-Zapadnogo Zabaikal'ya, Severnoi Mongolii i Tsentral'noi Gobi (Birds of Southwestern Trans-Baikal, Northern Mongolia, and Central Gobi). Leningrad.

Kozlova, E.V. 1932. Ptitsy vysokogornogo Khangaya (Birds of the Khangai Mountains). Leningrad.

Kozlova, E.V. 1946. Svyaz' mezhdu obrazom zhizni ptits i napravleniem razvitiya ikh letatel'nogo apparata (Relation between the lifestyles of birds and the evolution trends of flight organs). *Izvestiya AN SSSR, Seriya Biol.,* No. 4.

Kozlova, E.V. 1952. Avifauna Tibetskogo nagor'ya, ee rodstvennye svyazi i istoriya (Avifauna of the Tibetan plateau, its phylogenetic links and evolution). *Trudy Zool. In-ta AN SSSR,* vol. 9, No. 4.

Kozlova, E.V. 1960. K metodike izucheniya istorii regional' nykh faun (On the methodology of studying the history of regional fauna). *Trudy Problemnykh i Tematicheskikh Soveshchanii ZIN AN SSSR,* No. 9.

Kozlova, E.V. 1970. Resselenie fazana *Phasianus colchicus* L. v pustyni Tsentral'noi Azii (Distribution of pheasants *Phasianus colchicus* L. in the deserts of Central Asia). In: *Ornitologicheskii Sbornik k 100-Letiyu so Dnya Rozhdeniya P.P. Sushkina.* Leningrad.

Kozlova-Pushkareva, E.V. 1933. Ptitsy i promyslovye mlekopitayushchie Vostochnogo Kenteya (Birds and Game Mammals of the Eastern Kentei). Leningrad.

Krasavtsev, B.A. 1940. O pishche severo-kavkazskogo fazana (Food of North Caucasian Pheasants). *Priroda,* No. 9.

Krasovskii, L.I. and G.A. Troitskii. 1958. Osobennosti pitaniya ryabchika v god neurozhaya yagod (Feeding of Hazel Grouse in years with poor fruit crops). *Zool. Zhurn.,* vol. 37, No. 6.

Kravtsov, V.A. 1936. Zametka o sostave pishchi zakavkazskogo turacha (Notes on the food composition of Trans-Caucasian Francolin). *Boets-Okhotnik,* No. 4.

Krechmar, A.V. 1966. Ptitsy Zapadnogo Taimyra (Birds of western Taimyr). In: *Biologiya Ptits.*

Krishtofovich, A.N. 1959. Razvitie botaniko-geograficheskikh oblastei severnogo polushariya s nachala tretichnogo perioda (Development of phytogeographic regions in the northern hemisphere from the early Tertiary). *Izbrannye Trudy.* Moscow-Leningrad, vol. 1.

Krivitskii, I.A. 1962. O zimnei ornitofaune raiona Kurgal'dzhinskogo zapovednika (On the winter bird fauna of Kurgal'dzhinsk Preserve). *Ornitologiya,* No. 4.

Krivitskii, I.A. 1969. Ptitsy yuzhnykh stepei Tselinogradskoi oblasti (ekologo-faunisticheskii analiz) [Birds of the southern steppes of Tselinograd District (ecological and faunistic analysis)]. Avtoref. Kand. Diss., Khar'kov.

Krutovskaya, E.A. and E.V. Krutovskaya. 1953. Opyt poluvol'nogo razvedeniya glukharya (Semicaptive breeding of capercaillies). In: *Preobrazovanie Fauny Pozvonochnykh Nashei Strany.* Moscow.

Kuchin, A.P. 1963. K biologii perepela v verkhnem Priob'e (Biology of quail in the upper Ob' region). *Izvestiya Altaisk. Otd. Geogr. Ob-va Soyuza SSR,* No. 3.

Kurochkin, E.N. 1968. Lokomotsiya i morfologiya tazovykh konechnostei plavayushchikh i nyryayushchikh ptits (Locomotion and morphology of the hind limbs of swimming and diving birds). Avtoref. Kand. Diss., Moscow.

Kurochkin, E.N. 1971. Osnovnye voprosy izucheniya iskopaemykh ptits (Major problems in the study of bird fossils). In: *Itogi Nauki. Seriya Biologiya, "Zoologiya Pozvonochnykh,"* 1970. *Voprosy Ornitologii.* Moscow.

Kutovaya, T. 1976. Razvedenie glukharei v Berezinskom zapovednike (Breeding of capercaillies in Berezinsk Preserve). *Okhota i Okhotnich'e Khoz-vo,* No. 6.

Kutubidze, M.E. 1961. Ekologiya i rasprostranenie kavkazskogo tetereva (*Lyrurus mlokosiewiczi* Tacz.) v predelakh Gruzii [Ecology and distribution of Caucasian Blackcocks (*Lyrurus mlokosiewiczi* Tacz.) in Georgia]. *Trudy In-ta Zool. An GruzSSR,* vol. 18.

Kuz'mina, M.A. 1948. Ekologiya keklika (*Alectoris kakelik Falk.*) v Kazakhstane i ego rol' kak khozyaina iksodovykh kleshchei [Ecology of Chukar Partridges (*Alectoris kakelik* Falk.) in Kazakhstan and its role as a host of ixodid ticks]. *Izvestiya AN KazSSR, Seriya Parazitol.*, No. 5.

Kuz'mina, M.A. 1948. Materialy po avifaune predgorii Zapadnogo Altaya (On the bird fauna of the foothills of western Altai). *Izvestiya AN KazSSR, Seriya Zool.*, No. 7.

Kuz'mina, M.A. 1953. Materialy po ptitsam Zapadnogo Altaya (On the birds of western Altai). *Trudy In-ta Zool. AN KazSSR*, vol. 2.

Kuz'mina, M.A. 1955. Dzhungarskii keklik, ego ekologiya i khozyaistvennoe znachenie (The Dzhungar Chukar Partridge, its ecology, and economic importance). *Trudy In-ta Zool. AN KazSSR*, vol. 4.

Kuz'mina, M.A. 1955. Materialy po ekologii i morfologii temnobryukhogo ulara i keklika (On the ecology and morphology of Himalayan Snowcocks and Chukar). *Zool. Zhurn.*, vol. 34, No. 1.

Kuz'mina, M.A. 1959. Kharakter prisposobleniya teterevinykh i fazanovykh k pereneseniyu surovykh klimaticheskikh uslovii (Nature of adaptation of Tetraonidae and Phasianidae to severe climatic conditions). In: *Tezisy Dokl. 2-i Vsesoyuzn. Ornitol. Konfer.*, Moscow.

Kuz'mina, M.A. 1959. Nekotorye morfologicheskie osobennosti ryabchika (Some morphological features of Hazel Grouse). In: *Tezisy Dokl. 2-i Vsesoyuzn. Ornitol. Konfer.*, Moscow.

Kuz'mina, M.A. 1959. Rasprostranenie kurinykh i nekotorye voprosy istorii etogo otryada (Distribution of Galliformes and some aspects of the evolution of this order). *Trudy In-ta Zool. AN KazSSR*, vol. 10.

Kuz'mina, M.A. 1960. Materialy po razmnozheniyu, pitaniyu i lin'ke temnobryukhogo ulara (On the breeding, feeding, and molt of Himalayan Snowcocks). *Trudy In-ta Zool. AN KazSSR*, vol. 13.

Kuz'mina, M.A. 1961. Prisposoblenie teterevinykh i fazanovykh k osobennostyam klimaticheskikh uslovii (Adaptation of tetraonids and phasianids to climatic conditions). *Trudy In-ta Zool. AN KazSSR*, vol. 15.

Kuz'mina, M.A. 1962. Ekologo-morfologicheskie osobennosti ryabchika (Ecomorphological features of Hazel Grouse). *Ornitologiya*, No. 4.

Kuz'mina, M.A. 1962. Nekotorye adaptivnye osobennosti zadnikh konechnostei kurinykh ptits (Some adaptive features of the hind limbs of galliform birds). *Materialy 3-i Vsesoyuzn. Ornitol. Konfer. L'vov*, bk. 2.

Kuz'mina, M.A. 1962. Otryad kurinye—Galliformes (The order Galliformes). In: *Ptitsy Kazakhstana*. Alma-Ata, vol. 2.

Kuz'mina, M.A. 1964. Morfofunktsional'nye osobennosti zadnikh konechnostei kurinykh (Morphological and functional adaptations of the hind limbs in Galliformes). *Trudy In-ta Zool. AN KazSSR,* vol. 24.

Kuz'mina, M.A. 1965. Izuchenie pitaniya kurinykh (Studies on the food of Galliformes). In: *Novosti Ornitologii.* Alma-Ata.

Kuz'mina, M.A. 1968. Rasprostranenie i chislennost' teterevinykh ptits v Kazakhstane (Distribution and population of Tetraonidae in Kazakhstan). In: *Resursy Teterevinykh Ptits v SSSR.* Moscow.

Kuz'mina, M.A. 1968. Sravnitel'naya kharakteristika pitaniya teterevinykh i fazanovykh v SSSR (Comparative characteristics of Tetraonidae and Phasianidae feeding in the Soviet Union). *Trudy In-ta Zool. AN KazSSR,* vol. 29.

Kuz'mina, M.A. 1969. O morfologicheskikh adaptatsiyakh dvukh populyatsii keklika (Morphological adaptations of two populations of Chukar). In: *Ornitologiya v SSSR.* Ashkhabad, bk. 2.

Kuz'mina, M.A. 1974. Nekotorye razlichiya v letatel'nom apparate Altaiskogo i temnobryukhogo ularov (Certain differences in the flight organs of Altai and Himalayan snowcocks). *Ornitologiya,* No. 11.

Kuz'mina, M.A. 1974. Osobennosti letatel'nogo apparata kurinykh (Peculiarities of flight organs in Galliformes). In: *Materialy 6-i Vsesoyuzn. Ornitol. Konfer.,* Moscow, pt. I,

Kuz'mina, M.A. 1975. Kazakhstan. In: *Teterevinye Ptitsy.* Moscow.

Kuz'mina, M.A. and E.F. Savinov. 1953. Materialy po ekologii temnobryukhogo ulara (*Tetraogallus himalayensis sewertzowi* Zar.) v Zailiiskom Alatau [On the ecology of Himalayan Snowcocks (*Tetraogallus himalayensis sewertzowi* Zar.) in Trans-Ili Alatau]. *Zool. Zhurn.,* vol. 32, No. 6.

Kuznetsov, V.I. and E.I. Korenberg. 1963. K biologii tetereva, ryabchika i glukharya na yuge Kirovskoi oblasti (Biology of Northern Black Grouse, Hazel Grouse, and Western Capercaillies in the southern part of the Kirov District). *Ornitologiya,* No. 6.

Lambrecht, K. 1933. *Handbuch der Palaeornithologie.* Berlin.

Larin, I.V. 1950. Kormovye rasteniya senokosov i pastbishch SSSR (Fodder Plants and Pastures in the Soviet Union). Moscow-Leningrad, vol. 1.

Larionov, G.P. 1962. K ekologii tetereva Tsentral'noi Yakutii (Ecology of Northern Black Grouse in central Yakutia). *Uchen. Zap. Yakut. Un-ta,* No. 12.

Larionov, G.P. 1965. K biologii kamennogo glukharya i ryabchika v Tsentral'noi Yakutii (Biology of Black-billed Capercaillie and Hazel Grouse in Central Yakutia). *Uchen. Zap. Yakut. Un-ta,* No. 15.

Lavrenko, E.M. 1938. Istoriya flory i rastitel'nosti SSSR po dannym sovremennogo rasprostraneniya rastenii (History of the flora in the Soviet Union based on the present-day distribution of plants). In: *Rastitel'nost' SSSR.* Moscow-Leningrad, vol. 1.

Likhachev, G.N. 1957. Nekotorye dannye po pitaniyu ryabchika v Tul'skikh zasekakh (Some data on the feeding of Hazel Grouse in Tula Abatis). *Zool. Zhurn.*, vol. 36, No. 7.

Lisenka, P. 1929. Iz zhizni Sakhalinskogo ryabchika–dikushi (so slov Sakhalinskogo okhotnika A. Leble) [On the life of Sakhalin Hazel Grouse and Spruce Grouse (in the words of Sakhalin hunter, A. Leble)]. *Okhotnik i Rybak Sibiri,* No. 9.

Livron, A.R. 1938. Ptitsy Naurzumskikh stepei (Birds of the Naurzumsk steppe). *Trudy Naurzumskogo Zapovednika,* No. 1.

Lobachev, S.V. and F.A. Shcherbakov. 1933. Estestvennye korma tetereva-kosacha (Natural foods of Black Grouse). *Byul. Mosk. Ob-va Ispyt. Prirody,* vol. 42, No. 1.

Lobachev, S.V. and F.A. Shcherbakov. 1936. Estestvennye korma teterevaglukharya v godovom tsikle zhizni i v letnikh nagul'nykh stantsiyakh v svyazi s zadachami bonitirovki ugodii pri spetsial'nom okhotoustroistve (Natural foods of Black Grouse and capercaillies in its annual life cycle and in summer habitats relative to yields for special game management). *Zool. Zhurn.*, vol. 15, No. 2.

Logminas, V.V. 1959. Materialy po ekologii kurinykh Litovskoi SSR (On the ecology of Galliformes in the Lithuanian SSR). *Trudy 3-i Pribalt. Ornitol. Konfer.*

Logminas, V.V. 1962. Kurapku (*Perdix perdix* L.) Lietuvoje, *Liot TSR Mokslu Akad. Darbai* (27), 1.

Logminas, V.V. 1968. Seraya kuropatka v Litve (The Gray Partridge in Lithuania). *Okhota i Okhotnich'e Khoz-vo,* No. 12.

Lorents, F. 1884. Kavkazskii teterev (*Tetrao mlokosiewiczii* Tacz.) [The Caucasian Black Grouse (*Tetrao mlokosiewiczii* Tacz.)]. *Priroda i Okhota,* August.

Luchnik, Z.I. and S.A. Padetskii. 1938. Novye dannye po sostavu rastitel'nykh kormov dikikh zhivotnykh i promyslovykh ptits Suputinskogo zapovednika (New data on the composition of the plant food of wild animals and game birds of the Suputinsk Preserve). *Trudy Gornotaezhnoi Stantsii,* vol. 2.

Lüdicke, M. 1964. Das Integument der Vögel, *Studium Generale,* No. 6.

Lumsden, H. 1961. Displays of Spruce Grouse, *Canad. Field-Naturalist,* vol. 75, No. 3.

MacDonald, S.D. 1968. The courtship and territorial behavior of the Francolin race of the Spruce Grouse. *The Living Bird, 7th Annual of the Cornell Labor. of Ornithol.*

Mackworth-Pread, C.W. and C.B.H. Grant. 1952. *Birds of Eastern and Northeastern Africa.* London, vol. 1.

Mackworth-Pread, C.W. and C.B.H. Grant. 1962. *Birds of the Southern Third of Africa.* London, vol. 1, 2nd ed.

Madsen, H. and K.G. Wingstrand. 1958. Some behavioral reactions and structures enabling birds to endure the winter frost in Arctic regions, *Vid. Medd. Dansk. Naturhist., Kobenhayn,* No. 120.

Magnan, A. 1921. De la variation en poids des muscles abaisseurs et releveus de láile suivant l'étendue de la surface alarie chez les oiseaux, *Comptes Rend. d. L'ákad. d. Sci.,* V, vol. 172.

Makatsch, W. 1954. Die Vögel der Erde. Eine systematische Übersicht aller auf der Erde vorkommender Vögelordnungen mit kurzen tiergeographischen und fortpflanzungsbiologischen Angaben. Berlin.

Mal'chevskii, A.S. 1947. Rol'ptits v polezashchitnykh polosakh Zavolzh'ya (Role of birds in the wind-breaks of the Trans-Volga). *Vestnik LGU, Seriya Biol. Nauk,* No. 4.

Mal'chevskii, A.S. 1961. K biolcgii i okhrane perepela (*Coturnix coturnix* L.) [Biology and protection of Common Quail (*Coturnix coturnix* L.)]. *Vestnik LGU, Seriya Biol. Nauk,* No. 15.

Malinovskii, A.V. 1966. Okhotnich'e khozyaistvo v Chekhoslovakii (Game Management in Czechoslovakia). Moscow.

Markov, V.I. 1968. Ob osobennostyakh raśpredeleniya i chislennosti kamennogo glukharya na Vostochnoi Kamchatke (Distribution features and population of Black-billed Capercaillies in eastern Kamchatka). In: *Resursy Teterevinykh Ptits v USSR.* Moscow.

Mayr, E. 1946. History of the North American bird fauna, *Wilson Bull.,* vol. 58.

Meinertzhagen, R. 1954. *Birds of Arabia.* London.

Meklenburtsev, R.N. 1940. Novye dannye po raspredeleniyu i ekologii zeravshanskogo fazana (*Phasianus principalis* Tarn.) (New data on the distribution and ecology of Zeravshan Pheasants, *Phasianus principalis* Tarn.). *Byul. Mosk. Ob-va Ispyt. Prirody, Otd. Biol.,* vol. 9, Nos. 5–6.

Meklenburtsev, R.N. 1946. Zimuyushchie i proletnye ptitsy Vostochnogo Pamira (Wintering and migratory birds of the eastern Pamir). *Byul. Mosk. Ob-va Ispyt. Prirody, Otd. Biol.,* vol. 51, No. 1.

Meklenburtsev, R.N. 1948. Materialy po systematike i ekologii Amu-Dar'inskogo fazana *Phasianus colchicus zarudnyi* But.) [On the systematics and ecology of Amu-Darya Pheasants *Phasianus colchicus zarudnyi* But.]. *Izvestiya AN UzbSSR,* No. 1.

Meklenburtsev, R.N. 1958. Materialy po nazemnym pozvonochnym basseina r. Kashkadar'ya (Data on the land vertebrates of the Kashka-Dar'ya basin). *Trudy Sredneaz. Un-ta,* No. 130.

Meklenburtsev, R.N. 1959. Nablyudeniya po osennemy proletu perepela (*Coturnix coturnix coturnix*) v okrestnostyakh g. Tashkenta [Observations on the autumn migration of Quail (*Coturnix coturnix coturnix*) in the vicinity of Tashkent]. *Trudy Sredneaz. Un-ta,* No. 123.

Menzbir, M.A. 1900. Okhotnich'i i promyslovye ptitsy Evropeiskoi Rossii i Kavkaza (Game Birds of the European Part of Russian and the Caucasus). Moscow, vol. 2.

Menzbir, M.A. 1904–1909. Ptitsy (Birds). St Petersburg.

Mikhailovskii, B.A. 1972. Kurinye verkhov'ev r. Amur (Galliform birds in the upper reaches of the Amur River). In: *Sbornik Nauchno-Tekhnicheskoi Informatsii,* Kirov, Nos. 37–39.

Mikheev, A.V. 1939. Lin'ka i izmenchivost' beloi kuropatki (*Lagopus lagopus* L.) Vostochnoi Palearktiki [Molt and transformation of Willow Grouse (*Lagopus lagopus* L.) in the eastern Palearctic]. *Trudy Zool. Muzeya MGU,* No. 5.

Mikheev, A.V. 1948. Belaya kuropatka (The Willow Grouse). Moscow.

Mikheev, A.V. 1952. Rod Belye kuropatki—*Lagopus*, Rod Tetereva—*Lyrurus*, Rod Dikusha—*Falcipennis,* Rod Ryabchiki—*Tetrastes* (Genera *Lagopus, Lyrurus, Falcipennis,* and *Tetrastes*). In: *Ptitsy Sovetskogo Soyuza,* vol. 4.

Mikheev, A.V. 1956. Seraya kuropatka (The Rock Partridge). *Okhota i Okhotnich'e Khoz-vo,* No. 4.

Milne-Edwards, A. 1869–1871. *Recherches Anatomiques et paléontologiques pour servir à l'histoire des oiseous fossiles de la France.* Paris.

Minin, N.V. 1939. Nekotorye zakonomernosti v letnem pitanii keklika (*Alectoris graeca falki* Hart) na severnom sklone Turkestanskogo khrebta [Some patterns in the summer feeding of Chukar (*Alectoris graeca falki* Hart) on the northern slope of the Turkestan range]. *Voprosy Ekologii i Biotsenologii,* No. 4.

Mishin, I.P. 1959. O pitanii dikushi na Sakhaline (On the feeding of Siberian Spruce Grouse of Sakhalin). *Ornitologiya,* No. 2.

Mishin, I.P. 1960. K biologii teterevinykh ptits na Sakhaline (Biology of the Tetraonidae of Sakhalin). *Ornitologiya,* No. 3.

Mlokosevich, L. 1879. Zametka o rasprostranenii v Zakavkaz'e kurinykh ptits. Turach (*Francolinus vulgaris* Steph.), kavkazskii teterev (*Tetrao mlokosiewiczii* Tacz.) [Notes on the distribution of galliform birds in the Trans-Caucasus: Francolin (*Francolinus vulgaris* Steph.) and Caucasian Black Grouse (*Tetrao mlokosiewiczii* Tacz.)]. *Izvestiya Kavk. Ob-va Lyubit. Estestvozn. i Al'piiskogo Kluba,* No. 1.

Moreau, R.E. and P. Wayre. 1968. On Palearctic Quail, *Ardea,* vol. 56, Nos. 3-4.

Müller-Using, D. 1958. Einige Beobachtungen und Feststellungen beim Alpenschneehuhn (*Lagopus mutus helveticus* Thienemann), *Ornithol. Mitt.*, vol. 10, No. 3.

Nasimovich, A.A. 1947. Gornaya indeika v gorakh Kavkazskogo zapovednika (Mountain turkeys in the Caucasian Preserve). *Priroda,* No. 3.

Nasimovich, A.A. 1950. Prichiny dinamiki granits areala i chislennosti fazanov v SSSR (Reasons for fluctuations in the geographic range and population of pheasants in the Soviet Union). *Okhrana Prirody,* No. 12.

Naumann, J. 1897. *Naturgeschichte der Vögel Mitteleuropas,* band 6.

Naumov, S.P. 1931. Mlekopitayushchie i ptitsy Gydanskogo poluostrova (Mammals and birds of Gydansk Peninsula). *Trudy Polyarnoi Komissii AN SSSR,* No. 4.

Nechaev, V.A. 1968. Osobennosti pitaniya ryabchika v kedrovo-shirokolistvennykh lesakh Primor'ya i Nizhneogo Priamur'ya (Peculiarities of Hazel Grouse feeding in the cedar-deciduous forests of Primor'e and the lower Priamur). In: *Resursy Teterevinykh Ptits v SSSR.* Moscow.

Nechaev, V.A. 1975. Priamur'e i Primor'e (Priamur and Primor'e). In: *Teterevinye Ptitsy.* Moscow.

Nechaev, V.A. and I.G. Nikolaev. 1970. Borodataya kuropatka v Primor'e (The Daurian Partridge in Primor'e). *Byul. Mosk. Ob-va Ispyt. Prirody,* No. 3.

Nefedov, N.I. 1943. Sel'skokhozyaistvennoe znachenie seroi kuropatki v nizhnem Povolozh'e (Importance of agriculture to Gray Partridges in the Lower Volga region). *Zool. Zhurn.,* vol. 22, No. 1.

Neifel'd, I.A. 1958. Ob ornitofaune Yuzhnoi Karelii (On the avifauna of southern Karelia). *Trudy Zool. In-ta AN SSSR,* vol. 25.

Neifel'd, I.A. 1958. Pitanie nekotorykh lesnykh ptits Yuzhnoi Karelii (The feeding of some jungle birds of southern Karelia). *Zool. Zhurn.,* vol. 37, No. 2.

Nemtsev, V.V., V.V. Krinitskii, and E.K. Semenova. 1973. Razvedenie teterevinykh ptits v vol'erakh (Raising tetraonid birds in captivity). In: *Prirodnye Resursy Mologo-Sheksinskoi Niziny,* No. 11, Vologda.

Niethammer, G. 1942. Handbuch der deutschen Vögelkunde. In: *Auftrag der Deutschen Ornithologischen Gesellschaft,* band 3.

Niethammer, G. 1968. Die Könige unter den Feldhühnern: Königshühner, *Vogel-Kosmos,* vol. 5, No. 3.

Nikul'tsev, A.P. 1962. Ekologicheskie faktory statsionarnogo razmeshcheniya tetereva i perespektivy ego promyslovogo ispol'zovaniya (Ecological factors of the localized distribution of Northern Black Grouse and the prospects for its commercial exploitation). In: *Materialy 3-i Vsesoyuzn. Ornitol. Konfer.* L'vov, bk. 2.

Nikul'tsev, A.P. 1964. Urozhai muzhskikh serezhek na berezakh i kontsentratsiya teterevinykh v stai (Harvest of male catkins on birches and the flocking of tetraonids). In: *Geografiya Plodonosheniya Lesnykh Drevesnykh Porod Kustarnikov i Yagodnikov, Znachenie ikh Urozhaev v Narodnom Khozyaistve i Zhizhi Fauny.* Moscow.

Nikul'tsev, A.P. 1967. Materialy po chislennosti i okhotnich'e-promyslovomu ispol'zovaniyu tetereva v SSSR (The Black Grouse population and its game and commercial exploitation in the Soviet Union). *Trudy Vsesoyuzn. Nauchno-Issled. In-ta Zhivotnogo Syr'ya i Pushniny.*

Nikul'tsev, A.P. 1968. Materialy po pitaniyu tetereva v Kirovskoi oblasti (On the food of Northern Black Grouse in the Kirov District). In: *Resursy Teterevinykh Ptits v SSSR.* Moscow.

Nikul'tsev, A.P. 1968. Teterev-kosach (The Black Grouse). *Okhota i Okhotnich'e Khoz-vo,* No. 5.

Nikul'tsev, A.P. 1972. Osnovnye statsii tetereva i plotnost' ikh zaseleniya (Main habitats of Northern Black Grouse and its population density). *Trudy CNII Okhot. Khoz-va i Zverovodstva,* No. 24.

Noska, M. and V.R. Tschusi zu Schmidhoffen. 1895. Das kaukasische Birkhuhn (*Tetrao mlokosiewiczii*), *Ornith. Jb.,* vol. 6, No. 2.

Noska, M. and V.R. Tschusi zu Schmidhoffen. 1896. Das kaukasische Königshuhn (*Tetraogallus caucasicus*), *Ornith. Jb.,* vol. 7, No. 1.

Novikov, G.A. 1941. Promyslovo-okhotnich'ya fauna sev.-zap. Zabaikal'ya (Game fauna of northwestern Trans-Baikal). *Trudy SOPS AN SSSR, Seriya Vost.-Sibir.,* No. 4.

Novikov, G.A. 1952. Ekologiya ptits Khibinskikh gor (Kol'skii poluostrov) [Ecology of the birds of Khibinsk mountains (Kola Peninsula)]. *Trudy Zool., In-ta AN SSSR,* No. 9.

Novikov, G.A. 1952. Materialy po pitaniyu lesnykh ptits Kol'skogo poluostrova (Data on the food of the wild birds of the Kola Peninsula). *Trudy Zool. In-ta AN SSSR,* No. 9.

Novikov, G.A. 1956. Elovye less kak sreda obitaniya i rol' ikh v zhizni mlekopitayushchikh i ptits (Spruce forests as a habitat and their role in the life of mammals and birds). In: *Rol' Zhivotnykh v Zhizni Lesa.* Moscow.

Novikov, G.A. 1967. O nekotorykh adaptivnykh osobennostyakh ekologii i povedeniya mlekopitayushchikh i ptits v zimnikh usloviyakh obitaniya (Certain adaptive features of the ecology and the behavior of mammals and birds in their winter abodes). In: *Vidovye i Prirodno-Klimaticheskie Adaptatsii Organizma Zhivotnykh.* Novosibirsk.

Oehme, H. 1968. Physiologische und morphologische Aspekte der Muskelleistung fliegender Tiere, *Biologische Rundschau,* vol. 6, No. 5.

Ogilvie-Grant, W.R. 1893. *Catalog of Birds in the British Museum.* London, vol. 22.

Oko, Z. 1963. Pozywienie kuropatw doroslich (*P. perdix* L.) w cyklu rocznym na terenie województwa poznanskiego w latach 1960–1961, *Prace Komis Nauk Roln. i Lesn Poznan Towarz. Przyjaciól Nauk*, vol. 14, No. 1.

Oliger, I.M. 1973. Materialy po pitaniyu teterevinykh ptits lesnoi zony evropeiskoi chasti RSFSR (Data on the food of tetraonid birds of the forest zone in the European part of the RSFSR). In: *Prirodnye Resursy Mologo-Sheksinskoi Niziny*. Vologda, No. 11.

Orlov, S.I. 1930. O biologii severnoi kuropatki (Biology of Hungarian Partridges). *Okhotnik i Rybak Sibiri*, No. 4.

Osmolovskaya, V.I. 1950. Rol' ptits v unichtozhenii nasekomykh—vreditelei lesnykh posadok Stalingradskoi oblasti (Role of birds in the destruction of insect pests of forest stands of Stalingrad District). *Zool. Zhurn.*, vol. 29, No. 3.

Osmolovskaya, V.I. 1961. Nasekomoyadnye ptitsy Nizhnego Povolzh'ya i ikh znachenie v organichenii chislennosti vreditelei lesa (Insectivorous birds of the Lower Volga region and their importance in controlling forest pests). In: *Polezashchitnoe Lesorazvedenie na Kashtanovykh Pochvakh*. Moscow, No. 1.

Osmolovskaya, V.I. 1966. Chislennost' i raspredelenie seroi kuropatki v evropeiskoi chasti SSSR (Population and distribution of Gray Partridges in the European part of the USSR). *Zool. Zhurn.*, vol. 45, No. 1.

Osmolovskaya, V.I. 1966. Prichiny izmenenii chislennosti teterevinykh (Reasons for fluctuations in populations of Tetraonidae). *Byul. Mosk. Ob-va Ispyt. Prirody, Seriya Biol.*, vol. 71, No. 5.

Osmolovskaya, V.I. 1970. Raspredelenie i chislennost' beloi kuropatki v lesnoi polose evropeiskoi chasti Soyuza (Distribution and population of Willow Grouse in the forest belt of the European part of the USSR). *Byul. Mosk. Ob-va Ispyt. Prirody, Seriya Biol.*, vol. 55, No. 1.

Osmolovskaya, V.I. and A.N. Formozov. 1950. Ocherki ekologii nekotorykh poleznykh ptits lesa (An outline of the ecology of some useful forest birds). In: *Ptitsy i Vrediteli Lesa*. Moscow.

Ostapenko, M.M. 1958. K voprosu o pitanii kamennoi kuropatki v yugozapadnykh otrogakh Chatkal'skogo khrebta (The problem of food for Rock Partridges in the south-western ridges of the Chatkal'sk range). *Uzbeksk. Biol. Zhurn.*, No. 1.

Ostapenko, M.M. 1965. Materialy po ekologii kamennoi kuropatki v yugozapadnykh ostrogakh Chatkal'skogo khrebta (Data on the ecology of Rock Partridges in the southwestern ridges of Chatkal'sk range). *Trudy Chatkal'skogo Gornolesnogo Zapovednika*, No. 2.

Pavlov, B.M. 1971. O pitanii beloi i tundryanoi kuropatok na Taimyre v snezhnyi period (Diet of Willow Grouse and Rock Ptarmigans in Taimyr during the snow season). In: *Problemy Okhotnich'ego Khozyaistva Krasnoyarskogo Kraya*. Krasnoyarsk.

Pavlov, B.M. 1973. Dinamicheskaya kharakteristika inter'ernykh osobennostei beloi i tundryanoi kuropatok Taimyra (Dynamic characteristics of internal features of Willow Grouse and Rock Ptarmigans of Taimyr). *Ekologiya,* No. 4.

Pavlov, B.M. 1974. Belaya i tundryanaya kuropatka Taimyra (Willow Grouse and Rock Ptarmigans of Taimyr). Avtoref. Kand. Diss., Moscow.

Pavlov, E.I. 1948. Ptitsy yugo-vostochnoi chasti Vitimskogo ploskogor'ya v raione ozer Tasei, Ivan i Arakhlei (Birds of the southeastern part of the Vitim plateau in the regions of Lakes Tasei, Ivan, and Arakhlei). In: *Ptitsy i Zveri Chitinskoi Oblasti.* Chita.

Pavlov, N.V. 1931. Kompleksnye stepi i pustyni Karsakpaya (The Steppe and Barren Land Complex of Karsakpai). Moscow.

Pavlov, N.V. 1948. Nekotorye biogeograficheskie otnosheniya v lesostepi Kazakhstana (Some biogeographic relations in the forest steppes of Kazakhstan). *Vestnik AN KazSSR,* No. 10

Pek, L.V. and T.F. Fedyanina. 1961. Pishcha ptits Kirgižii (Food of birds of Kirgizia). In: *Ptitsy Kirgizii.* Frunze, vol. 3.

Perfil'ev, V.I. 1975. Yakutiya (Yakutia). In: *Teterevinye Ptitsy.* Moscow.

Peters, Y.L. 1934. *Check-List of Birds of the World.* Cambridge, vol. 2.

Petrov, P. and S. Donchev. 1969. Prouchvaniya v'rkhu p'dp'd'ka (*Coturnix coturnix* L. v B'lgariya. *Izvestiya na Zoologich. In-ut s Muzei B'lg. AN,* bk. 30.

Pil'man, E.E. 1961. Oblepikha kak ekologicheskii faktor v zhizni ptits Selenginskoi Daurii (Sallow-thorn as an ecological factor in the life of birds of Selenginsk Dauriya). In: *Kraevedcheskii Sbornik Buryat. Filiala Geog. Ob-va SSSR,* No. 6.

Polivanov, V.M. and N.I. Polivanova. 1971. K. voprosu o sootnoshenii vidovoi spetsializatsii i ekologicheskoi plastichnosti u ptits (On the problem of the relation of species specialization with ecological adaptability in birds). In: *Ekologiya i Fauna Ptits Yuga Dal'nego Vostoka.* Vladivostok.

Popov, A.V. 1959. Ptitsy Gissaro-Karategina (Birds of Gissar and Karategin). Stalinabad.

Popov, A.V. 1960. Migratsii keklika v Tadzhikistane (Migration of Chukar in Tadzhikistan). *Trudy Problemnykh i Tematicheskikh Soveshchanii ZIN AN SSSR,* No. 9.

Popov, L. 1969. V zashchitu perepela (Protection of Common Quail). *Okhota i Okhotnich'e Khoz-vo,* No. 8.

Popov, Yu.K. and V.D. Shklyaev. 1967. Materialy po pitaniyu ryabchika v Udmurtskoi ASSR (Data on the food of the Hazel Grouse in Udmurtsk ASSR). *Uchen. Zap. Perm. Ped. In-ta,* No. 41.

Portenko, L.A. 1939. Fauna Anadyrskogo kraya. Ch. II (Fauna of the Anadyr region, pt. II). *Trudy Nauchno-Issled. In-ta Polyarnogo Zemledeliya i Promysl. Khoz-va,* No. 6.

*Portenko, L.A. 1972. Ptitsy Chukotskogo p-va i ostrova Vrangelya (Birds of the Chukchi Peninsula and Wrangel Island). Leningrad, vol. 1.

Porter, R.D. 1955. The Hungarian Partridge in Utah, *J. Wildlife Manag.,* vol. 19, No. 1.

Potapov, R.L. 1966. Ptitsy Pamira (Birds of Pamir). In: *Biologiya Ptits.* Moscow-Leningrad.

Potapov, R.L. 1969. O funktsional'nom znachenii rogovoi bakhromy na pal'tsakh teterevinykh ptits (Functional importance of the horny fringe on the toes of tetraonids). *Zool. Zhurn.*, vol. 48, No. 9.

Potapov, R.L. 1969. Tokovanie aziatskoi dikushi (*Falcipennis falcipennis*) [Breeding habits of Siberian Spruce Grouse (*Falcipennis falcipennis*)]. *Zool. Zhurn.*, vol. 48, No. 6.

Potapov, R.L. 1970. Sravnitel'nyi obzor dikush (rody *Falcipennis i Canachites*, Tetraonidae) Azii i Severnoi Ameriki [Comparative review of grouse (genera *Falcipennis* and *Canachites*, Tetraonidae) of Asia and North America]. In: *Ornitologicheskii Sbornik k 100-Letiyu so Dnya Rozhdeniya P.P. Sushkina.* Leningrad.

Potapov, R.L. 1971. Nekotorye momenty evolyutsii i rasseleniya glukharei (*Tetrao urogallus*) v antropogene [Some stages in the evolution and dispersal of capercaillies (*Tetrao urogallus*) in the Anthropogene]. *Zool. Zhurn.*, vol. 50, No. 1.

Potapov, R.L. 1974. Adaptatsii semeistva Tetraonidae k zimnemu sezonu (Adaptation of the family Tetraonidae to the winter season). In: *Issledovaniya po Biologii Ptitis.* Leningrad.

Potapov, R.L. and A.V. Andreev. 1973. K bioenergetike tetereva *Lyrurus tetrix v* zimnii period [Bioenergetics of Black Grouse (*Lyrurus tetrix*) in winter]. *Dokl. AN SSSR,* Vol. 210, No. 2.

Poyarkov, D.V. 1955. K ekologii seroi kuropatki v stepnykh raionakh evropeiskoi chasti SSSR (Ecology of Gray Partridges in the steppe regions of European USSR). *Uchen. Zap. Mosk. Gor. Ped. In-ta*, vol. 38, No. 3.

Poznanin, L.P. 1962. O znachenii razlichnykh sposobov peredvizheniya v evolyutsii ptits (On the importance of different modes of locomotion in the evolution of birds). In: *Ekologiya i Migratsiya Ptits Pribaltiki.* Riga.

Priroda i zhivotnyya Tsentral'noi Azii (iz puteshestvii Przheval'skogo) [Nature and the animals of central Asia (from Przewalski's expeditions)]. In: *Sbornik Priroda*, St. Petersburg, 1875, bk. 2.

*Translated into English and published by Amerind Publishing Company Pvt. Ltd., New Delhi, 1984.

Przheval'skii [Przewalski], N.M. 1870. Puteshestvie v Ussuriiskom krae v 1867–1869 (Expeditions through the Ussuriisk Region in 1867–1869). St. Petersburg.

Przheval'skii [Przewalski], N.M. 1875-1876. Mongoliya i strana tangutov (Mongolia and the Tangutov Country). St. Petersburg, vol. 2.

Pukinskii, Yu. and A. Nikanorov. 1974. Dikusha (Spruce Grouse). *Okhota i Okhotnich'e Khoz-vo*, No. 7

Pulliainen, E. 1966. Peltopyyn syksyisestä ja talvisesta ravinnosta, *Suomen Riista*, No. 18.

Pulliainen, E. 1966. Peltopyyn Talvisesta käyttäytymisestä, *Suomen Riista*, No. 18.

Pulliainen, E., L. Paloneimo, and L. Syrjälä. 1968. Digestibility of blueberry (bilberry) stems (*Vaccinium myrtillus*) and cowberries (*Vaccinium vitis-idae*) in Willow Grouse (*Lagopus lagopus*), *Ann. Acad. Sci. Fennicae S. A. Biologia*, No. 4.

Pynnönen, A. 1954. Beiträge zur Kenntnis der Lebensweise des Hasselhuhns *Tetrastes bonasia* (L.), *Riistatieteellisia Julkaisujä*. Helsinki, vol. 12.

Radde, G.I. 1884. Ornitologicheskaya fauna Kavkaza (*Ornis caucasica*). Sistematicheskoe i biologogeograficheskoe opisanie Kavkazskikh ptits [Bird fauna of the Caucasus (*Ornis caucasica*). Systematics, Biology, and Geographic Distribution of Caucasian Birds). Tiflis.

Rakhilin, V.K. 1968. Materialy po ekologii ryabchika na vostochnykh sklonakh Tsentral'nogo Sikhote-Alinya (Data on the ecology of Hazel Grouse on the eastern slopes of central Sikhote-Alin'). In: *Resursy Teterevinykh Ptits v SSSR*. Moscow.

Rashkevich, N.A. 1962. Zimnyaya ornitofauna tugaev Kara-Kalpakii (Winter bird fauna of Kara-Kalpak). *Ornitologiya*, No. 4.

Raus, L. 1969. Skol'ko yagod s'edayut zhivotnye (How many berries do animals consume?). *Okhota i Okhotnich'e Khoz-vo,* No. 5.

Raus, L.K. 1973. Dikorastushchie yagodniki v pitanii nekotorykh zverei i ptits (Wild berries in the food of some animals and birds). In: *Sbornik Nauchno-Tekhnicheskoi Informatsii (Okhota, Pushnina, Dich')*. Kirov, Nos. 40–41.

Razevig, V.A. 1880. Glukhar' v chernoles'e (Capercaillies of the deciduous forests). *Priroda i Okhota*, vol. 2, April.

Rikhter, G.D. 1948. Rol' snezhnogo pokrova v fizikogeograficheskom protsesse (Role of snow cover in physical and geographic processes). *Trudy In-ta Geografii AN SSSR*.

Rodahl, K. 1944. Content of vitamin C (1-ascorbic acid) in Arctic plants, *Trans. Proc. Bot. Soc. Edinb.*, vol. 34, part 1.

Rodionov, M.A. 1960. Osobennosti biologii kurinykh ptits v Leningradskoi oblasti (Features of the biology of Galliformes in the Leningrad

District). *Trudy Problemnykh i Tematicheskikh Soveshchanii ZIN AN SSSR*, No. 9.

Rodionov, M.A. 1963. K biologii ryabchika (*Tetrastes bonasia* L.) Leningradskoi oblasti [Biology of Hazel Grouse (*Tetrastes bonasia* L.) in the Leningrad District]. In: *Ekologiya Pozvonochnykh Zhivotnykh Leningradskoi Oblasti*. Leningrad.

Rodionov, M.A. 1963. Materialy po biologii glukharya (*Tetrao urogallus* L.) v Leningradskoi oblasti [Data on the biology of Western Capercaillies (*Tetrao urogallus* L.) in the Leningrad District]. In: *Ekologiya Pozvonochnykh Zhivotnykh Leningradskoi Oblasti*. Leningrad.

Rodionov, M.A. 1964. Plody i semena v pitanii teterevinykh ptits (Fruits and seeds in the food of tetraonids). In: *Geografiya Plodonosheniya Lesnykh Drevesnykh Porod Kustarnikov i Yagodnikov, Znachenie ikh Urozhaev v Narodnom Khozyaistve i Zhizni Fauny*. Moscow.

Rodionov, M.A. and V.V. Nemtsev. 1954. K biologii seroi kuropatki (Biology of Gray Partridge). *Priroda*, No. 12.

Rodionova, T.V. 1940. Morfobiologicheskii analiz myshechnoi sistemy nogi kurinykh (Galliformes) [Morphological and biological analysis of the muscular system of the legs of Galliformes]. *Trudy Nauchno-Issled. Zool. Biol. In-ta, Khar'kov, Gos. Un-ta,* vols. 8–9.

Romanov, A.N. 1960. Nekotorye cherty ekologii glukharya v svyazi s vyrubkami v severnoi taige (Some ecological characteristics of Western Capercaillies associated with felling in the northern taiga). *Trudy Komi Fil. AN SSSR,* No. 9.

Romanov, A.N. 1963. Ocherk ekologii ryabchika v Komi ASSR (An outline of the ecology of Hazel Grouse in Komi ASSR). *Trudy Komi Fil. AN SSSR*, vol. 24, No. 12.

Romanov, S. 1969. Vid poleznyi i perspektivnyi (o razvedenii seroi kuropatki) [Useful and promising species (breeding and raising Gray Partridges)]. *Okhota i Okhotnich'e Khoz-vo*, No. 4.

Rusakov, O.S. 1963. Chislennost', pitanie i statsial'nye razmeshcheniya teterevinykh v Leningradskoi oblasti (Population, feeding, and local migrations of tetraonids in Leningrad District). In: *Promyslovaya Fauna i Okhotnich'e Khozyaistvo Severo-Zapada RSFSR.* Leningrad.

Rustamov, A.K. 1945. K biologii keklika (*Alectoris graeca* Meisner) v Turkmenii [Biology of Chukar (*Alectoris graeca* Meisner) in Turkmenia]. *Izvestiya Turkm. Fil. AN SSSR*, No. 1.

Rustamov, A.K. and Sh. I. Kogan. 1954. Zametki o pitanii pustynnoi kuropatki (Notes on the food of Sand Partridge). *Izvestiya AN TurkmSSR,* No. 4.

Sabaneev, L.P. 1876. Glukhoi teterev (The Capercaillie). *Priroda,* bks. 3–4.

Sabaneev, L.P. 1876. Ryabchik (Hazel Grouse). *Priroda,* bk. 4.
Sabaneev, L.P. 1876. Teterev-kosach (The Black Grouse). *Zhurnal Okhoty. Prilozhenie*, vols. 3-4.
Salikhbaev, Kh.S., D.Yu. Kashkarov, and A. Sharipov. 1970. Ptitsy, mlekopitayushchie (Birds and mammals). In: *Ekologiya Pozvonochnykh Zhivotnykh Khrebta Nuratau.* Tashkent.
Salikhbaev, Kh. S. and M.M Ostapenko. 1964. Ptitsy (Birds). In: *Ekologiya i Khozyaistvennoe Znachenie Pozvonochnykh Zhivotnykh Yuga Uzbekistana.* Tashkent.
Savile, D. 1957. Adaptive evolution in the avian wing, *Evolution*, vol. 11, No. 2.
Schäfer, E. 1938. Ornithologische Ergebnisse zweier Forschungsreisen nach Tibet, *J. fur Ornithol.*, vol. 88.
Scholander, P.F., V. Walters, R. Hock, and L. Irving. 1950. Body insulation of some Arctic and tropical mammals and birds, *Biolog. Bull.*, No. 90.
Sdobnikov, V.M. 1957. Sibirskaya tundrovaya kuropatka *(Lagopus mutus kellogae* Grin.) na severnom predele svoego areala [The Siberian Rock Ptarmigan (*Lagopus mutus kellogae* Grin.) in the northern limits of its range]. *Trudy Arktichesk. In-ta*, vol. 205.
Sedova, I.A. 1950. Morfo-biologicheskii analiz rotovogo apparata ptits v svyazi s ikh pitaniem (Morphological and biological analysis of the oral apparatus of birds in relation to their feeding). Avtoref. Kand. Diss., Moscow.
Seiskari, P. 1957. Riekon talvisesta ruokailusta, *Suomen Riista,* No. 11.
Seiskari, P. 1957. Teeren talvisen ruokailupaikan rakenteesta, *Suomen Riista*, No. 11.
Seiskari, P. 1962. On the winter ecology of the Capercaillie (*Tetrao urogallus*) and the Black Grouse (*Lyrurus tetrix*) in Finland. S.L.S., riistanhoitosäätio 1962, *Suomen Kirjakappalehti*, vol. 65, No. 19.
Semenov-Tyan-Shanskii, O.I. 1938. Ekologiya borovoi dichi Laplandskogo zapovednika (Ecology of the game birds of pine forests in the Lapland Preserve). *Trudy Lapl. Zapovednika*, No. 1.
Semenov-Tyan-Shanskii, O.I. 1939. Ekologiya tundryanoi kuropatki v Laplandskom zapovednike (Ecology of Rock Ptarmigan in the Lapland Preserve). *Voprosy Ekologii Biotsenologii*, No. 4.
Semenov-Tyan-Shanskii, O.I. 1959. Ekologiya teterevinykh ptits (Ecology of Tetraonidae). Moscow.
Sergeeva, N.A. and E.B. Sumina. 1963. Popytki akklimatizatsii borodatykh i reakklimatizatsiya belykh kuropatok v srednei polose (Attempts to introduce Daurian Partridge and the reintroduction of Willow Grouse in temperate latitudes). *Ornitologiya,* No. 6.
Severtsov, N.A. 1873. Vertikal'noe i gorizontal'noe rasprostranenie Turkestanskikh zhivotnykh (Latitudinal and altitudinal distribution of

the animals of Turkestan). *Izvestiya Ob-va Lyubitelei Estestvozn., Antropologii i Etnografii,* vol. 8, No. 2.

Severtsov, N.A. 1947. Puteshestvie po Turkestanskomu kraya i issledovanie gornoi strany Tyan'-Shanya, sovershennoi po porucheniyu Russk. geogr. ob-va (Expedition to Turkestan and a Study of the Hilly Regions of Tien-Shan Commissioned by the Russian Geographic Society). Moscow, 2nd ed.

Shaposhnikov, L. 1936. O biologii i promyslovom znachenii Syr-dar'inskogo fazana (Biology and commercial value of the Syr-Darya Pheasant). *Boets-Okhotnik,* No. 12.

Shaposhnikov, L.V. 1936. O migratsii seroi kuropatki v Kazakhstane (Migration of Gray Partridge to Kazakhstan). *Boets-Okhotnik,* No. 3.

Shatas, Ya.F. 1963. Fazan del'ty Volgi i faktory, opredelyayushchie dinamiku ego chislennosti (The pheasants of the Volga delta and factors determining their population dynamics). *Trudy Astrakhanskogo Zapovednika,* No. 8.

Shcherbakov, A.F. 1932. Teterev--glukhar', ego rasprostranenie i zapas (The Capercaillie—its distribution and population). In: *Verkhne-Vychegodskaya Ekspeditsiya.* Moscow.

Shcherbina, M.N. 1915. Gruppovoi tok glukharei (*Tetrao urogallus uralensis* L.) na zemle [Group breeding of Western Capercaillies (*Tetrao urogallus uralensis* L.)on land]. *Ptitsevedenie i Ptitsevodstvo,* vol. 6, No. 4.

Shcherbina, M.N. 1916. Na glukharei (*Tetrao urogallus* L.) [The Capercaillie (*Tetrao urogallus* L.)]. *Ptitsevedenie i Ptitsevodstvo*, vol. 7, Nos. 1-2.

Shestakova, G.S. 1950. Znachenie differentsiatsii poleta i evolyutsii ptits (Significance of flight differences and the evolution of birds). *Byul. Mosk. Ob-va Ispyt. Priorody, Otd. Biol.*, vol. 55, No. 3.

Shestakova, G.S. 1953. Morfo-biologicheskii analiz kak metod izucheniya poleta ptits (Morphological and biological analysis as a method of studying the flight of birds). *Trudy In-ta Morfologii Zhivotnykh im. A.N. Severtsova*, No. 9.

Shestakova, G.S. 1971. Stroenie kryl'ev i mekhanika poleta ptits (Structure of the Wings and Flight Mechanisms of Birds). Moscow.

Shestoperov, E.L. 1929. Materialy dlya ornitologicheskoi fauny Iliiskogo kraya (Data on the bird fauna of the Ili region). *Byul. Mosk. Ob-va Ispyt. Prirody, Otd. Biol.,* vol. 38, Nos. 3-4.

Shevareva, T.P. 1965. Izuchenie sezonnogo razmeshcheniya i migratsii ptits v SSSR s pomeshch'yu kol' tsevaniya (Study on the seasonal dispersal and migration of birds in the Soviet Union by ringing). In: *Migratsiya Ptits i Mlekopitayushchikh.* Moscow.

Shevchenko, V.V. 1960. Surovaya zima 1944-1945 gg. i ee vliyanie na ptits predgorii Talasskogo Alatau (The severe winter of 1944-1945 and its effects on birds of the foothills of Talassk Alatau). *Okhrana Prirody*, No. 6.

Shilo, A.A. 1974. Perepel v Barabinskoi lesostepi (The Common Quail in the Barabinsk forest-steppe). In: *Materialy 6-i Vsesoyuzn. Ornitol. Konfer,* Moscow, pt. 2.

Shilov, I.A. 1968. Regulyatsiya teploobmena u ptits (ekologo-fiziologicheskii ocherk) [Thermoregulation in Birds (Ecological and Physiological Aspects)]. Moscow.

Shinkin, N.A. 1967. Kurinye yugo-vostochnoi chasti Zapadnoi Sibiri, ikh khozyaistvennoe ispol'zovanie i okhrana (Galliform birds of the southeastern part of western Siberia—their economic value and conservation). Avtoref. Kand. Diss., Tomsk.

Shlyakhtin, A. 1896. Gornaya kuropatka (The Rock Ptarmigan). *Okhotnich'ya Gazeta,* No. 9.

Shnitnikov, V.N. 1949. Ptitsy Semirech'ya (Birds of Semirech'e). Moscow-Leningrad.

Short, L.J. 1967. A review of the genera of grouse (Aves, Tetraoninae), *Amer. Museum Novit.,* No. 2289.

Shtegman, B.K. 1931. O proiskhozhdenii ornitofauny taigi (Origin of the taiga avifauna). *Dokl. AN SSSR. Seriya A,* No. 31.

Shtegman, B.K. 1934. Ptitsy Kokchetavskikh borov (Birds of the Kochetavsk pine forests). *Trudy Kazakhstan. Bazy AN SSSR,* No. 1.

Shtegman, B.K. 1936. Problema bylykh kontinental'nykh svyazei mezhdu Starym i Novym svetom v ornitologicheskom osveshchenii (The problem of former continental bridges between the Old and New Worlds in light of ornithological records). *Priroda,* No. 7.

Shtegman, B.K. 1938. Osnovy ornitogeograficheskogo deleniya Palearktiki (Principles of ornithological and geographic division of the Palearctic). In: *Fauna SSSR. Ptitsy.* Moscow-Leningrad, vol. 1, No. 2.

Shtegman, B.K. 1945-1946. K ekologii fazana v del'te r. Ili (Ecology of the pheasants of the Ili River delta). *Vestnik Kaz. Fil. AN SSSR,* 1945, No. 5 (8); 1946, No. 1 (10).

Shtegman, B.K. 1946. Osnovnye tipy avifauny Tyan'-Shanya i ikh genezis (Main types of avifauna in Tien-Shan and their genesis). *Vestnik AN KazSSR,* Nos. 7–8.

Shtegman, B.K. 1950. Issledovaniya o polete ptits. O letnykh sposobnostyakh kurinykh ptits (Studies of bird flight. Flight capability of Galliformes). In: *Sbornik Pamyati Akademika P.P. Sushkina.* Moscow-Leningrad.

Shtegman, B.K. 1953. Osobennosti letnykh kachestv seroi i kamennoi kuropatok (Flight characteristics of Gray and Rock partridges). *Zool. Zhurn.,* vol. 32, No. 4.

Shtegman, B.K. 1954. O ptitsakh vysokogornoi zony Zailiiskogo Alatau (Birds of the alpine Trans-Ilian Alatau). *Trudy Leningr. Ob-va Estestvoispyt.,* vol. 72, No. 4.

Shtegman, B.K. 1954. Osobennosti letnykh kachestv voronovykh ptits (Flight characteristics of Corvidae). *Zool. Zhurn.*, vol. 33, No. 3.

Shtegman, B.K. 1958. O nekotorykh osobennostyakh stroeniya skeleta golubei i ryabkov (Some peculiarities in the skeletal structure of pigeons and grouse). *Uchen. Zpl. MGU. Ornitologiya,* No. 197.

Shtegman, B.K. 1960. Osobennosti stroeniya zadnikh konechnostei u golubei i ryabkov i nekotorye cherty prisposobleniya ptits k peredvizheniyu po zemle (Structural features of the hind limbs of pigeons and grouse and some adaptive features of birds to locomotion on land). *Trudy Problemnykh i Tematicheskikh Soveshchanii ZIN AN SSSR*, No. 9.

Shtegman, B.K. 1966. O zakonomernostyakh v rasprostranenii avifauny taigi (Distribution patterns of avifauna in the taiga). *Izvestiya AN EstSSR*, vol. 15, No. 2.

Shul'pin, L.M. 1936. Promyslovye, okhotnich'i i khishchnye ptitsy Primor'ya (Game and Predatory Birds of Primor'e). Vladivostok.

Shul'pin, L.M. 1939. Ekologicheskii ocherk ptits Alma-Atinskogo gosudarstvennogo zapovednika (An ecological account of birds of the Alma-Ata State Preserve). *Trudy Alma-Atinskogo Zapovednika*, No. 1.

Shul'pin, L.M. 1940. Ornitologiya (Ornithology). Leningrad.

Shvarts, S.S. 1949. Novye dannye po otnositel'nomu vasu serdtsa i pecheni ptits (New data on the relative weight of the heart and liver of birds). *Zool. Zhurn.*, No. 4.

Shvarts, S.S. 1961. Puti prisposobleniya nazemnykh pozvonochnykh (preimushchestvenno mlekopitayushchikh) k usloviyam Subarktiki [Adaptations of land vertebrates (mostly mammals) to subarctic conditions]. In: *Problemy Severa*. Moscow, No. 4.

Shvarts, S.S., V.S. Smirnov, and L.N. Dobrinskii. 1968. Metod morfofiziologicheskikh indikatorov v ekologii nazemnykh pozvonochnykh (The Morphological and Physiological Method of Indicators in the Ecology of Land Vertebrates). Sverdlovsk.

Sirakhulidze, Z.D. 1974. K biologii kavkazskogo tetereva (Biology of Caucasian Blackcocks). *Ornitologiya*, No. 11.

Skripchinskii, K.K. 1962. K voprosu akklimatizatsii borodatoi kuropatki na srednem i yuzhnom Urale (The problem of the naturalization of Daurian Partridge in the central and southern Urals). In: *Materialy 3-i Vsesoyuzn. Ornitol. Konfer.* L'vov, bk. 2.

Sleptsov, M.M. and N.N. Gorchakovskaya. 1952. K ekologii Ussuriiskogo fazana (Ecology of Ussurian Pheasants). *Byul. Mosk. Ob-va Ispyt. Prirody, Otd. Biologii*, vol. 57, No. 3.

Sludskii, A.A. 1965. Rol' stikhiinykh bedstvii v dinamike chislennosti ptits v Kazakhstane (Role of natural calamities in the population dynamics of the birds of Kazakhstan). In: *Novosti Ornitologii*. Alma-Ata.

Smirnov, A. 1967. V dzhunglyakh Khamar-Dabana (In the Jungles of Khamar-Daban). Irkutsk.

Smythyes, B.E. 1953. *The Birds of Burma*. Edinburgh-London.

Snigirevskii, S.I. 1946. Teterevinye SSSR (Tetraonidae of the Soviet Union). Kand. Diss., ZIN AN SSSR, Leningrad.

Snigirevskii, S.I. 1950. Lin'ka teterevinykh (sem. Tetraonidae, otr. Galliformes) [Molting in tetraonids (family Tetraonidae, order Galliformes)]. In: *Sbornik Pamyati Akademika P.P. Sushkina*. Moscow-Leningrad.

Sokolov, E.A. 1949. Okhotnich'i zhivotnye korma i pitanie promyslovykh zverei i ptits (Game Animals: Food and Nutrition of Game Animals and Birds). Moscow.

Sokolov, N.O. 1960. O kuropatkakh Bol'shezemel'skoi tundry (Partridges of the Bol'she-Zemlya tundra). *Okhota i Okhotnich'e Khoz-vo*, No. 2.

Sokolov, O. 1895. O prolete kuropatok. Otvet g. S. Buturlinu (Migration of galliform birds. A rejoinder to S. Buturlin). *Okhotnich'ya Gazeta*, No. 49.

Spangenberg, E.P. 1948. Kratkie dopolneniya k faune ptits Armenii (Brief supplement to the avian fauna of Armenia). *Izvestiya AN ArmSSR. Biol. i S.-Kh. Nauki*, vol. 1, No. 3.

Spangenberg, E.P. 1948. Ob osennem perelete perepelov v Krymu (Autumn migration of quail to Crimea). *Okhrana Prirody*, No. 4.

Spangenberg, E.P. 1951. Zapiski naturalista. Perepela (A Naturalist's Diary. Quail). Moscow.

Spangenberg, E.P. 1965. Ptitsy basseina r. Imana (Birds of the Iman River basin). *Trudy Zool. Muzeya MGU*, No. 9.

Spangenberg, E.P. and G.A. Feigin. 1930. Okhotnich'i ptitsy Kzyl-Ordinskogo uezda Syr-Dar'inskoi oblasti (Kazakhstan) [Game birds of the Kzyl-Orda District of the Syr-Darya region (Kazakhstan)]. *Trudy po Lesnomu Opytnomu Delu*, No. 7.

Spangenberg, E.P. and A.M. Sudilovskaya. 1959. Materialy po biologii i vertikal' nomu razmeshcheniyu ptits v Kirgizskom Alatau (On the biology and vertical migration of birds of Kirgiz Alatau). *Uchen. Zap. Mosk. Obl. Ped. In-ta*, vol. 71, No. 4.

Starosivil'skii. 1897. Turach. Sposoby okhoty za nim i ego istreblenie v Zakavkaz'e (Francolins. Methods of hunting them and their extermination in Trans-Caucasus). *Priroda i Okhota*, March.

Stegmann [Shtegman], B.K. 1932. Die geographischen Formen des Birkhuns (*Lyrurus tetrix*). *J. fur Ornithol.*, vol. 80, No. 3.

Stepanyan, L.S. 1958. O proiskhozhdenii nekotorykh elementov vysokogornoi avifauny Turkestana (Origin of some elements of the alpine avifauna of Turkestan). In: *Problemy a Zoogeografi Sushi.* L'vov.

Stepanyan, L.S. 1959. Ptitsy khrebta Terskei-Alatau (Tyan'-Shan') [Birds of the Terskei-Alatau (Tien-Shan)]. *Uchen. Zap. Mosk. Obl. Ped. In-ta*, vol. 71, No. 4.

Stepanyan, L.S. 1960. Razmeshchenie avifaunisticheskikh kompleksov v Tsentral'nom Tyan'-Shane (Distribution of avifaunal complexes in central Tien-Shan). *Ornitologiya*, No. 4.

Stepanyan, L.S. 1960. Zamechaniya o roli lednikovogo perioda v istorii vysokogornoi avifauny Palearktiki (Notes on the role of the glacial period in the history of the alpine avifauna of the Palearctic). In: *Materialy k Konferentsii po Voprosam Zoogeografii Sushi, 15-21 Avgusta 1960 g.* Alma-Ata.

Stepanyan, L.S. 1962. Sistematicheskie vzaimootnosheniya dikushi i Kanadskogo ryabchika (Taxonomic relationships between Siberian Spruce Grouse and Canadian Hazel Grouse). *Ornitologiya*, No. 5.

Stepanyan, L.S. 1975. Sostav i raspredelenie ptits fauny SSSR. Nevorob'inye Non-Passeriformes (Composition and Distribution of Birds in the Fauna of the Soviet Union—Non-Passeriformes). Moscow.

Storer, J.H. 1948. The flight of birds, *Cranbrook Institute of Science Bull.* No. 28.

Strakhov, N.M. 1948. Osnovy istoricheskoi geologii (Principles of Historic Geology). Moscow-Leningrad, pt. II.

Strautman, F.I. 1954. Ptitsy Sovetskikh Karpat (Birds of the Soviet Carpathians). Kiev.

Strautman, F.I. 1963. Ptitsy Zapadnykh oblastei USSR (Birds of the western districts of the Ukrainian SSR). L'vov.

Strelkov, E.I. 1961. K ekologii ryabchika v taezhnoi zone Tomskoi oblasti (Ecology of Hazel Grouse in the taiga zone of the Tom District). *Uchen. Zap. Tomsk. Ped. In-ta,* No. 1.

Stresemann, E. 1927-1954. *Handbuch der Zoologie*, vol. 7, 2 Hälfte. *Sauropsida Aves*. Berlin-Leipzig.

Stresemann, E. 1959. The status of avian systematics and its unsolved problems. *Auk.*, vol. 76, No. 3.

Stroganov, S.U. 1946. Gibel' fazana ot beskormitsy v Gissarskoi doline (Death of pheasants from starvation in the Gissar Valley). *Priroda*, No. 3.

Stroganov, S.U. 1957. Fazan v Tadzhikistane (Pheasants of Tadzhikistan). *Okhota i Okhotnich'e Khoz-vo,* No. 4.

Sukachev, V.N. 1938. Istoriya rastitel'nosti SSSR vo vremya pleistotsena (History of the flora of the Soviet Union in the Pleistocene). In: *Rastitel'nost' SSSR.* Leningrad, pt. I.

Sumina, E.B. 1962. Nekotorye rezul'taty akklimatizatsii i reakklimatizatsii borodatykh i belykh kuropatok (po materialam kol'tsevaniya) [Some results of the introduction and reintroduction of Daurian Partridge and Willow Grouse (based on data obtained by ringing)]. In: *Materialy 3-i Vsesoyuzn. Ornitol. Konfer.*, L'vov.

Sungurov, A.N. 1959. Morfofiziologicheskie i ekologicheskie osobennosti teterevinykh i fazanovykh (Morphological, physiological, and ecological peculiarities of Tetraonidae and Phasianidae). In: *Materialy 2-i Vsesoyuzn. Ornitol. Konfer.*, Moscow.

Sushkin, P.P. 1908. Ptitsy srednei Kirgizskoi stepi (Birds of the central Kirgiz steppes). In: *Materialy k Poznaniyu Fauny i Flory Rossiiskoi Imperii, Otd. Zool.*, Moscow, No. 8.

Sushkin, P.P. 1914. Ptitsy Minusinskogo kraya, Zapadnogo Sayana i Uryankhaiskoi zemli (Birds of the Minusinsk area, western Sayan, and Uryankhai land). In: *Materialy k Poznaniyu Fauny i Flory Rossiiskoi Imperii, Otd. Zool.* No. 13.

Sushkin, P.P. 1938. Ptitsy Sovetskogo Altaya in prilezhashchikh chastei severo-zapadnoi Mongolii (Birds of the Soviet Altai and Adjacent Areas of Northwestern Mongolia). Leningrad.

Sviridov, A. 1876. Kamennaya kuropatka (The Rock Partridge). *Priroda*, bk. 2.

Syroechkovskii, E.E. 1960. Izmeneniya arealov ptits v Srednei Sibiri v rezul'tate potepleniya klimata i vozdeistviya cheloveka (Changes in the geographic ranges of birds of central Siberia due to the warming climate and anthropogenic activity). *Ornitologiya*, No. 3.

Syroechkovskii, E.E. 1967. Biologo-khozyaistvennoe raionirovanie Eniseiskogo severa kak geograficheskaya osnova dlya razmeshcheniya i organizatsii severnogo khozyaistva (Biological and economic zoning of the northern Yenisey as a geographic basis for the distribution and organization of economy in the North). In: *Problemy Severa.* Moscow, No. 11.

Szederjei, A. 1958. Die Bedeutung der Deckung fur Rebhüner und Fasanen, *Z. Jagdwiss,* vol. 4, No. 2.

Taka-Tsukasa. 1932–1939. *The Birds of Nippon.* London-Tokyo, vol. 1.

Tarasov, M.P. 1960. Materialy po ekologii borovoi dichi Zapadnogo Khamar-Dabana (Data on the ecology of pine forest game birds of western Khamar-Daban). *Izvestiya Irkutsk S.-Kh. In-ta*, No. 18.

Telepnev, V.G. 1972. Nekotorye osobennosti v pitanii yuzhno-sibirskogo glukharya (Some peculiarities in the feeding of southern Siberian Capercaillies). In: *Sbornik Nauchno-Tekhnicheskoi Informatsii (Okhota, Pushnina, Dich').* Kirov, Nos. 37–39.

Teplov, V.P. 1947. Glukhar' v Pechorsko-Ilychskom zapovednike (The Capercaillie in the Pechora-Ilych Preserve). *Trudy Pechorsko-Ilychskogo Zapovednika*, No. 4, pt. I.

Teplov, V.P. 1947. K ekologii borovoi dichi Pechorsko-Ilychskogo zapovednika (Ecology of the pine forest game birds of Pechora-Ilych Preserve). *Trudy Pechorsko-Ilychskogo Zapovednika*, No. 4, pt. I.

Teplov, V.P. 1960. Dinamika chislennosti i godovye izmeneniya v ekologii promyslovykh zhivotnykh Pechorskoi taigi (Population dynamics and annual changes in the ecology of the game animals of the Pechora taiga). *Trudy Pechoro-Ilychskogo Zapovednika*, No. 8.

Teplov, E.N. 1957. Ptitsy raiona Pechoro-Ilychskogo zapovednika (Birds of the Pechora-Ilych Preserve). *Trudy Pechoro-Ilychskogo Zapovednika*, No. 6.

Teplova, E.N. and V.P. Teplov. 1947. Znachenie zimnego pokrova v biologii mlekopitayushchikh i ptits Pechorsko-Ilychskogo zapovednika (Importance of winter cover in the biology of mammals and birds of the Pechora-Ilych Preserve). *Trudy Pechorsko-Ilychskogo Gos. Zapovednika*, No. 5.

Teterevinye ptitsy, Promyslovye zhivotnye i sreda ikh obitaniya (Tetraonids. Game Animals and Their Habitats). 1975. Moscow.

Tikhomirov, B.A. 1946. Puti formirovaniya rastitel' nogo pokrova Arkticheskoi Evrazii v chetvertichnom periode (Courses of the formation of the plant canopy of Arctic Eurasia in the Quaternary period). *Sovetskaya Botanika*, vol. 14, No. 5.

Tikhomirov, B.A. 1959. Vzaimosvyazi zhivotnogo mira i rastitel'nogo pokrova tundry (Interrelations of the Animal World and Tundra Vegetation). Moscow-Leningrad.

Tinbergen, N. 1969. Povedenie zhivotnykh (Animal Behavior). Translated from English. Moscow.

Tkachenko, V.I. 1966. Ekologiya kurinykh ptits vysokogornoi oblasti Severo-Zapadnogo Kavkaza (Ecology of Galliformes in the alpine regions of the northwestern Caucasus). *Trudy Teberdinskogo Zapovednika*, No. 6.

Tolmachev, A.I. 1927. K avifaune o-va Kolgueva (Avifauna of Kolguev). *Ezhegodn. Zool. Muzeya Akad. Nauk.* Leningrad, vol. 28.

Tolmachev, A.I. 1954. K istorii vozniknoveniya i razvitiya temnokhvoinoi taigi (History of the Origin and Growth of Boreal Forests). Moscow-Leningrad.

Tolstoi, A.K. 1964. Dva dnya v Kirgizskoi stepi (Two days in a Kirgiz steppe). *Sobr. Soch.* Moscow, vol. 3.

Tomme, M.F. 1964. Korma SSSR (Food in the Soviet Union). Moscow.

Tsvelen'ev, L.A. 1938. Materialy po pitaniyu ryabchika na Altae (Data on the feeding of Hazel Grouse in Altai). *Trudy Altaiskogo Zapovednika*, No. 1.

Tsyurupa, V.D. 1933. Vetochnyi korm (Twig Fodder). Moscow-Leningrad.

Tugarinov, A. Ya. 1929. O proiskhozhdenii fauny Arktiki (On the origin of arctic fauna). *Priroda*, Nos. 7-8.

Tugarinov, A. Ya. 1932. K kharakteristike chetvertichnoi ornitofauny Sibiri (Characteristics of the Quaternary bird fauna of Siberia). *Trudy Komissii po Izucheniyu Chetvertichnogo Perioda*, No. 1.

Tugarinov, A. Ya. 1940. Novye nakhodki pliotsenovoi ornitofauny Odessy (New discoveries of the Pleistocene avifauna of Odessa). *Dokl. AN SSSR*, vol. 26, No. 2.

Tugarinov, A. Ya. and E.V. Kozlova. 1945. Tadzhikskii fazan (Pheasants of Tadzhikistan). *Izvestiya Tadzhik, Fil. AN SSSR*, No. 6.

Tyurekhodzhaev, Zh. M. 1967. Ekologiya fazana na Syr-Dar'e (Ecology of pheasants in Syr-Darya). Avtoref. Kand. Diss., Alma-Ata.

Tyurin, P.S. 1957. Teterev (*Lyrurus tetrix*) v Kirgizii Northern Black Grouse (*Lyrurus tetrix*) in Kirgizia]. *Trudy In-ta Zool. i Paraz. AN Kirg-SSR*, No. 6.

Ul'yanin, N.S. 1939. K biologii beloi kuropatki *Lagopus lagopus major* Lor. Severnogo Kazakhstana (po nablyudeniyam v Naurzumskom zapovednike 1935-1936 gg.) [Biology of Willow Grouse *Lagopus lagopus major* Lor. in northern Kazakhstan (based on observations made in Naurzum Preserve in 1935-1936)]. *Trudy Zool. Muzeya MGU*, No. 5.

Ul'yanin, N.S. 1941. Materialy k ekologii perepela Severnogo Kazakhstana (Data on the ecology of quail in northern Kazakhstan). *Trudy Zool. Muzeya MGU*, No. 6.

Ul'yanin, N.S. 1949. K ekologii tetereva, beloi i seroi kuropatok Severnogo Kazakhstana (Ecology of Black Grouse, Willow Grouse, and Gray Partridges in northern Kazakhstan). *Trudy Naurzumskogo Zapovednika*, No. 2.

Ushakov, S.L. 1949. Nekotorye periodicheskie yavleniya v zhizni ptits v Il'menskom zapovednika (Some cyclic phenomena in the life of birds in the Il'mensk Preserve). *Trudy Il'menskogo Zapovednika*, No. 4.

Uspenskii, S.M. 1965. Materialy po faune ptits severa Anabarskikh tundr (Data on the avifauna in the northern Anabarsk tundra). *Trudy Zool. Muzeya MGU*, No. 9.

Uspenskii, S.M. 1969. Zhizn' v vysokikh shirotakh na primere ptits (Bird Life in High Altitudes). Moscow.

Uspenskii, S.M., R.L. Beme, S.G. Priklonskii, and V.I. Vekhov. 1962. Ptitsy Severo-Vostoka Yakutii (Birds of northeastern Yakutia). *Ornitologiya*, No. 5.

Vasilenko, V.G. and A.A. Tkachenko. 1970. Materialy po ekologii teterevinykh ptits Yuzhnogo Urala (Data on the ecology of Tetraonidae in the southern Urals). *Nauchn. Trudy Tul'skogo Ped. In-ta. Biologiya i Sel'skoe Khozyaistvo*, No. 4.

Vasil'ev, V.N. 1944. Neskol'ko slov k voprosu o proiskhozhdenii taigi (On the origin of taiga). *Priroda*, No. 3.

Verbolovich, P.A. 1961. Mioglobin i ego rol' v fiziologii i patologii zhivotnykh i chesloveka (Myoglobin and Its Role in the Physiology and Pathology of Animal and Man). Moscow.

Vereshchagin, N.K. 1931. Fazan v srednem techenii Syr-Dar'i (Pheasants in the middle reaches of the Syr-Darya). *Okhotnik*, June.

Vereshchagin, N.K. 1947. Okhotnich'ki i promyslovye zhivotnye Kavkaza (Game and Commercial Animals of the Caucasus). Baku.

Vereshchagin, N.K. 1960. Osobennosti rasprostraneniya turacha *Francolinus francolinus* L. v Azerbaidzhane i perspektivy rasshireniya ego areala [Peculiarities in the distribution of francolins (*Francolinus francolinus* L.) in Azerbaidzhan and prospects for expanding their geographic range]. *Okhrana Prirody i Ozelenenie*, No. 2.

Vereshchagin, N.K. and A.G. Dyunin. 1949. Povedenie ptits i mlekopitayushchikh pri sil'nykh zamorozkakh v Azerbaidzhane (Behavior of birds and mammals during severe frosts in Azerbaidzhan). *Izvestiya AN AzerbSSR,* No. 7.

Verheyen, R. 1956. Contribution à l'anatomie et à le systematique des Galliformes. *Bull. Inst. Roy. Sci. Natur Belgigue*, vol. 32, No. 42.

Verheyen, R. 1961. A new classification of nonpasserine birds of the world. *Bull. Inst. Roy. Sci. Natur Belgique*, vol. 37, No. 27.

Vishnyakov, S.V. 1962. K voprosu o khozyaistvennoi roli seroi kuropatki v lesostepnykh biotsenozakh (The problem of the economic role of Gray Partridges in forest steppe biocenoses). In: *Materialy III Vsesoyuzn. Ornitol. Konfer*. L'vov, pt. I.

Voinstvenskii, M.A. 1959. Novye dannye o predkakh domashnei kuritsy (New information on the ancestors of domestic fowl). In: *Vtoraya Vsesoyuzn. Ornitol. Konfer.* Moscow.

Volkov, N.I. 1968. Izmenenie upitannosti ryabchika v svyazi s sezonnymi osobennostyami ego ekologii (Change in the well being of Hazel Grouse depending on the seasonal characteristics of its ecology). In: *Resursy Teterevinykh Ptits v SSSR*. Moscow.

Volkov, N.I. 1968. Snezhnyi pokrov i teterevinye ptitsy (Snow cover and tetraonids). *Okhota i Okhotnich'e Khoz-vo*, No. 3.

Volkov, N.I. 1968a. Eksperimental'noe izucheni temperaturnykh uslovii v snezhnykh norakh teterevinykh ptits (Experimental studies of temperature conditions in the snow holes of tetraonids). *Zool. Zhurn.*, vol. 47, No. 2.

Volkov, N.I. 1969. K voprosu o vliyanii potepleniya klimata na mnogoletnie izmeneniya chislennosti teterevinykh ptits (The influence of climate warming on long-time changes in the tetraonid population). In: *Ornitologiya v SSSR*. Ashkhabad, pt. 2.

Volkov, N.I. 1970. Ob odnom uslovii pitaniya belykh kuropatok v zimnii period (One of the conditions of Willow Grouse feeding in winter). *Byul. Mosk. Ob-va Ispyt. Prirody*, vol. 75, No. 1.

Vorob'ev, K.A. 1946. Materialy k biologii turacha (*Francolinus francolinus* L.) i ego rasprostranenie v Yuzhnoi Turkmenii [Data on the biology of Francolins (*Francolinus francolinus* L.) and their distribution in southern Turkmenia]. *Byul. Mosk. Ob-va Ispyt. Prirody*, vol. 51, No. 1.

Vorob'ev, K.A. 1954. Ptitsy Ussuriiskogo kraya (Birds of the Ussuriisk region). Moscow.

Vorob'ev, K.A. 1963. Ptitsy Yakutii (Birds of Yakutia). Moscow.

Vorob'ev, K.A. 1966. Dikusha (The Spruce Grouse). *Okhota i Okhotnich'e Khoz-vo,* No. 7.

Vorob'ev, K.A. 1967. Turach (The Francolin). *Okhota i Okhotnich'e Khoz-vo*, No. 5.

Vorob'ev, N. 1896. K voprosu o prolete kuropatok (Migration of partridges). *Okhotnich'ya Gazeta*, No. 2.

Voronin, R.N. 1969. Kriterii optimal'noi emkosti zimnikh ugodii dlya beloi kuropatki na evropeiskom severe (The criterion of the optimal carrying capacity of winter abodes for Willow Grouse in the European North). In: *Proizvoditel'nost' i Produktivnost' Okhotnich'ikh Ugodii SSSR.* Kirov, pt. I.

Voronin, R.N. 1972. Ekologo-morfologicheskii analiz populyatsii beloi kuropatki Bol'shezemel'skoi tundry (Ecomorphological analysis of Willow Grouse populations in the tundra of Bol'shezemlya). Avtoref. Kand. Diss., Sverdlovsk.

Voronov, A.G. 1963. Biogeografiya (Biogeography). Moscow.

Vul'f [Wolff], E.V. 1944. Istoricheskaya geografiya rastenii (Evolutionary Geography of Plants). Moscow-Leningrad.

Ward, E.C. 1955. The Golden Pheasant. *Wyoming Wildlife*, vol. 19, No. 11.

Watson, A. 1972. A behavior of the Ptarmigan. *Brit. Birds*, vol. 65, No. 3.

West, G.C. and M.S. Meng. 1966. Nutrition of the Willow Ptarmigan in northern Alaska. *Auk*, vol. 83, No. 4.

Westerskov, K. 1956. History and distribution of the Hungarian Partridge in Ohio, 1909—1948. *Ohio Sci. J.*, vol. 56, No. 2.

Westerskov, K. 1965. Winter ecology of the Common Partridge (*Perdix perdix*) in the Canadian prairie. *Proc. NZ Ecol. Soc.*, No. 12.

Westerskov, K. 1966. Winter food and feeding habits of Common Partridge (*Perdix perdix*) in the Canadian prairie. *Canad. J. Zool.*, vol. 44, No. 2.

Wetmore, A. 1930. Two fossil birds from the Miocene of Nebraska. *Condor*, vol. 32.

Wetmore, A. 1956. A check-list of the fossil and prehistoric birds of North America and the West Indies. *Smithson. Misc. Collect.*, vol. 131, No. 5.

Wetmore, A. 1960. A classification for the birds of the world. *Smithson. Misc. Collect.*, vol. 139, No. 11.

Whistler, H. 1949. *Popular Handbook of Indian Birds*. London.

Wolfson, A. 1955. Origin of the North American bird fauna: critique and reinterpretation from the standpoint of continental drift. *Amer. Midland Naturalist*, vol. 53, No. 3.

Yakobi, V.E. 1960. Morfo-funktsional'nye issledovaniya poleta ptits semeistva yastrebinykh (Morphological and functional studies of the flight of birds of the family Accipitridae). *Trudy In-ta Morfol. Zhivotnykh AN SSSR,* No. 32.

Yamaschina, Y. 1939. Note sur le *Tetras falcipenne* de Siberia, *L'Oiseau et la Rev. Franc. Ornithol., Nouv. Ser.*, vol. 9.

Yanushevich, A.I. 1952. Fauna pozvonochnykh Tuvinskoi oblasti (Vertebrate Fauna of the Tuva Region). Novosibirsk.

Yanushevich, A.I., A.S. Tyurin, I.D. Yakovleva, A. Kydyraliev, and N.I. Semenova. 1959. Ptitsy Kirgizii (Birds of Kirgizia). Frunze, vol. 1.

Yudakov, A.G. 1967. Dikusha v verkhnem Priamur'e (The Siberian Spruce Grouse in the upper Amur region). In: *Okhrana, Ratsional'noe Ispol'zovanie i Vosproizvodstvo Estestv. Resursov Priamur'ya.* Khabarov.

Yudakov, A.G. 1972. Biologiya dikushi (*Falcipennis falcipennis*) v Amurskoi oblasti [Biology of Siberian (Spruce Grouse) *Falcipennis falcipennis* in the Amur District]. *Zool. Zhurn.*, vol. 51, No. 4.

Yudin, K.A. 1950. Morfologicheskie adaptatsii sem. Falconidae v svyazi s voprosami sistematiki (Morphological adaptations of the family Falconidae from the viewpoint of systematics). In: *Sbornik Pamyati Akademika P.P. Sushkina*. Moscow-Leningrad.

Yudin, K.A. 1957. O zadachakh morfologicheskikh issledovanii v oblasti ornitologii (Morphological problems in the field of ornithology). *Zool. Zhurn.*, vol. 36, No. 1.

Yudin, K.A. 1970. O nekotorykh printsipal'nykh i metodicheskikh voprosakh nadvidovoi sistematiki ptits (Some principal and methodological problems in avian supraspecific taxonomy). *Zool. Zhurn.*, vol. 49, No. 4.

Yunnatov, A.A. 1954. Kormovye rasteniya pastbishch i senokosov Mongol'skoi Narodnoi Respubliki (Fodder plants of the pastures and hayfields of the Mongolian People's Republic). *Trudy Mongol'sk Komissii*, No. 56, Moscow-Leningrad.

Yurgenson, P.B. 1964. Rezul'taty iskusstvennogo rasseleniya pernatoi dichi kak metoda vosproizvodstva v okhotnich'em khozyaistve (Results of the artificial distribution of wild birds as a game management method to increase their population). In: *Problemy Ornitologii*, L'vov.

Yurlov, K.T. 1956. K ekologii teterevinykh (Tetraonidae) lesostepi Zapadnoi Sibiri (Ecology of Tetraonidae of the forest steppes of western Siberia). Avtoref. Kand. Diss., Novosibirsk.

Yurlov, K.T. 1956. Materialy po pitaniyu beloi kuropatki v Barabe i Kulunde (Data on the nutrition of Willow Grouse in Baraba and Kulunda). *Trudy Biol. In-ta Zap. Sib. Fil. Akad. Nauk SSSR*, No. 1.

Yurlov, K.T. 1960. Materialy po ekologii beloi kuropatki i tetereva v Barabe i Kulunde (Data on the ecology of Willow and Northern Black Grouse in Baraba and Kulunda). *Trudy Biol. In-ta Zap. Sib. Otd. Akad. Nauk SSSR*, No. 6.

Zarudnyi, N.A. 1896. Ornitologicheskaya fauna Zakaspiiskogo kraya (Severnoi Persii, Zakaspiiskoi oblasti, Khivinskogo oazisa i ravninnoi Bukhary) [Avian fauna of the Trans-Caspian region (northern Persia, Trans-Caucasian Districts, Khivinsk oasis, and plains of Bukhara)]. In: *Materialy k Poznaniyu Fauny i Flory Ros. Imperii, Otd. Zool.*, No. 2.

Zarudnyi, N.A. 1903. Ptitsy Vostochnoi Persii. Ornitologicheskie rezul'taty ekskursii po Vostochnoi Persii v 1898 g. (Birds of eastern Persia. Ornithological results of an expedition through eastern Persia in 1898). *Zap. RGO po Obshch. Geogr.*, vol. 36.

Zhirnov, L.V. and A.A. Vinokurov, 1959. Materialy po ekologii fazana v verkhov'yakh r. Tekes (Data on the ecology of pheasants in the upper reaches of the Tekes River). *Uchen. Zap. Mosk. Gos. Ped. In-ta*, No. 104.

Zhitkov, B.M. and S.A. Buturlin. 1906. Materialy dlya ornitofauny Simbirskoi gubernii (Data on the *Ornithofauna* of Simbirsk Province). *Zap. RGO po Obshch. Geogr.*, vol. 41, No. 2.

Zimina, R.P. 1956. Materialy k zimnei ekologii gimalaiskogo ulara (Data on the winter ecology of Himalayan Snowcocks). *Trudy In-ta Zool. i Parazit. AN KirgSSR*, No. 5.

Zimina, R.P. 1959. K biologii gimalaiskogo ulara (On the biology of Himalayan Snowcocks) In: *Tezisy Dokl. 2-i Vsesoyuzn. Ornitol. Konfer.*, Moscow, pt. 3.

Znamenskaya, L.A. 1961. Khimicheskii sostav nekotorykh rastenii Krainego Severa Yakutii (Chemical composition of some plants of far northern Yakutia). In: *Materialy po Rastitel'nosti Yakutii*. Leningrad, No. 1.

Zverev, M.D. 1962. K ekologii Tyan-Shan'skogo tetereva (Zailiiskii Alatau) [Ecology of Tien-Shan Black Grouse (Trans-Ilian Alatau)]. *Ornitologiya*, No. 5.

Zwickel, Fred C. 1966. Winter food habits of the Capercaillie in northeast Scotland. *Brit. Birds*, vol. 59, No. 8.

Index of Plant Names

1	2	3	4
Beech, oriental	Buk	*Fagus orientalis*	5, 84
Bent	Polevitsa	Gen. *Agrostis*	112
Bergenia	Badan	Gen. *Bergenia*	56
Bilberry	Chernika	*Vaccinium myrtillus*	9, 25, 71, 78, 81, 88, 89, 90, 91
Bilberry, Caucasian	Chernika kavkazskaya	*Vaccinium arctostaphylos*	82, 84
Bindweed, greater	Povoi zabornyi	*Calystegia sepium*	106
Bindweed, black	Gorets v'yunkovyi	*Polygonum convolvulus*	97
Bindweed, lesser	V'yunok polevoi	*Calystegia sepium*	22, 88, 106, 114
Birch, common	Bereza borodavchataya	*Betula verrucosa*	81, 82
Birch, costate	Bereza rebristaya	*Betula costata*	88
Birch, dwarf	Bereza karlikovaya	*Betula nana*	9, 56, 72
Birch, Erman's	Bereza kamennaya	*Betula ermanii*	15, 25, 76
Birch, ground	Bereza kruglolistnaya	*Betula rotundifolia*	10
Birch, white	Bereza pushistaya	*Betula pubescens*	81, 82
Birthwort	Khokhlatka	Gen. *Corydalis*	89
Bittersweet	Paslen sladko-gorkii	*Solanum persicum*	93, 112, 114
Blackthorn	Tern	Gen. *Prunus*	20
Bluegrass, annual	Myatlik odnoletnyi	*Poa annua*	109
Bluegrass, Kentucky	Myatlik lugovoi	*Poa pratensis*	109
Borages	Burachnikovye	Fam. Boraginaceae	93, 101, 118
Box-thorn, Turkmenian	Litsii	*Lycium turcomanicum*	113
Bramble, stone	Kostyanika	*Rubs saxatilis*	72, 83, 86, 90
Bristle-grass, green	Shchetinnik zelenyi	*Setaria viridis*	93, 94

1	2	3	4
Bristle-grass, yellow	Shchetinnik sizyi	*Setaria glauca*	94
Burnet, ground	Krovokhlebka	*Sanguisorba officinalis*	104
Butterbur	Podbel	*Andromeda polifolia*	82
Buttercup	Lyutikovye	Fam. Ranunculaceae	74, 82, 83, 89, 91, 109, 122
Buttercup, corn	Lyutik polevoi	*Ranunculus arvensis*	89
Callianthemum	Kalliantemum	Gen. *Callianthemum*	110
Camel's thorn	Verblyuzh'ya kolyuchka	*Alhagi pseudalhagi*	53
Campanula	Kolokol'chik	Gen. *Campanula*	84
Caraway	Tmin polevoi	*Carum carvi*	97
Caryophyllaceae	Gvozdichnye	Fam. Caryophyllaceae	73, 103
Catmint	Kotovnik kokandskii	*Nepeta kokanica*	108
Cherry, ground	Vishnya stepnaya	*Cerasus chamaecerasus*	48, 73, 83
Cherry, small-fruited	Vishnya melkoplodnaya	*Cerasus microcarpa*	104
Cherry, Tien-Shan	Vishnya Tyan'-shanskaya	*Cerasus tianschanica*	103
Chestnut, sweet	Kashtan	*Castanea sativa*	5
Chokecherry	Cheremukha	*Padus racemosa*	20, 38, 91, 99
Chokecherry, Amur	Cheremukha maaka	*Padus maackii*	88, 90, 91
Cinquefoil, pitted	Lapchatka vyemchataya	*Potentilla sericea*	71
Clematis, oriental	Lomonos vostochnyi	*Clematis orientalis*	112
Cloudberry	Moroshka	*Rubus chamaemorus*	25, 78, 86
Clover	Klever	Gen. *Trifolium*	78, 82, 90, 106
Clover, white	Klever polzuchii	*Trifolium repens*	108

1	2	3	4
Cobresia	Kobereziya	*Cobresia humilis*	109
Cotoneaster	Kizil'nik	Gen. *Cotoneaster*	83, 98, 105
Columbine	Akvilegiya	Gen. *Aquilegia*	10, 56
Cornflower	Vasilek	Gen. *Centauria*	98, 104
Cottongrass	Pushitsa	Gen. *Eriophorum*	71, 82
Couchgrass	Pyrei polzuchii	*Agropyrum repens*	83, 110, 112, 114
Crabapple	Yablonya dikaya	Gen. *Malus*	48
Cranberries	Brusnichnye	Vassiniaceae subfam.	73, 79, 82, 83
Cranesbill, tuberous	Geran'	*Geranium tuberosum*	102, 104
Crocus	Krokus	Gen. *Crocus*	110
Crowberries	Shikshievye	Fam. Empetraceae	121
Crowberry	Voronika	*Empetrum nigrum*	25, 71, 86, 90
Crowberry, black	Shiksha	*Empetrum nigrum*	72, 77
Crucifers	Krestotsvetnye	Fam. Cruciferae	83, 101
Crupina	Krupina	*Crupina pauciflora*	104
Currant, black	Smorodina chernaya	*Ribes nigrum*	20
Currant, red	Smorodina krasnaya	*Ribes gladellum*	79, 88
Cypress, prostrate summer	Kokhiya stelyushchayasya	*Kochia prostrata*	112, 113
Cypress, summer	Kokhiya	Gen. *Kochia*	112
Daisies or Composites	Slozhnotsvetnye	Fam. Compositae	72, 74, 83, 97, 104
Dandelion	Oduvanchik	Gen. *Taraxacum*	84, 104
Desert candle	E'remurus	*Eremurus spectabilis*	105
Dewberry	Ezhevika	*Rubus caesicus*	45, 53, 111, 112

1	2	3	4
Dock, sorrel	Shchavel' shchavelkovidnyi	*Rumex acetosella*	106, 112
Dropwort, fine-leaved water	Omezhnik	*Oenanthe aquatica*	112, 114
Dryas	Driada	*Dryas punctata*	10, 56
Dryas, punctate	Kuropach'ya trava	*Dryas punctata*	72
Ebelek	E'belek	Gen. *Ceratocarpus*	102
Elaeagnus	Lokhovye	Fam. Elaeagnaceae	116
Elder, red-berried	Buzina	*Sambucus racemosa*	91
Elm	Vyaz	Gen. *Ulmus*	5, 20
Elm, Manchurian	Il'm lopastnyi	*Ulmus laciniata*	88
Ephedra	Khvoinik	*Ephedra equisetina*	102
Ephedra	E'fedra	*Ephedra equisetina*	48
Ericaceae	Vereskovye	Fam. Ericaceae	121
Euonymus, Sakhalin	Bersklet sakhalinskii	*Euonymus sachalinensis*	88
Euonymous, sparse-flowered	Bersklet malotsvetnyi	*Euonymus pauciflora*	88
Fescue, celestial	Ovsyanitsa podnebesnaya	*Festuca coelestis*	84, 108
Fescue, sheep's	Tipchak	*Festuca sulcata*	58, 108
Fescue, Tien-Shan sheep's	Tipchak Tyan'-shanskii	*Festuca tianschanica*	109
Figwort	Pogremok	Gen. *Alectorolophus*	84
Filbert, Manchurian	Leshchina man'chzhurskaya	*Corylus manshurica*	88
Filbert, Siberian	Leshchina raznolistnaya	*Corylus heterophylla*	88
Fir, Manchurian	Pikhta belokoraya	*Abies holophylla*	25, 86, 227
Fir, Nordmann	Pikhta kavkazskaya	*Abies nordmaniana*	82

1	2	3	4
Garlic	Luk polevoi	Gen. *Allium*	102
Garlic, wooly	Luk sherstistyi	*Allium trachyscordum*	103
Gladiolus, Turkmenian	Shpazhnik turkmenskii	*Gladiolus turkmenorum*	104, 105
Globeflower, Jungarian	Kupal'nitsa dzhungarskaya	*Troslius dschungaricus*	89, 110
Goat's beard	Kozloborodnik	*Tragopogon rutenicus*	98
Goldbachia, smooth	Gol'dbakhiya gladkaya	*Goldbachia laevigata*	101
Gooseroots	Marevye	Fam. Chenopodiaceae	95, 97, 103, 104
Grapevine	Vinograd	Gen. *Vitis*	106
Grapevine, Amur	Vinograd amurskii	*Vitis amurensis*	88, 90
Grass, alpine meadow	Myatlik al'piiskii	*Poa alpina*	109
Grass, bulbous meadow	Myatlik lukovichnyi	*Poa bulbosa*	98, 101, 102
Grass, Brome	Koster	Gen. *Bromus*	102
Grass, cockspur	Kurinoe proso	*Panicum crus galli*	112
Grass, Litvinov's meadow	Myatlik litvinova	*Poa litivinowiana*	108
Grass, meadow	Myatlik al'berta	*Poa alberti*	109, 110
Grass, spiked oat	Trishchetinnik	*Trisetum spicatum*	109
Grass, timothy	Timofeevka	Gen. *Phleum*	112
Grass, wavy hair	Lugovnik	*Deschampsia flexuosa*	72
Grass, whitlow	Krupka prodolgovataya	*Draba oblongifolia*	74
Grass, willow	Grechikha	Gen. *Polygonum*	84, 92, 93, 114
Grasses	Zlaki	Fam. Gramineae	71, 89, 96, 97, 100
Gromwell, corn	Vorobeinik polevoi	*Lithospermum arvense*	93, 94, 101
Gagea	Luk gusinyi	Gen. *Gagea*	89, 104
Gypsophila, prostrate	Kachim-rasprostertyi	*Gypsophila forrigens*	101

1	2	3	4
Hawkweed	Yastrebinka	Gen. *Hieracium*	78, 98
Hawthorn	Boyaryshnik	Gen. *Crataegus*	20, 48, 91
Hawthorn, Maximovich	Boyaryshnik maksimovicha	*Crataegus maximowiczii*	88, 91
Heath	Veresk	Gen. *Erica*	70
Hemp, wild	Konoplya dikaya	*Cannabis ruderalis*	93, 94
Hempnettle	Pikul'nik	Gen. *Galeopsis*	73, 94
Herbaceous seablite	Sveda	Gen. *Suaeda*	112
Honeysuckle	Zhimolost'	Gen. *Lonicera*	15, 77, 98, 99
Honeysuckle	Kopeechnik	*Hedysarum obscurum*	106
Honeysuckle, Maack's	Zhimolost' maaka	*Lonicera maackii*	88
Honeysuckle, Persian	Zhimolost' persidskaya	*Lonicera persica*	104
Horbeam, heartleaf	Grab serdtselistnyi	*Carpinus cordata*	88
Horn-wort	Rogolistnik	Gen. *Ceratophyllum*	112
Horsetail	Khvoshch	Gen. *Equisetum*	77, 82, 86, 98
Iron-tree	Zheleznoe derevo	*Celtis caucasica*	104
Ixiolirion, Tatar	Iksiolirion	*Ixiolirion tataricum*	102, 103
Juniper, common	Mozhzhevel'nik	*Juniperus communis*	19, 20, 21, 44, 76, 82, 90
Juniper, dwarf	Mozhzhevel'nik karlikovyi	*Juniperus depressa*	38
Juniper, low	Mozhzhevelovoi stlanets	*Juniperus sibirica*	77
Knotgrass	Grechishnye	Fam. Polygonaceae	72, 83, 89, 93, 95
Knotgrass, slender	Gorets malyi	*Polygonum alpinum*	106

1	2	3	4
Larch, Daurian	Listvennitsa daurskaya	*Larix dahurica*	76, 86
Larch, Siberian	Listvennitsa sibirskaya	*Larix sibirica*	15, 16, 19, 77, 81, 86
Legumes	Bobovye	Fam. Leguminosae	72, 74, 83, 95, 101, 115
Lemon tree	Limonnik	*Schizandra chinensis*	91
Lettuce	Latuk	*Mulgedium tataricum*	98, 99
Lilies	Lileinye	Fam. Liliaceae	97, 116
Lily	Liliya	Gen. *Lilium*	114
Lily, May	Mainik	*Majanthenum bifolium*	78, 91
Liquorice, Spanish	Solodka	*Glycyrrhiza glabra*	98, 112
Lungwort	Medunitsa	Gen. *Pulmonaria*	84
Madder	Morena	*Rubia florida*	104
Maple, bearded	Klen borodatyi	*Acer barbinerve*	88
Maple, Manchurian	Klen man'chzhurskii	*Acer mandschuricum*	88
Maple, Mono	Klen melkolistnyi	*Acer mono*	88
Marigold, three-lobed bur	Chereda	*Bidens tripartita*	99
Marsh tea	Bagul'nik	*Ledum palustre*	9
Marshwort	Klyukva	*Vaccinium oxycoccus*	9, 71, 79
Meadowsweet	Labaznik	*Filipendula ulmaria*	71, 90
Melilot, dentate	Donnik	*Melilotus dentatus*	131
Mercury	Proleska	Gen. *Mercurialis*	89
Moschatel	Adoksa muskusnaya	*Adoxa moschatellina*	89

1	2	3	4
Moss, haircap	Kukushkin len	*Polytrichum juniperinum*	72
Moss, Reindeer	Yagel'	Gen. *Cladonia* sp. [*C. rangifera*]	86
Nightshade, black-berried	Paslen chernyi	*Solanum nigrum*	93, 104, 106, 114
Oak, pedunculate	Dub	*Quercus robur*	77
Oak, Mongolia	Dub manchzhurskii	*Quercus mongolica*	29, 65, 112
Oat, common wild	Ovsyug	*Avena fatua*	83, 99
Olive, Russian	Dzhida, Lokh	*Elaeagnus angustifolia*	20, 53, 63, 65, 111, 112, 113
Orache	Lebeda	*Atriplex nitens*	96, 99
Oxytropis	Ostrolodochnik	*Oxytropis lapponica*	104
Oxytropis, pitted	Ostrolodochnik uglublennyi	*Oxytropis immersa*	108
Parsnip, cow	Borshchevik	*Borszczowia aralocaspica*	96
Pasque-flower	Prostrel	*Pulsatilla campanella*	82, 110
Pasque-flower, spreading	Son-tsvety	*Pulsatilla patens*	77, 89
Pea, meadow	China lugovaya	*Lathyrus pratensis*	71, 78
Pea-shrub	Karagana	Gen. *Caragana*	20
Pepper, Guinea	Perets v'yunkovyi	*Capsicum annuum*	94
Petrosimonia, Siberian	Petrosimoniya	*Petrosimonia sibirica*	112
Pine, mountain	Kedrovoi stlanets	*Pinus pumila*	10, 15
Pine, Scots	Sosna	*Pinus sylvestris*	14, 19, 81
Pine, Swiss mountain	Sosna gornaya	*Pinus montana*	19

1	2	3	4
Pine, Siberian stone	Kedr	*Pinus sibirica*	14, 76
Plantain, water	Chastukha	*Alisma plantago aquatica*	112, 114
Pondweed	Rdest	Gen. *Potamogetan*	114
Poplar	Topol’	Gen. *Populus*	29, 37
Poplar, Asiatic	Turanga	Populus diversifolia	63, 111
Poplar, black	Osokor’	*Populus nigra*	37
Poplar, white	Topol’ belyi	*Populus alba*	20
Poppies	Makovye	Fam. Papaveraceae	74
Poppy, polar	Mak polyarnyi	*Papaver polare*	71
Potentilla	Driadotsvet	*Dryadanthe tetranda*	109
Primrose	Primula	Gen. *Primula*	56
Prune	Uryuk	Gen. *Armeniaca*	48
Ptarmigan berry, alpine	Toloknyaka alpiiskaya	*Arctous alpinus*	71
Ragwort	Krestovnik lugovoi	*Senecio jacobaea*	112, 114
Raspberry	Malina	*Rubus idaeus*	78, 86, 90
Raspberry, hawthorn	Malina boyarshnikolistnaya	*Rubus crataegifolius*	90
Rattle, red	Mytnik	*Pedicularis palustris*	84
Reed, common	Trostnik	*Phragmites communus*	53, 112
Reedmace	Rogoz	Gen. *Typha*	106
Rhododendron	Rododendron	Gen. *Rhododendron*	28, 58
Rhododendron, Caucasian	Rododendron kavkazskii	*Rhododendron caucasicum*	82
Rose, dog	Shipovnik	Gen. *Rosa*	20, 38, 63, 72, 81, 88, 91, 104

1	2	3	4
Rose, Guelder	Kalina	*Viburnum opulus*	91
Roses	Rozotsvetnye	Fam. Rosaceae	83, 88, 106, 109
Rushes	Sitnikovye	Fam. Juncaceae	53, 74
Sallow-thorn	Oblepikha	*Hippophae rhamnoides*	38, 53, 63, 65, 91, 101, 111
Salt tree	Chingil	*Halimodendron halodendron*	37, 63, 112
Saltwort	Selitryanka	*Nitraria schoberi*	53, 106, 111
Saltwort, scrub	Boyalych	*Salsola arbuscula*	48
Saltwort, wooly	Solyanka sherstistaya	*Salsola lanata*	112
Saussurea	Sossyureya lednikovaya	*Sauscurea glacialis*	108
Saxaul	Saksaul	Gen. *Haloxylon*	37, 48
Saxifragaceae fam.	Kamnelomkovye	Fam. Saxifragaceae	71
Saxifrage, alternate leaved golden	Selezenochnik	*Chrysosplenium alternifolium*	89
Saxifrage, marsh	Kamnelomka bolotnaya	*Saxifraga hirculus*	74, 108
Saxifrage, tufted	Kamnelomka dernistaya	*Saxifraga caespitosa*	71
Scaligeria	Skaligeriya	*Scaligeria setacea*	103, 104
Schrenkia	Shrenkiaya	*Schrenkia ugamica*	102
Scirpus, sea	Kamysh	*Scirpus maritimus*	37, 112
Sedge	Osoka	Gen. *Carex*	37, 56, 72, 74, 86
Sedges	Osokovye	Fam. Cyperaceae	72, 74
Sorrel	Andromeda	Gen. *Andromeda*	9
Sorrel, sheep's	Shchavelek	*Rumex*	111

1	2	3	4
Sorrel, wood	Kislitsa	*Oxalis acetosella*	84, 88, 89
Spiraea	Tavolga	Gen. *Spiraea*	48, 56, 58, 71
Spruce, Schrenk	El' Tyan-shanskaya	*Picea schrenkiana*	81
Spruce, Yeddo	El' ayanskaya	*Picea jezoensis*	25, 86, 227
Spurge	Volchnik	Gen. *Daphne*	84
Spurge	Molochaisereyushchii	*Euphorbia canescens*	101
Statice	Kermek	Gen. *Statice*	98
Stitchwort, wood	Zvezdchatka	*Stellaria nemorum*	108
Stickseed	Lipuchka	*Lappula patula*	99
Strawberry	Klubnika	*Fragaria collina*	83, 90
Strawberry, common	Zemlyanika	*Frageria vesca*	20, 72, 79, 88, 90
Storksbill	Zhuravel'nik ostronosnyi	*Erodium cicutarium*	101
Tamarisk	Grebenshchik	Gen. *Tamarix*	37, 45
Tamarisk	Tamariks	*Tamarix* sp.	53
Thermopsis	Termopsis	*Thermopsis alpina*	109
Tulip	Tyul'pan	Gen. *Tulipa*	98, 104
Ungernia	Ungerniya	*Ungernia minor*	104
Vetch	Astragal	Gen. *Astragalus*	84, 106
Vetch	Vika	Gen. *Vicia*	97
Vetch, tufted	Goroshek myshinyi	*Vicia cracca*	71, 94
Violet, wonder	Fialka	*Viola mirabilis*	91

1	2	3	4
Watercress, common	Zherukha	*Nasturtium officinale*	112, 114
Water-lily	Nimfeinik	*Limnanthemum nymphoides*	112
Wheat, cow	Mar'yannik	Gen. *Melampyrum*	91
Whortle-berry, bog	Golubika	*Vaccinium uliginosum*	9, 25, 71, 77, 88
Whortle-berry, red	Brusnika	*Vaccinium vitis-idaea*	25, 71, 77, 82, 84, 88, 91
Willow, dwarf	Iva karlikovaya	*Salix herbacea*	71
Willow, polar	Iva polyarnaya	*Salix polaris*	72
Willow, shrub	Iva kustarnikovaya	*Salix arbuscula*	19
Willow, Thunberg's	Iva tunberga	*Salix thunbergiana*	88
Winterfat, European	Teresken	*Eurotia ceratoides*	112
Wormwood	Polyn'	Gen. *Artemisia*	97

Index of Animal Names (Excluding Birds)

English Name	Russian Name	Latin Name	Pages
1	2	3	4
Ants	Murav'i	Fam. Formicidae	69, 91, 104, 106
Austrian bug	Cherepashka vrednaya	*Eurygaster austriacus, E. integriceps*	92, 100, 103
Bees and wasps	Pereponchatokryle	Ord. Hymenoptera	83, 105
Beetles, click	Shchelkuny	Fam. Elateridae	93
Beetles, darkling	Chernotelki	Fam. Tanebrionidae	92
Beetles, ground	Zhuzhelitsy	Fam. Carabidae	93
Beetles, jewel	Zlatki	Fam. Buprestidae	96
Beetles, ladybird	Bozh'i korovki	Fam. Coccinellidae	83, 90
Beetles, leaf	Listoedy	Fam. Chrysomelidae	84, 86, 93
Beele, long-horned	Usachi	Fam. Cerambycidae	86
Beetle, powder-post	Medlyak peschanyi'	*Opatrum sabulosum*	92
Blackflies	Moshki	Fam. Simulidae	84, 90
Bugs	Klopy	Ord. Heteroptera	83, 92, 94, 96
Bugs	Poluzhestokrylye	Ord. Heteroptera	105
Bush crickets	Kuznechiki	Fam. Tettigoniidae	93, 94, 108
Butterflies and moths	Cheshuekrylye	Ord. Lepidoptera	86, 90, 104, 105
Centipedes	Mnogonozhki	Class Chilopoda	95, 116

1	2	3	4
Dragonflies	Strekozy	Ord. Odonata	105, 107
Earthworm	Dozhdevoi cherv'	*Lumbricus terrestris*	69
Earwigs	Ukhovertka	Ord. Dermaptera	105
Flies	Mukhi	Fam. Muscidae	69, 104
Flies	Dvukrylye	Ord. Diptera	105
Grasshopper	Pryamokrylye	Ord. Orthoptera, *Saltatoria*	83, 86, 90, 92, 94, 96, 98, 105
Grasshoppers, acridoid	Saranchovye	Superfam. Acridoidea	84, 92
Horseflies	Slepni	Fam. Tabanidae	93
Inchworms	Pyadenitsy	Fam. Geometridae	90
Lacewings	Setchatokrylye	Ord. Neuroptera	116
Leafhoppers	Tsikadka	Fam. Cicadellidae	99, 104
Lizard	Yasheritsa zhivorodyashchaya	*Lacerta vivipara*	83
Louse, wood	Mokritsy	*Hemilapistes elegans*	105
Mites	Kleshchiki	Gen. *Aticula*	73
Mollusks	Mollyuski	Phylum Mollusca	83, 92, 93, 95, 99, 108, 111

1	2	3	4
Mosquitoes	Komary	Fam. Culicidae	69, 84, 90
Moth, brindled beauty	Pyadenitsa-shelkopryad buropolosaya	*Biston hirtaria* [= *Lycia hirtaria*]	100
Moth, clearwing	Streklyannitsy	Fam. Aegiriidae	90
Moths, owlet	Sovki	Fam. Noctuidae	90
Moth, pine beauty	Sovka sosnovaya	*Panolis flammea*	100
Moth, sphinx	Brazhniki	Fam. Sphingidae	90
Mouse, harvest	Mysh' malyutka	*Mus minutus* [= *Micromys minutus*]	111
Notodontid, plumed	Peristousaya khokhlatka	*Ptilophora plumigera*	100
Sawflies	Pilil'shchiki	Fam. Tenthredinidae	90, 93
Slugs	Slizni	Gen. *Arion*	73
Spiders	Pauki	Superfam. Arachnoidea	69, 84, 86, 95, 105
Spider, karakut	Karakury	*Lathrodectus tredecimguttatus*	111
Springtail, elm	Nogokhvost il'movyi	*Exaereta ulmi*	100
Ticks, argasine	Kleshchi argasovye	Subfam. Odonthophorinae	111
Ticks, ixodid	Kleshchi iksodovye	Fam. Ixodidae	111
Weevils	Dolgonosiki	Fam. Curculionidae	84, 86, 92, 93, 100, 108
Weevils	Sloniki	Fam. Curculionidae	83